Invasive Species and Human Health

CABI INVASIVES SERIES

Invasive species are plants, animals or microorganisms not native to an ecosystem, whose introduction has threatened biodiversity, food security, health or economic development. Many ecosystems are affected by invasive species and they pose one of the biggest threats to biodiversity worldwide. Globalization through increased trade, transport, travel and tourism will inevitably increase the intentional or accidental introduction of organisms to new environments, and it is widely predicted that climate change will further increase the threat posed by invasive species. To help control and mitigate the effects of invasive species, scientists need access to information that not only provides an overview of and background to the field, but also keeps them up to date with the latest research findings.

This series addresses all topics relating to invasive species, including biosecurity surveillance, mapping and modelling, economics of invasive species and species interactions in plant invasions. Aimed at researchers, upper-level students and policy makers, titles in the series provide international coverage of topics related to invasive species, including both a synthesis of facts and discussions of future research perspectives and possible solutions.

Titles Available

Invasive Species and Human Health

_____

Edited by

Giuseppe Mazza

CREA Research Centre for Plant Protection and Certification, Florence, Italy
Department of Biology, University of Florence, Italy

and

Elena Tricarico

Department of Biology, University of Florence, Italy

CABI is a trading name of CAB International

CABI
Nosworthy Way
Wallingford
Oxfordshire OX10 8DE
UK

CABI
745 Atlantic Avenue
8th Floor
Boston, MA 02111
USA

Tel: +44 (0)1491 832111
Fax: +44 (0)1491 833508
E-mail: info@cabi.org
Website: www.cabi.org

T: +1 (617)682-9015
E-mail: cabi-nao@cabi.org

A catalogue record for this book is available from the British Library, London, UK.

Library of Congress Cataloging-in-Publication Data

Names: Mazza, Giuseppe (Biologist), editor.
Title: Invasive species and human health / [edited by] Giuseppe Mazza, PhD, CREA Research Centre for Plant Protection and Certification, Florence, Italy, Department of Biology, University of Florence, Florence, Italy and Elena Tricarico, PhD, Department of Biology, University of Florence, Florence, Italy.
Description: Wallingford, Oxfordshire, UK ; Boston, MA : CABI, [2018] | Series: CABI invasives series ; 10 | Includes bibliographical references and index.
Identifiers: LCCN 2018002402 (print) | LCCN 2018003542 (ebook) | ISBN 9781786390998 (ePDF) | ISBN 9781786391001 (ePub) | ISBN 9781786390981 (hbk : alk. paper)
Subjects: LCSH: Introduced organisms--Health aspects. | Biological invasions-- Health aspects.
Classification: LCC QH353 (ebook) | LCC QH353 .I582697 2018 (print) | DDC 578.6/2--dc23
LC record available at https://lccn.loc.gov/2018002402

ISBN-13: 978 1 78639 098 1

Commissioning editor: David Hemming
Editorial assistant: Emma McCann
Production editor: Tim Kapp

Typeset by AMA DataSet Ltd, Preston, UK.
Printed and bound in the UK by Antony Rowe, CPI Group (UK) Ltd.

Contents

Contributors

Pedro M. Anastácio, MARE – Marine and Environmental Sciences Centre, Departamento de Paisagem, Ambiente e Ordenamento, Universidade de Évora, Rua Romão Ramalho 59, 7000-671 Évora, Portugal. E-mail: anast@uevora.pt

Leonardo Ancillotto, Wildlife Research Unit, Laboratorio di Ecologia Applicata, Dipartimento di Agraria, Università degli Studi di Napoli Federico II, Portici, Naples, Italy. E-mail: leonardo.ancillotto@unina.it

Marie-Anne Auger-Rozenberg, INRA Zoologie Forestière, Orléans, France. E-mail: marie-anne.auger-rozenberg@inra.fr

Sylvie Augustin, INRA Zoologie Forestière, Orléans, France. E-mail: sylvie.augustin@inra.fr

Daniela Boccolini, Vector Borne Diseases Unit, Department of Infectious Diseases, Istituto Superiore di Sanità, Rome, Italy. E-mail: daniela.boccolini@iss.it

Giuseppe Brundu, Department of Agriculture, University of Sassari, 07100 Sassari, Italy. E-mail: gbrundu@uniss.it

Dario Capizzi, Latium Region, Environment and Natural Systems, via del Pescaccio 96, 00166 Rome, Italy. E-mail: dcapizzi@regione.lazio.it

Lucilla Carnevali, ISPRA Institute for Environmental Protection and Research, Via V. Brancati 48, 00144 Rome, Italy. E-mail: lucilla.carnevali@isprambiente.it

Marco Di Luca, Vector Borne Diseases Unit, Department of Infectious Diseases, Istituto Superiore di Sanità, Rome, Italy. E-mail: marco.diluca@iss.it

Franz Essl, Division of Conservation Biology, Vegetation Ecology and Landscape Ecology, University of Vienna, Rennweg 14, 1030 Vienna, Austria. E-mail: franz.essl@univie.ac.at

Bella Galil, The Steinhardt Museum of Natural History, Israel National Center for Biodiversity Studies, Tel Aviv University, Tel Aviv, Israel. E-mail: galil@post.tau.ac.il

Piero Genovesi, ISPRA Institute for Environmental Protection and Research, Via V. Brancati 48, 00144 Rome, Italy. E-mail: piero.genovesi@isprambiente.it

Giulio Grandi, Department of Biomedical Sciences and Veterinary Public Health, Swedish University of Agricultural Sciences (SLU), Uppsala, Sweden; Department of Microbiology, National Veterinary Institute (SVA), Uppsala, Sweden. E-mail: giulio.grandi@slu.se

Lorenzo Lazzaro, Department of Biology, University of Florence, Via G. La Pira, 4, 50121 Florence, Italy. E-mail: lorenzo.lazzaro@unifi.it

Antonella Lugliè, Department of Architecture, Design and Urban Planning, University of Sassari, 07100 Sassari, Italy. E-mail: luglie@uniss.it

Angeliki F. Martinou, Joint Services Health Unit (Cyprus), BFC RAF Akrotiri BFPO 57, Cyprus. E-mail: BFC-JSHU-HQ-Entomologist@mod.uk

Giuseppe Mazza, CREA Research Centre for Plant Protection and Certification, Florence, Italy; Department of Biology, University of Florence, Italy. E-mail: giuseppe.mazza@unifi.it

Jolyon M. Medlock, Medical Entomology Group, Emergency Response Department, Public Health England, Porton Down, Salisbury, United Kingdom. E-mail: jolyon.medlock@phe.gov.uk

Saverio Meini, Centro Veterinario Cimarosa, Livorno, Italy. E-mail: saveriomeini@hotmail.com

Mattia Menchetti, Department of Biology, University of Florence, Sesto Fiorentino, Florence, Italy. E-mail: mattiamen@gmail.com

Andrea Monaco, Latium Region, Environment and Natural Systems, via del Pescaccio 96, 00166 Rome, Italy. E-mail: amonaco@regione.lazio.it

Emiliano Mori, Research Unit of Behavioural Ecology, Ethology and Wildlife Management, Dipartimento di Scienze della Vita, Università di Siena, Siena, Italy. E-mail: moriemiliano@tiscali.it

Wolfgang Nentwig, Institute of Ecology and Evolution, University of Bern, Baltzerstrasse 6, 3012 Bern, Switzerland. E-mail: wolfgang.nentwig@iee.unibe.ch

Bachisio Mario Padedda, Department of Architecture, Design and Urban Planning, University of Sassari, 07100 Sassari, Italy. E-mail: bmpadedda@uniss.it

Nikola Pantchev, IDEXX Laboratories, Ludwigsburg, Germany. E-mail: nikola-pantchev@idexx.com

Olivier S.G. Pauwels, Institut Royal des Sciences Naturelles de Belgique, Brussels, Belgium. E-mail: osgpauwels@yahoo.fr

Cristina Preda, Ovidius University of Constanţa, Constanţa, Romania. E-mail: cristina.preda@univ-ovidius.ro

Petr Pyšek, Department of Invasion Ecology, Institute of Botany, The Czech Academy of Sciences, CZ-252 43 Průhonice, Czech Republic; Department of Ecology, Faculty of Science, Charles University, Viničná 7, CZ-128 44 Prague, Czech Republic. E-mail: pysek@ibot.cas.cz

Wolfgang Rabitsch, Department of Biodiversity and Nature Conservation, Environment Agency Austria, Spittelauer Lände 5, 1090 Vienna, Austria. E-mail: wolfgang.rabitsch@umweltbundesamt.at

Julian Reynolds, Emeritus, Trinity College, University of Dublin, Dublin, Ireland. E-mail: jrynolds@tcd.ie

Roberto Romi, Vector Borne Diseases Unit, Department of Infectious Diseases, Istituto Superiore di Sanità, Rome, Italy. E-mail: roberto.romi@iss.it

Alain Roques, INRA Zoologie Forestière, Orléans, France. E-mail: alain.roques@inra.fr

Helen E. Roy, Centre for Ecology & Hydrology, Benson Lane, Wallingford OX10 8BB, United Kingdom. E-mail: hele@ceh.ac.uk

Riccardo Scalera, IUCN SSC Invasive Species Specialist Group, Via Mazzola 38, 00142 Rome, Italy. E-mail: scalera.riccardo@gmail.com

Francis Schaffner, National Centre for Vector Entomology, Institute of Parasitology, University of Zurich, Zurich, Switzerland; Francis Schaffner Consultancy, Riehen, Switzerland. E-mail: francis.schaffner@uzh.ch

Stefan Schindler, Department of Biodiversity and Nature Conservation, Environment Agency Austria, Spittelauer Lände 5, 1090 Vienna, Austria. E-mail: Stefan.Schindler@umweltbundesamt.at

Francesco Severini, Vector Borne Diseases Unit, Department of Infectious Diseases, Istituto Superiore di Sanità, Rome, Italy. E-mail: francesco.severini@iss.it

Sauro Simoni, CREA Research Centre for Plant Protection and Certification, Florence, Italy. E-mail: sauro.simoni@crea.gov.it

Catherine Souty-Grosset, Laboratoire Ecologie & Biologie des Interactions, Equipe Ecologie Evolution Symbiose, Université de Poitiers, UMR CNRS 7267, F-86073 Poitiers, France. E-mail: catherine.souty@univ-poitiers.fr

Paolo Sposimo, NEMO s.r.l., Florence, Italy. E-mail: sposimo@nemoambiente.com

Diederik Strubbe, Terrestrial Ecology Unit, Ghent University, Ghent, Belgium. E-mail: diederik.strubbe@uantwerpen.be

Luciano Toma, Vector Borne Diseases Unit, Department of Infectious Diseases, Istituto Superiore di Sanità, Rome, Italy. E-mail: luciano.toma@iss.it

Elena Tricarico, Department of Biology, University of Florence, via Romana 17, I-50125 Florence, Italy. E-mail: elena.tricarico@unifi.it

To Francesca Gherardi

Introduction
From Local Strategy to Global Frameworks: Effects of Invasive Alien Species on Health and Well-being

Angeliki F. Martinou[1]* and Helen E. Roy[2]

[1]*Joint Services Health Unit, Cyprus;* [2]*Centre for Ecology & Hydrology, Wallingford, UK*

Introduction

The rate of introduction of species by humans to regions in which they did not exist previously is increasing, with no evidence of saturation (Seebens *et al.*, 2017). A proportion of these alien species are widely accepted to have an adverse effect on biodiversity, economies and society and are termed invasive alien species. Traditionally, ecologists and invasion biologists have placed focus on the impacts of invasive alien species on biodiversity and ecosystem functions. The adoption of a more anthropocentric approach, that of ecosystem services (Vilà and Hulme, 2017) where biodiversity is viewed as a central provider to human well-being, has highlighted the need to consider human health impacts of invasive alien species alongside environmental impacts. The impacts of invasive alien species on human health vary from psychological effects, discomfort, nuisance and phobias to skin irritations, allergies, poisoning, disease and even death.

Some taxonomic groups represent particular pressure to human health, perhaps over and above the impacts they pose to biodiversity and ecosystems. There has been extensive research on species of plant that are highly allergenic, such as *Ambrosia artemisiifolia*, ragweed (Chapman *et al.*, 2016; Lazzaro *et al.*, Chapter 2, this volume), but it is the Culicidae or mosquitoes that are the most notable in terms of medical importance because of their capacity to vector human diseases (Romi *et al.*, Chapter 6, this volume). However, 43 of the 1418 documented alien species of insect in Europe have been noted as having demonstrable biting, urticating or allergenic properties (Roques *et al.*, Chapter 5, this volume), including cockroaches, many ants, the Asian hornet, some true bugs and mosquitoes. After insects, spiders are the most diverse group of terrestrial invertebrates and are notoriously associated with biting, in some cases venomously, but can also be urticarious (Nentwig, Chapter 3, this volume). There are a number of venomous and poisonous alien species within the marine environment (Galil, Chapter 1, this volume), including fish such as *Plotosus lineatus*, striped eel catfish, which is spreading through the Mediterranean and has venomous glands situated along the dorsal and pectoral fins. Alien species of urchin and jellyfish also pose a threat

* E-mail: BFC-JSHU-HQ-Entomologist@mod.uk

© CAB International 2018. *Invasive Species and Human Health*
(eds G. Mazza and E. Tricarico)

to human health through their spines and stings respectively (Galil, Chapter 1, this volume).

Freshwater environments have also been invaded by a number of invertebrate alien species that directly affect human health, including some that cause allergic reactions either through contact or consumption but also others that can exert physical damage because of their sharp shells. The shells of *Dreissena polymorpha*, zebra mussel, can cause injuries to people using water bodies for a range of purposes, for example recreational swimming and commercial fishing (Souty-Grosset *et al.*, Chapter 7, this volume). Alien species of freshwater crustaceans can harbour pathogens and cause food-poisoning, while other toxic effects can be exerted through the proliferation of alien bryozoans or the concentration of pesticides, herbicides and heavy metals within aquatic invaders (Souty-Grosset *et al.*, Chapter 7, this volume). Perhaps the greatest threat to human health in freshwater environments is disease transmission mediated by crustaceans and fish (Souty-Grosset *et al.*, Chapter 7, this volume). Amphibians and reptiles also harbour zoonoses, including emerging pathogens of medical importance such as *Brucella* sp. (Pauwels and Pantchev, Chapter 8, this volume). Of course, some of the alien species of reptile are able to inflict bites that warrant medical concern (Pauwels and Pantchev, Chapter 8, this volume), but it should be noted that there is a low probability of direct contact between people and these species. Bites and transmission of pathogens are perhaps more likely from alien mammal species, rather than reptile species. It is noted that alien species of mammal and bird can introduce new pathogens to human populations, alter the epidemiology of resident pathogens and become reservoir hosts so increasing disease risk for humans (Capizzi *et al.*, Chapter 10, this volume; Mori *et al.*, Chapter 9, this volume). However, although high-profile pandemics have received considerable attention, such as avian influenza, there is still a lack of empirical data on the risk of disease transmission to humans by alien species. Furthermore, it is likely that the impacts of alien species on human health may increase as climate change (and other environmental change) further facilitates invasions (Schindler *et al.*, Chapter 11, this volume). It is also important to note that both native and alien species have the potential to affect human health and well-being, but that climate warming is perhaps likely to exacerbate the contribution from alien species to a greater extent than native species. Indeed, a semi-systematic review highlighted the alien species that are likely to be increasingly implicated as of nuisance to humans as the climate warms in Britain, including *Lasius neglectus* (invasive garden ant), *Thaumetopoea processionea* (oak processionary moth), *Linepithema humile* (Argentine ant), *Reticulitermes grassei* (Mediterranean termite) and mosquitoes, but also noted the propensity for native nuisance insects including Diptera such as *Musca domestica* (house fly) to respond positively to climate warming (Roy *et al.*, 2009). However, the need for further evidence across environments and taxonomic groups is apparent.

Evidence underpinning the impacts of invasive alien species is growing, but there is still a need for empirical approaches to inform risk assessment. Encouraging recent developments have included refinements to risk assessments (Roy *et al.*, 2017a), development of environmental impact assessments (Blackburn *et al.*, 2014) and socio-economic impact assessments (Bacher *et al.*, 2017). It is recognized that an invasive alien species can exert multiple impacts, spanning environmental and socio-economic contexts, but adverse effects are often considered in isolation. The fragmented nature of approaches to understanding the impacts of alien species is not only evident in terms of the context of risk and impact assessments but also in taxonomic perspectives; pathogens are generally not included in inventories of non-native species and the subsequent classification of their impacts (Roy *et al.*, 2017b). This is particularly problematic when considering the effects of invasive alien species on human health and the need for prioritizing management. Here, we provide an introduction to the invasive alien species and their effect on health. We also highlight the knowledge gaps and the importance of addressing these to meet the demands of the UN Sustainable Development Goals (Griggs *et al.*, 2013).

Embracing the Framework of the UN Development Agenda

It is widely accepted that the major drivers of change identified through the Millennium Ecosystem Assessment will interact. Climate change and the arrival of invasive alien species were described as a 'deadly duo' at the Nagoya Biodiversity Summit (2010) by Sarah Simons, Executive Director of the Global Invasive Species Programme (GISP). She stated that:

> The dangers posed by this 'deadly duo' cannot be overestimated. Each driver poses an enormous threat to biodiversity and human livelihoods but now evidence is rapidly emerging which shows that climate change is compounding the already devastating effects of invasive species, resulting in a downward spiral with increasingly dire consequences.

The 17 UN Sustainable Development Goals depend on functioning ecosystems, economies and societies and, as such, the threats posed by invasive alien species present a challenge to meeting the UN Development Agenda (Table 1).

From Global Frameworks to Local Strategy: Insect Vectors on Cyprus as a Case Study

Among all invasive alien species that cause health impacts, mosquitoes are the most notorious. Mosquitoes were first associated with deadly diseases 120 years ago. Sir Ronald Ross found the malaria parasite in the stomach tissue of an anopheline mosquito that had previously fed on a malarious patient and went on to prove the role of *Anopheles* mosquitoes in the transmission of malaria parasites in humans. A day after his Nobel-Prize-winning discovery on 21 August 1897, Ross wrote a poem with the title 'In Exile, Reply – What Ails the Solitude?':

> . . . seeking His secret deeds with tears and toiling breath I find thy cunning seeds,
> O million-murdering Death.
> I know this little thing a myriad men will save.
> O Death, where is thy sting, thy victory, O Grave!
>
> (After Baton and Ranford-Cartwright, 2005)

Since Ross' remarkable discovery significant advances have been made for humanity in terms of mosquito- and vector-borne diseases surveillance (European Centre for Disease Prevention Control [ECDC], 2012), integrated vector and resistance management as well as medical advances. However, despite all the advances, invasive mosquito vectors still remain at the top of the news headlines as threat to human life and well-being globally. Invasive mosquito species of the genera *Aedes* and *Culex* are the protagonists. With their amazing talent in cruising and hitchhiking, they have managed to reach some of the most remote places on Earth, such as small oceanic islands (Toto *et al.*, 2003; Almeida *et al.*, 2007; Bullivant and Martinou, 2017). The adaptation of mosquitoes to life with humans has established them as some of the most successful invaders globally (Romi *et al.*, Chapter 6, this volume). Mosquitoes are universally unpopular, many people desire their demise and research activity is responding through innovative approaches to control, but yet they are still spreading globally. So why are people failing in their quest to eliminate mosquitoes?

As for any other invasive alien species, there are three steps that will determine the success of an invasive mosquito: introduction, establishment and spread. Globalization, international trade and climate change have a pivotal role in all three steps. Guidelines at the European or international scale exist (ECDC, 2012), aiming mainly at surveillance, suggesting early warning systems. However, mosquitoes are introduced at the local scale (for example, at a local port or an airport) and therefore it is essential that European or international guidelines are interpreted and implemented promptly at the local scale. Adapting the

Table 1. Examples of ways in which invasive alien species pose a threat to meeting the UN Sustainable Development Goals.

Sustainable Development Goals	Example	Reference
1. End poverty in all its forms everywhere	Alien species of plants adversely affect catchments and availability of water. Water scarcity in developing countries is closely linked to the prevalence of poverty, hunger and disease.	Turton and Henwood, 2002
2. End hunger, achieve food security and improved nutrition and promote sustainable agriculture	There are many examples of how alien species have adversely affected sustainable agriculture. There are also examples where invasive alien species have been integrated into local livelihoods, as managed species as well as through the exploitation of wild invasive alien species, but adverse consequences have been documented as species spread into the broader landscape away from the control of the local communities, and so ultimately again undermine livelihoods. Indeed, Witt (2017) recognizes that alien species can provide benefits to poor communities, but there are also costs associated with these introductions when species escape cultivation or culturing and establish populations in the wild, noting 'the unfortunate reality is that many donors and development agencies have failed to recognise or acknowledge that cultured organisms can have significant impacts on ecosystems and human health should they escape and establish invasive populations'. It is acknowledged that although there is minimal disagreement when environmental and socio-economic impacts of an alien species are generally considered either negative or positive, there are species that evoke controversy when the assessment of impacts results in diverging priorities by different stakeholders. For example, the alien species of plant *Echium plantagineum* causes detrimental impacts to Australian agriculture because of its toxicity to livestock but provides benefits to beekeepers because it represents a floral resource for bees.	Reaser *et al.*, 2007; Shackleton *et al.*, 2007; Essl *et al.*, 2017; Witt, 2017
3. Ensure healthy lives and promote well-being for all at all ages	The impacts of invasive non-native species on human health vary from psychological effects, discomfort, nuisance and phobias to skin irritations, allergies, poisoning, disease and even death.	Vilà and Hulme, 2017
4. Ensure inclusive and equitable quality education and promote lifelong learning opportunities for all	How this manifests in the context of invasive species is not clear, but it could result in women having less awareness of the problem of invasive species, and of how to manage them effectively.	Fish, 2010

5. Achieve gender equality and empower all women and girls	It is acknowledged that there are gender differences in the context of natural resource management, including differences in ownership and access to assets and resources, division of labour, access to education, knowledge and information. Therefore, it is anticipated that invasive alien species will be have differing effects on the well-being of men and women.	Fish, 2010
	Loss of biodiversity has many direct and indirect impacts on human health and well-being, but gender differences occur when the services or disservices are accrued unequally. Therefore, for example, in some communities, women collect and use medicinal herbs more than men and so invasive alien species that reduce biodiversity would more immediately impact women.	
	The impacts of trade restrictions imposed to prevent the introduction of invasive alien species can have differential effects on gender. Pesticides are commonly used as a management strategy for invasive alien species and poisonings by pesticides are common. In Africa, women do much of the labour within high-value non-traditional crops, where pesticide use is greatest.	
6. Ensure availability and sustainable management of water and sanitation for all	Alien species of plants have a measurable negative effect on stream flow and it has been estimated that the total incremental water use of invading alien species of plants (i.e. the additional water use compared with the natural vegetation) is about 3300 million m^3 of water per year. Current loss of usable water due to invasive alien plants in primary catchments in the Western Cape has been estimated at 695 million m^3, equivalent to 4% of the total registered water use.	Turpie et al., 2008
7. Ensure access to affordable, reliable, sustainable and modern energy for all	Alien species of plant can be used to contribute to climate-change mitigation, but while the use of fast-growing alien plants for biofuel production to reduce greenhouse-gas emissions is seen as important by some stakeholders, others consider the risks of detrimental impacts through invasion of the landscape an unacceptable risk.	Fish, 2010; Essl et al., 2017
	Alien species of aquatic plants can block hydroelectric dams, and the fine hairs from water hyacinth roots have been shown to clog filters in turbine inlets.	
8. Promote sustained, inclusive and sustainable economic growth, full and productive employment and decent work for all	The cost of invasive alien species to economies has been estimated for a number of countries and regions. In Britain, a recent study estimated the direct costs conservatively to be £1.7 billion.	Williams et al., 2010
9. Build resilient infrastructure, promote inclusive and sustainable industrialization and foster innovation	Alien species can cause damage to hard infrastructure, affecting buildings, transportation, water and energy supplies globally. In Britain, in 2010 an estimate of the direct cost of alien species to infrastructure was approximately £310 million per annum, comprising 18% of the overall cost of alien species to Britain.	Booy et al., 2017

continued

Table 1. *Continued*

Sustainable Development Goals	Example	Reference
10. Reduce inequality within and among countries	The threat from invasive alien species will remain high (and show no signs of saturation) in wealthy countries, but worryingly invasive alien species are predicted to increasingly threaten human livelihoods in poor and developing countries. Globally, policies are under-equipped to address emerging threats from invasive alien species but this is particularly the case throughout Africa and the eastern hemisphere.	Early *et al.*, 2016; Seebens *et al.*, 2017
11. Make cities and human settlements inclusive, safe, resilient and sustainable	Invasive alien species can reduce the resilience of urban environments through adversely impacting infrastructure, but also by reducing ecosystem services. However, the aesthetics of urban environments can also be adversely affected. From a cultural perspective, people can take a 'sense of place' from the species around them, but this can come from both native species and novel ecosystems (dominated by alien species).	Booy *et al.*, 2017; Kueffer and Kull, 2017
12. Ensure sustainable consumption and production patterns	Invasive alien species can affect high-nature-value farming systems in Europe, for example in grasslands in Serbia or carob groves (due to the intensive use of rodenticides for black rat *Rattus rattus* control) in Cyprus.	Cooper *et al.*, 2010; Martinou, pers. comm.
13. Take urgent action to combat climate change and its impacts	Climate change facilitates the establishment of invasive alien species globally.	Early *et al.*, 2016
14. Conserve and sustainably use the oceans, seas and marine resources for sustainable development	Invasive alien species are considered one of the major threats to global marine ecosystems.	Galil, Chapter 1, this volume
15. Protect, restore and promote sustainable use of terrestrial ecosystems, sustainably manage forests, combat desertification, and halt and reverse land degradation and halt biodiversity loss	One-sixth of the global land surface is highly vulnerable to invasion, including substantial areas in developing economies and biodiversity hotspots.	Early *et al.*, 2016

16. Promote peaceful and inclusive societies for sustainable development, provide access to justice for all and build effective, accountable and inclusive institutions at all levels	The Zika epidemic in Brazil caused global panic during the Olympic games of Brazil in 2016, which are a worldwide celebration of peace between nations and athletic spirit, despite the predictions of scientists that the chance of visitors getting infected was low. Some athletes even withdrew from the games.	Grubaugh and Andersen, 2016; Petersen et al., 2016
17. Strengthen the means of implementation and revitalize the global partnership for sustainable development	Global collaborations on invasive alien species are inspiring and recognized through partnerships in research, conservation and strategy. A recent initiative, InvasivesNet, exemplifies global partnerships on alien species. The EU has adopted a Regulation (1143/2014) on invasive alien species emphasizing the need for prevention, early warning and rapid response by invasion biologists to address this problem. The involvement of wider society is seen as critical to develop more responsible behaviours to underpin this regulatory approach. There is no doubt that invasive alien species cause conflict with differing viewpoints projected by stakeholders with different priorities, but the need for global collaborations to achieve sustainable development is unequivocal. The Intergovernmental Platform on Biodiversity and Ecosystem Services (IPBES) addresses the major challenges facing biodiversity, including alien species. International meetings such as the Convention on Biological Diversity (CBD) now explicitly include invasive alien species through, for example, the 2020 Aichi targets. The EU has recently enforced a Regulation to deal with the threat from invasive alien species (although this focuses on biodiversity and ecosystems, human health impacts are of course recognized as important).	Genovesi et al., 2015; Lucy et al., 2016; Nentwig et al., 2017; Russell and Blackburn, 2017

guidelines at the local scale is not easy, as each local mosquito management programme has its own needs, resources and particularities. The success or failure of managing an initial introduction at the local scale could be sufficient to determine the fate for establishment and spread at the national or regional level of an invasive mosquito species. There are three pillars on which the success of local integrated mosquito management will depend:

- Collaboration (between public health authorities, local governments, research institutes and relevant stakeholders, including those responsible for urban development and the public).
- Advanced planning (including the development of a framework and common policy, appointing roles and allocating resources, horizon scanning, establishing early warning rapid response systems, developing codes of practice for integrated management of invasive mosquitoes considering environmental and legal issues, acquiring funding, addressing personnel training issues, and developing an evaluation scheme to assess the effectiveness of the adopted mosquito management programme).
- Raising awareness (among stakeholders and public, and developing initiatives such as citizen science projects that will support surveillance).

All of the above conditions need to be met to achieve success for an integrated vector management programme and none of them are simple to achieve.

Mosquito Control on Cyprus

The following case study aims to highlight some of the issues that can impede local integrated vector management programmes based on the three pillars mentioned above. Cyprus is an island of the Mediterranean, located at the cross-roads of three continents. Early pioneers settled on the island (10,500–9000 BP) from somewhere in the northern Levant and transported with them the full complement of economically important fauna (Zeder, 2008). Since then, 664 alien species have been introduced (CYIAS database). A recent horizon-scanning exercise looking at invasive species that are likely to arrive and establish within Cyprus in the next ten years and cause health impacts shows invasive mosquitoes at the top of the list (www.ris-ky.eu). Despite all the alien fauna and flora, Cyprus has high levels of endemism (Sparrow and John, 2016) and some recent threats to the island's biodiversity other than biological invasions are urbanization (for both the resident population and tourism), land abandonment, desertification and agricultural intensification – all processes that often take place in the most biodiverse sites of the island. Urban development at biodiverse wetland sites, such as the areas surrounding the salt lake of the Akrotiri peninsula, is expected to have an impact not only on the desired endemic flora and fauna but also on the undesired resident mosquito species, the salt marsh mosquito *Ochlerotatus detritus*. Developers and the local population who buy property in close proximity to the marshes put pressure on the relevant public health authorities to adopt intensive control methods based on synthetic chemicals, which could have detrimental impacts on the environment. This pressure will increase further if *Aedes albopictus* and *Aedes aegypti*, not currently present on the island, are introduced and manage to establish. A code of practice for the management of mosquitoes is currently lacking, but it is imperative this is agreed and acted on in order to provide local authorities, organizations and individuals involved in mosquito management with reasonable and practicable measures that minimize negative environmental impacts from mosquito management activities. There are a number strengths and weaknesses in the local mosquito management programmes currently operating in Cyprus (Table 2). Some of the strengths are that public health authorities are responsible for both surveillance and control, which can simplify rapid response in case of a mosquito invasion because arranging

Table 2. Summary of strengths and weaknesses of local integrated mosquito management programmes in Cyprus.

	Strengths	Weaknesses
Collaboration	Collaborative efforts between public health authorities and environment and conservation departments in the southern part of the island	No collaboration or communication between the southern and northern parts of the island regarding vector management No communication between urban development planners and public health authorities
Planning in advance	Horizon scanning	No common policy or agreement for management practices among authorities under different scenarios (introduction, establishment or spread of invasive mosquitoes)
	Surveillance system in place	Limitations in coverage of current surveillance system
	Appointed roles between health authorities in the southern part of the island	No local code of practice considering any impacts of management decisions to the environment
	Surveillance and management systems in place from public health authorities with no external contracting for vector control	No agreement currently on a rapid response system and the practices that should be adopted Limited training of vector control personnel, need for a better understanding of environmental implications of their management practices, e.g. application of chemicals, use of biocontrol agents such as fish for mosquito control No current evaluation scheme of the collaborative work and effectiveness of implemented programmes Lack of expertise in vector ecology and entomology among the pest control teams of the public health agencies
Raising awareness	Some communication between public health authorities and local stakeholders and members of the public	No citizen science or student science initiatives that could aid mosquito surveillance Lack of educational programmes for the public regarding mosquitoes

external contracts for control can be time-consuming. In the southern part of the island, there is also collaboration between public health and environmental authorities, which was demonstrated during a false alarm for *Aedes albopictus* in 2015 (Martinou *et al.*, 2016). A major impediment is the lack of communication and collaboration between the southern and northern parts of the island for political reasons, which could have detrimental effects if the surveillance programme of either side fails to identify invasive mosquitoes quickly. Some authorities have invested a lot of resources in terms of workforce and have dedicated surveillance teams which is encouraging because there are examples from other countries of increased management costs when cutting down budgets for surveillance and research (Vazquez-Prokopec *et al.*, 2010). However, advanced training of the workforce and pest control personnel on mosquito surveillance and management practices is needed. Some of the pest control teams on the island fail to comply with recommendations such as not to release the invasive fish *Gambusia* spp. for mosquito control and are not aware of the adverse environmental impacts of using synthetic chemical pesticides. A common local code of practice with a legal status could help the adoption of safe mosquito management approaches while safeguarding the environment. Lack of island-wide funding for local collaborations is also currently missing and research trials and collaborative management projects need to rely on international funding. Raising awareness is also a missing component from the local mosquito management programmes. Although brochures are distributed about personal protection measures that citizens could adopt to protect themselves from mosquitoes and on avoiding creating breeding sites around their residences, there is a need for citizens to become more involved and gain a better understanding of the components of local integrated vector management programmes. Initiatives such as the Global Mosquito Alert, developed by the United Nations, could greatly help local surveillance schemes. Other initiatives with incentives for citizens to recycle and not abandon their used tyres could greatly help to eliminate mosquito breeding sites.

Overall, while there are many strengths in current local integrated mosquito management programmes across Cyprus, there is still much to be achieved to ensure the weaknesses are addressed (Table 2). Failure to promptly address all issues could result in the failure of mosquito management programmes with detrimental impacts to human well-being, human health and the environment.

Advancing Towards One Health

Invasive alien species have the potential to severely affect the functioning of ecosystems and the well-being of the people that depend on them. However, perhaps the greatest direct threat to human health from invasive alien species is that of emerging diseases. One Health Initiatives (e.g. www.onehealthinitiative.com) aim to bring an interdisciplinary approach to the health of people, domestic animals and wildlife. As such, One Health Initiatives provide an excellent framework in which to address the potential impacts of invasive alien species within the context of human health, while accepting the multiple interacting effects that such species can exert. Coordinated and interdisciplinary approaches are key to understanding, detecting and managing the emergence of invasive alien species and their impacts on humans at various scales (Roy *et al.*, 2017b).

Acknowledgements

We acknowledge support from COST Action TD1209 ALIEN Challenge for providing networking opportunities. We are also grateful to the Darwin Initiative (Defra) project, which enabled us to pursue discussions on this topic. Finally, we thank Elena Tricarico for the invitation to contribute this chapter.

References

Almeida, A., Gonçalves, Y., Novo, M., Sousa, C., Melim, M. and Gracio, A. (2007) Vector monitoring of *Aedes aegypti* in the Autonomous Region of Madeira, Portugal. *Eurosurveillance* 12, E071115.

Bacher, S., Blackburn, T., Essl, F., Genovesi, P., Heikkilä, H., Jeschke, J.M., Jones, G., Keller, R., Kenis, M., Kueffer, C., Martinou, A.F., Nentwig, W., Pergl, J., Pyšek, P., Rabitsch, W., Richardson, D.M., Roy, H.E., Saul, W.-C., Scalera, R., Vilà, M., Wilson, J.R.U. and Kumschick, S. (2017) Socio-economic impact classification of alien taxa (SEICAT). *Methods in Ecology and Evolution* 10.1111/2041-210X.12844.

Baton, L.A. and Ranford-Cartwright, L.C. (2005) Spreading the seeds of million-murdering death: metamorphoses of malaria in the mosquito. *Trends in Parasitology* 21, 573–580.

Blackburn, T.M., Essl, F., Evans, T., Hulme, P.E., Jeschke, J.M., Kühn, I., Kumschick, S., Marková, Z., Mrugała, A., Nentwig, W., Pergl, J., Pyšek, P., Rabitsch, W., Ricciardi, A., Richardson, D.M., Sendek, A., Vilà, M., Wilson, J.R.U., Winter, M., Genovesi, P. and Bacher, S. (2014) A unified classification of alien species based on the magnitude of their environmental impacts. *PLoS Biology* 12(5), e1001850.

Booy, O., Cornwell, L., Parrott, D., Sutton-Croft, M. and Williams, F. (2017) Impact of biological invasions on infrastructure. In: Vilà, M. and Hulme, P.E. (eds) *Impact of Biological Invasions on Ecosystem Services.* Springer, Cham, Switzerland, pp. 235–247.

Bullivant, G. and Martinou, A.F. (2017) Ascension Island: a survey to assess the presence of Zika virus vectors. *Journal of the Royal Army Medical Corps* jramc-2016-000730.

Chapman, D.S., Makra, L., Albertini, R., Bonini, M., Páldy, A., Rodinkova, V., Šikoparija, B., Weryszko-Chmielewska, E. and Bullock, J.M. (2016) Modelling the introduction and spread of non-native species: international trade and climate change drive ragweed invasion. *Global Change Biology* 22, 3067–3079.

Cooper, T., Pezold, T. (eds), Keenleyside, C., Đorđević-Milošević, S., Hart, K., Ivanov, S., Redman, M. and Vidojević, D. (2010) *Developing a National Agri-Environment Programme for Serbia.* IUCN Programme Office for South-Eastern Europe, Gland, Switzerland and Belgrade, Serbia.

Early, R., Bradley, B.A., Dukes, J.S., Lawler, J.J., Olden, J.D., Blumenthal, D.M., Gonzalez, P., Grosholz, E.D., Ibañez, I., Miller, L.P., Sorte, C.J.B. and Tatem, A.J. (2016) Global threats from invasive alien species in the twenty-first century and national response capacities. *Nature Communications* 7, 12485.

Essl, F., Hulme, P.E., Jeschke, J.M., Keller, R., Pyšek, P., Richardson, D.M., Saul, W.-C., Bacher, S., Dullinger, S., Estévez, R.A., Kueffer, C., Roy, H.E., Seebens, H. and Rabitsch, W. (2017) Scientific and normative foundations for the valuation of alien species impacts: thirteen core principles. *BioScience* 67, 166–178.

European Centre for Disease Prevention Control (ECDC) (2012) *Guidelines for the Surveillance of Invasive Mosquitoes in Europe.* ECDC, Stockholm.

Fish, J. (2010) *Mainstreaming Gender into Prevention and Management.* Global Invasive Species Programme (GISP), Washington, DC.

Genovesi, P., Carboneras, C., Vilà, M. and Walton, P. (2015) EU adopts innovative legislation on invasive species: a step towards a global response to biological invasions? *Biological Invasions* 17, 1307–1311.

Griggs, D., Stafford-Smith, M., Gaffney, O., Ohman, M.C., Shyamsundar, P., Steffen, W., Glaser, G., Kanie, N. and Noble, I. (2013) Policy: Sustainable development goals for people and planet. *Nature* 495, 305–307.

Grubaugh, N.D. and Andersen, K.G. (2016) Navigating the Zika panic. *F1000Research* 5.

Kueffer, C. and Kull, C.A. (2017) Non-native species and the aesthetics of nature. In: Vilà, M. and Hulme, P.E. (eds) *Impact of Biological Invasions on Ecosystem Services.* Springer, Cham, Switzerland, pp. 311–324.

Lucy, F.E., Roy, H.E., Simpson, A., Carlton, J.T., Hanson, J.M., Magellan, K., Campbell, M.L., Costello, M.J., Pagad, S., Hewitt, C.L., *et al.* (2016) INVASIVESNET towards an international association for open knowledge on invasive alien species. *Management of Biological Invasions* 7, 131–139.

Martinou, A.F., Vaux, A.G., Bullivant, G., Charilaou, P., Hadjistyllis, H., Shawcross, K., Violaris, M., Schaffner, F. and Medlock, J. (2016) Rediscovery of *Aedes cretinus* (Edwards, 1921) (Diptera; Culicidae) in Cyprus, 66 years after the first and unique report. *Journal of the European Mosquito Control Association* 34, 10–13.

Nentwig, W., Mebs, D. and Vilà, M. (2017) Impact of non-native animals and plants on human health. In: Vilà, M. and Hulme, P.E. (eds) *Impact of Biological Invasions on Ecosystem Services.* Springer, Cham, Switzerland, pp. 277–293.

Petersen, E., Wilson, M.E., Touch, S., McCloskey, B., Mwaba, P., Bates, M., Dar, O., Mattes, F., Kidd, M., Ippolito, G., Azhar, E.I. and Zumla, A. (2016) Rapid spread of Zika virus in the Americas: implications for public health preparedness for mass gatherings at the 2016 Brazil Olympic Games. *International Journal of Infectious Diseases* 44, 11–15.

Reaser, J.K., Meyerson, L.A., Cronk, Q., De Poorter, M., Eldrege, L.G., Green, E., Kairo, M., Latasi, P., Mack, R.N., Mauremootoo, J., O'Dowd, D.J., Orapa, W., Sastroutomo, S., *et al.* (2007) Ecological and socioeconomic impacts of invasive alien species in island ecosystems. *Environmental Conservation* 34, 98–111.

Roy, H.E., Beckmann, B.C., Comont, R.F., Hails, R.S., Harrington, R., Medlock, J., Purse, B. and Shortall, C.R. (2009) *An Investigation into the Potential for New and Existing Species of Insect with the Potential to Cause Statutory Nuisance to Occur in the UK as a Result of Current and Predicted Climate Change.* DEFRA, London.

Roy, H.E., Rabitsch, W., Scalera, R., Stewart, A., Gallardo, B., Genovesi, P., Essl, F., Adriaens, T., Bacher, S., Booy, O., *et al.* (2017a) Developing a framework of minimum standards for the risk assessment of alien species. *Journal of Applied Ecology* doi: 10.1111/1365-2664.13025.

Roy, H.E., Hesketh, H., Purse, B.V., Eilenberg, J., Santini, A., Scalera, R., Stentiford, G.D., Adriaens, T., Bacela-Spychalska, K., Bass, D., *et al.* (2017b) Alien pathogens on the horizon: opportunities for predicting their threat to wildlife. *Conservation Letters* 10, 477–484.

Russell, J.C. and Blackburn, T.M. (2017) The rise of invasive species denialism. *Trends in Ecology & Evolution* 32, 3–6.

Seebens, H., Blackburn, T.M., Dyer, E.E., Genovesi, P., Hulme, P.E., Jeschke, J.M., Pagad, S., Pyšek, P., Winter, M., Arianoutsou, M., *et al.* (2017) No saturation in the accumulation of alien species worldwide. *Nature Communications* 8, 14435.

Shackleton, C.M., McGarry, D., Fourie, S., Gambiza, J., Shackleton, S.E. and Fabricius C. (2007) Assessing the effects of invasive alien species on rural livelihoods: case examples and a framework from South Africa. *Human Ecology* 35, 113–127.

Sparrow, D. and John, E. (eds) (2016) *An Introduction to the Wildlife of Cyprus*. Terra Cypria, Lefkosia, Cyprus.

Toto, J.C., Abaga, S., Carnevale, P. and Simard, F. (2003) First report of the oriental mosquito *Aedes albopictus* on the West African island of Bioko, Equatorial Guinea. *Medical and Veterinary Entomology* 17, 343–346.

Turpie, J., Marais, C. and Blignaut, J.N. (2008) The working for water programme: evolution of a payments for ecosystem services mechanism that addresses both poverty and ecosystem service delivery in South Africa. *Ecological Economics* 65, 788–798.

Turton, A. and Henwood, R. (2002) *Hydropolitics in the Developing World: A Southern African Perspective*. IWMI, Colombo, Sri Lanka.

Vazquez-Prokopec, G.M., Chaves, L.F., Ritchie, S.A., Davis, J. and Kitron, U. (2010) Unforeseen costs of cutting mosquito surveillance budgets. *PLoS Neglected Tropical Diseases* 4, e858.

Vilà, M. and Hulme, P.E. (2017) *Impact of Biological Invasions on Ecosystem Services*. Springer, Cham, Switzerland.

Williams, F., Eschen, R., Harris, A., *et al.* (2010) The economic cost of invasive non-native species on Great Britain. CABI Proj No VM10066, 1–199.

Witt, A.B. (2017) Use of non-native species for poverty alleviation in developing economies. In: Vilà, M. and Hulme, P.E. (eds) *Impact of Biological Invasions on Ecosystem Services*. Springer, Cham, Switzerland, pp. 295–310.

Zeder, M.A. (2008) Domestication and early agriculture in the Mediterranean Basin: origins, diffusion, and impact. *Proceedings of the National Academy of Sciences* 105, 11597–11604.

1

Poisonous and Venomous: Marine Alien Species in the Mediterranean Sea and Human Health

Bella Galil*

The Steinhardt Museum of Natural History, Tel Aviv University, Israel

Abstract

The Suez Canal is the main pathway of introduction of alien species into the Mediterranean Sea. Its successive enlargements left the entire sea prone to colonization by highly impacting invasive alien species, including poisonous and venomous ones. The temporal and spatial extent of occurrence in the Mediterranean Sea of nine species (fish, sea-urchin, jellyfish and stinging hydroid), the evidence of their impacts on human health in their native range, the frequency and severity of human health impacts in their introduced range are described, as well as management measures. This chapter aims to acquaint and forewarn the public, stakeholders and decision makers, and to urge the latter to take the necessary steps to control the pathways and vectors of introduction and prepare themselves for these new health hazards.

1.1 Introduction

Invasive alien species are considered one of the major threats to global marine ecosystems for impacting their structure, function and services. A small number of marine invasive alien species engender human health impacts. Intensification of anthropogenic activities, coupled with fast-increasing coastal urbanization, drive complex and fundamental changes in the relatively small, landlocked Mediterranean Sea, including increases in alien species. Some of these alien venomous and poisonous species have drawn the attention of scientists, managers, the media and the public for their conspicuous human health impacts. The main alien venomous (V) and poisonous (P) species are listed, their regions of origin and introduction pathways, the temporal and spatial extent of their occurrence in the Mediterranean Sea, the evidence of their impacts on human health in their native range, the frequency and severity of human health impacts in their introduced range are described, as well as management measures.

Human health hazards of invasive alien species are expected to worsen, benefitting from climate change and the greatly enlarged Suez Canal – the main pathway of invasive alien species introduction into the Mediterranean Sea (Galil *et al.*, 2018, 2017). In conjunction, these will enable the spread of thermophilic alien species to yet uncolonized regions,

* E-mail: galil@post.tau.ac.il

© CAB International 2018. *Invasive Species and Human Health*
(eds G. Mazza and E. Tricarico)

and admit entrance to additional species from the Indian Ocean and the Red Sea. The chapter aims to acquaint and forewarn the public, stakeholders and decision makers, and to urge the latter to take the necessary steps to control the pathways and vectors of introduction.

Actinopterygii Klein, 1885
Plotosidae Bleeker, 1858
Plotosus lineatus (Thunberg, 1787) V

The striped eel catfish *Plotosus lineatus* is widely distributed in the tropical Indo-west Pacific Ocean, including the Red Sea and the Suez Canal (Goren and Dor, 1994). It is found in marine, brackish and even freshwaters. Juveniles form closely packed ball-shaped schools in shallow waters, whereas adults are solitary or occur in small groups and shelter under overhangs in daytime. In 2002 it was recorded off the coast of Israel (Golani, 2002). Its population increased greatly within a couple of years, occurring on sandy and muddy bottoms to depths of 80 m, as well as rocky ridges. Recently it has been recorded off the Sinai Coast, Egypt (Temraz and Ben Souissi, 2013), Syria (Ali *et al.*, 2015) and Iskenderun Bay, Turkey (Doğdu *et al.*, 2016). Along the southern Levant it forms a sizeable portion of the shallow shelf trawl discards.

Halstead (1988) considered *P. lineatus* one of the most dangerous venomous fishes known, causing fatal envenomations. Toyoshima (1918) described its neurotoxic and haemotoxic properties (as *P. anguillaris*). Venom glands are associated with the serrate spines on the dorsal and pectoral fins and skin secretions contain proteinaceous toxins: at least one haemolysin, two lethal factors and two oedema-forming factors (Shiomi *et al.*, 1986, 1988). Once the spine penetrates the victim's skin, the thin sheath surrounding the venom gland cells is torn, releasing venom into the wound. Catfish venom is rarely fatal. Injury causes immediate throbbing pain, followed by cyanosis, numbness and swelling. Erythema, muscle fasciculations, severe lymphadenopathy and fever are also common. Tissue necrosis and gangrene are rare, but are more likely when a large amount of venom is delivered. Secondary bacterial infections may occur if the wound is unattended. The hand is the most common site of wounds and result from handling the fish after it has been caught.

A survey of injuries from marine organisms conducted among Israeli professional fishermen in 2003–2004 showed that 10% of fish-related injuries in Israel were caused by *P. lineatus* (Gweta *et al.*, 2008). Considering that in those years its population was much smaller than at present, the proportion of injuries it causes is rather high. Most injuries were incurred when sorting trawl hauls.

Scorpaenidae Risso, 1827
Pterois miles (Bennett, 1803) V

The lionfish *Pterois miles* (Fig. 1.1) is widely distributed in the tropical Indo-west Pacific Ocean, including the Red Sea. The Indo-Pacific lionfish was introduced into the western Atlantic in 1985 (Schofield, 2009). In 1991 a single specimen was recorded off the coast of Israel (Golani and Sonin, 1992). Recent records in Lebanon, Turkey, Cyprus, Greece, and as far west as Tunisia and Sicily, Italy, follow a major enlargement of the Suez Canal in 2010 (Bariche *et al.*, 2013; Turan *et al.*, 2014; Turan and Ozturk, 2015; Oray *et al.*, 2015; Iglésias and Frotté, 2015; Crocetta *et al.*, 2015; Kletou *et al.*, 2016; Dailianis *et al.*, 2016; Azzurro *et al.*, 2017). A couple of years ago this expansion of records across the Levant would have been inconceivable. A study combining remote sensing and computer modelling (Johnston and Purkis, 2014) had assured that this was unlikely to occur and argued that connectivity

Fig. 1.1. *Pterois miles* (Bennett, 1803). (O. Klein.)

among potential lionfish habitats in the Mediterranean was low and unfavourable to wide dispersion of its larvae. Alas, the lionfish was unaware of this and the authors of this study have been proved spectacularly wrong.

The venom apparatus consists of 13 dorsal spines, 3 anal spines, 2 pelvic grooved spines and their venom glands (Halstead, 1988). Envenomation produces intense pain and swelling, which may continue for several hours, depending on the amount of venom. A first-hand account of a *P. miles* sting inflicted by a juvenile specimen 10-cm long (Steinitz, 1959, as *P. volitans*) relates in precise detail the intense pain ('just short of driving oneself completely mad') and swelling that lasted 3 h.

No lionfish injuries have been reported from the Mediterranean Sea so far. However, their presence along coastlines popular with tourists is a marine health hazard.

Siganidae Richardson, 1837
Siganus luridus (Rüppell, 1829) V
Siganus rivulatus Forsskål & Niebuhr, 1775 V

The two species of rabbitfish, *Siganus rivulatus* and *S. luridus*, originate in the western Indian Ocean, where they inhabit rocky, coral reefs and algal-covered sandy bottoms. They entered the Mediterranean through the Suez Canal and were first recorded off the coast of Israel in

1924 and 1955, respectively (Steinitz, 1927; Ben-Tuvia, 1964). The species are found as far west as France and Tunisia (Ktari-Chakroun and Bahloul, 1971; Ktari and Ktari, 1974; Daniel et al., 2009). The schooling, herbivorous fishes form thriving populations in the Levant Sea, where 'millions of young abound over rocky outcropping' (George and Athanassiou, 1967).

The venom apparatus consists of 13 dorsal spines, 7 anal spines, 4 pelvic grooved spines and their venom glands which rupture and release their contents on penetration (Halstead, 1988). Care must be taken during fishing and cleaning, as rabbitfishes will use their venomous spines in defence. Their venom is not life-threatening to adult humans, but causes severe pain. The stings may cause pain and swelling lasting several hours.

Herzberg (1973) related two cases of poisoning following ingestion of *S. luridus* fished in Haifa Bay, Israel. The symptoms, which persisted for 24 h, comprised 'burning pain in the throat, followed by tingling and numbness of the muscles, some sensation reminding of electrical shock'. Similarly, Raikhlin-Eisenkraft and Bentur (2002) examined several patients who complained that after a dinner of rabbitfish caught in Haifa Bay, they suffered tremors, muscle cramps, nightmares, hallucinations, agitation, anxiety and nausea of varying severity. These symptoms lasted between 12 and 30 h and resolved completely. In all cases the intoxication occurred in the summer months.

Synanceiidae Swainson, 1839
Synanceia verrucosa Bloch and Schneider, 1801 V

The reef stonefish *Synanceia verrucosa* is widely distributed in the tropical Indo-west Pacific Ocean, including the Red Sea (Froese and Pauly, 2016). A solitary species, it occurs in shallow waters on rubble bottoms near rocky ridges or coral reefs. It was first recorded in the Mediterranean Sea in 2010 off the coast of Israel (Edelist et al., 2011) and in quick succession in the Gulf of Iskenderun, Turkey (Bilecenoğlu, 2012) and Tyr, Lebanon (Crocetta et al., 2015).

The reef stonefish is considered to be one of the most dangerous venomous marine fish – its stings may be fatal. Their venom apparatus consists of 13 dorsal spines, 3 anal spines and 2 pelvic spines. Venom is released from paired venom glands placed in lateral grooves at the base of each spine. The lethal fraction is a potent hypotensive agent which has myotoxic and neurotoxic activity. Envenomation results in excruciating pain to the point of losing consciousness and gross oedema, which may involve the entire extremity and regional lymph nodes, peaking around 60 to 90 min and lasting up to 12 h if untreated. Hypotension appears to be the primary cause of death. Systemic effects may include pallor, diaphoresis, nausea, muscle weakness, dyspnoea, headaches and delirium (Lee et al., 2004). Late complications include cutaneous abscesses, necrosis associated with painful oedema and lymphangitis necessitating surgery (Louis-Francois et al., 2003).

A first-hand account of *S. verrucosa* envenomation by the ichthyologist J.L.B. Smith (1951) describes in lucid, grim detail the succession of symptoms. Seconds after receiving a stab to the thumb from a specimen 12-cm long, the hand swelled and intense pain shot up the arm to the neck, spasms reaching to the neck, head and shoulder. An injection of Novocain did little to sooth the searing agony. Immersion in hot water diminished the pain after treatment by a hastily summoned native healer ('witch doctor') and a morphine injection failed. The hand and forearm remained swollen and intensely painful to the touch and large yellow blisters released a serous fluid for 6 days. On the sixth day inflammation and pain increased alarmingly and necessitated an injection of 1,000,000 units of penicillin. Similar symptoms arose a week later and subsided after another penicillin injection. After 14 days the hand and thumb were still swollen, painful and unusable. After 80 days the hand was still weak and the thumb slightly swollen and painful. His scientific interest was heightened by the experience and Smith (1957) related several fatalities resulting from *S. verrucosa* envenomations in the Seychelles and Mozambique.

No stonefish injuries have been reported in the Mediterranean Sea. The camouflage and habits of stonefish make them difficult for the unwary to detect and avoid: many reported victims are tourists (Louis-Francois *et al.*, 2003; Lee *et al.*, 2004; Lion, 2004; Prentice *et al.*, 2008; Ngo *et al.*, 2009; Diaz, 2015). If their numbers increase, their presence along coast-lines popular with tourists will be a marine health hazard.

Tetraodontidae Bonaparte, 1831
Lagocephalus sceleratus (Gmelin 1789) P

The silverstripe blaasop, *Lagocephalus sceleratus* (Fig. 1.2), a type of pufferfish, is widely dis-tributed in the tropical Indo-west Pacific Ocean, including the Red Sea (Goren and Dor, 1994). In 2003 a single specimen, considered to have been introduced through the Suez Canal, was collected along the south-western coast of Turkey (Akyol *et al.*, 2005). Soon after, its Levantine populations increased greatly and spread across the Mediterranean Sea to Spain (Katsanevakis *et al.*, 2014) and Algeria (Kara *et al.*, 2015), and into the Black Sea (Boltachev *et al.*, 2014). In the Mediterranean Sea, *L. sceleratus* has been recorded from waters as shallow as 3 m (Bilecenoğlu *et al.*, 2006) to 80 m (Halim and Rizkalla, 2011). The largest adult specimens recorded in the Mediterranean measure 83 cm in length.

Tetradotoxin (TTX), a potent neurotoxin that inhibits voltage-gated sodium channels, is produced by bacteria present in gonads, gastrointestinal tract, liver, muscle and skin of pufferfish. It is one of the most potent, non-protein poisons known. Symptoms include peri-oral paraesthesia (numbness or tingling of lips, tongue, around the mouth), nausea and vomiting, dizziness, headache, abdominal pain and progressive muscular paralysis, eventu-ally causing death due to respiratory paralysis. Coma has been reported in severe cases of TTX poisoning and in the final stages before death. TTX poisoning caused by ingestion of the fish is not uncommon in East Africa and South-east Asia (Chopra, 1967; Kan *et al.*, 1987;

Fig. 1.2. *Lagocephalus sceleratus* (Gmelin 1789). (O. Klein.)

Kanchanapongkul, 2001; Ahasan *et al.*, 2004; Chowdhury *et al.*, 2007). A TTX poisoning of epidemic proportions occurred in 2008 in Bangladesh after the native inland population were exposed to unfamiliar fish marketed locally. Of the 141 patients who had consumed these fish, 17 died from respiratory arrest (Islam *et al.*, 2011).

Following incidents of pufferfish poisonings in Egypt, mainly in the Suez Gulf, 45 specimens of *L. sceleratus* (34.5–65 cm in length) were collected in the Gulf of Suez and toxins were extracted from their gonads, liver, digestive tract, muscles and skin (El-Sayed *et al.*, 2003, as *Pleuranacanthus sceleratus*). The toxicity of the gonads was highest, but potentially harmful amounts were detected in the musculature as well. A subsequent study involving specimens from the Gulf of Suez (18.5–78.5 cm in length) provided evidence that the highest gonad toxicity for both sexes was recorded in late spring, before summer spawning (Sabrah *et al.*, 2006). Katikou *et al.* (2009) found that all tissues of a large specimen (length 66 cm) collected in the Aegean Sea were toxic. The gonads and liver of that specimen contained 17 lethal doses of TTX and consumption of 200 g of its flesh would be fatal.

Following the establishment of *L. sceleratus* in the Mediterranean Sea, evidence of severe TTX poisoning was noted. Between 2005 and 2008, 13 victims of TTX poisoning (aged 26–70 years) were hospitalized in Israel (Bentur *et al.*, 2008; Eisenman *et al.*, 2008). Onset of toxicity was within 1 h of consuming liver and gonads of *L. sceleratus*. The main clinical manifestations were paraesthesia, muscle weakness, hypertension, tachycardia, dyspnoea, ataxia, vomiting and diarrhoea. The two most severely affected victims ingested the entire liver of the fish. Other severe cases followed: after consuming the liver and gonads a recreational fisher experienced perioral paraesthesia with worsening limb muscle weakness. Shortly after admission to the hospital he developed acute respiratory failure with bradypnoea, accompanied by bradycardia, which quickly deteriorated to cardiac arrest. Complete paralysis with absence of motor responses and lack of pupil reactions necessitated treatment with cholinesterase inhibitors (Kheifets *et al.*, 2012). Similar cases were reported from Lebanon: a woman who had eaten the liver suffered proximal limb weakness and dyspnoea (Chamandi *et al.*, 2009); a man reported perioral tingling, dysarthria, became quadriparetic, developed respiratory and haemodynamic failure and within 3–4 h suffered deep nonreactive coma with absence of all brainstem reflexes (Awada *et al.*, 2010). The latter authors mention additional cases of poisoning that went unreported. Indeed, many cases are, at best, published locally, thus the full extent of the phenomenon is unknown (Ben Souissi *et al.*, 2014).

Lagocephalus sceleratus is widespread in the Mediterranean Sea and quite abundant along the Levant. Despite official prohibitions it is being sought by recreational fishers and occasionally commercially marketed (Nader *et al.*, 2012; Beköz *et al.*, 2013; Ben Souissi *et al.*, 2014; Farrag *et al.*, 2015, 2016). The Turkish Ministry of Food, Agriculture and Livestock issued a ban on the fishery of *L. sceleratus* in 2012. However, the results of a survey conducted by Beköz *et al.* (2013) examining awareness among fishermen, fish mongers, restaurant staff, medical staff and the public in Antalya, Turkey on the danger of consuming the locally abundant *L. sceleratus* are shocking. Fishermen, though aware that the fish is poisonous, admitted selling it to local markets and hotels, providing themselves with a good income. Restaurant managers and fish mongers denied selling the pufferfish, though specimens were in fact displayed for sale. Most customers had not heard of the pufferfish, were unaware of its lethality and admitted to relying on the sellers' advice. Of the physicians surveyed, 76% were not aware that pufferfish are poisonous and 98% were unfamiliar with the symptoms of pufferfish poisoning and believed an antidote to TTX exists. It is clear that both native local and tourist consumers are at serious risk given the high level of toxicity of *L. sceleratus*.

Echinoidea Leske, 1778
Diadematidae Gray, 1855
Diadema setosum (Leske, 1778) V

The long-spined urchin *Diadema setosum* is a native of the tropical Indo-west Pacific Ocean, including the Red Sea, and quite common in the northern Gulf of Suez (but see Lessios *et al.*, 2001). It is commonly observed on coral reefs, rocky ridges and coralligenous formations, but also in sand flats and seagrass beds (Coppard and Campbell, 2006). Commonly clustered in small groups, it may occasionally form large aggregations, especially in anthropogenically altered and eutrophic environments. The first record of *D. setosum* in the Mediterranean Sea was noted off Kaş peninsula, Turkey (Yokes and Galil, 2006). It has since spread along the Mediterranean coast of Turkey from Antakya to Gökova Bay at the south-east Aegean Sea (Turan *et al.*, 2011; Katsanevakis *et al.*, 2014; B. Yokes, 2016, personal communication) and to the adjacent Greek islands of Rhodes and Kastelorizo (Tsiamis *et al.*, 2015; Crocetta *et al.*, 2015), and southwards to Lebanese coastal waters (Nader and Indary, 2011).

Diadema setosum inflicts painful injuries on unwary swimmers, divers and fishers. The brittle spines may inflict deep penetrating wounds and break off easily to become embedded in the tissue (Halstead, 1988). Their venom is mild and may cause inflammation, swelling and acute pain, which gradually declines after a few hours. There is usually no residual disability and the injuries rarely come to the attention of physicians. However, in the minority of cases they may cause severe, irreversible tissue damage. A scuba diver with multiple penetrations of spines into his left hand experienced pain, discomfort and stiffness. The spines were absorbed within a few days, leaving one finger swollen and painful, with markedly reduced joint flexion and hand gripping power. The swelling reduced and hand strength was recovered within 6 weeks, yet residual joint swelling and range of flexion restriction persisted after 30 months. MRI findings were consistent with chronic, active synovial inflammation with surrounding fibrosis (Liram *et al.*, 2000).

No long-spined sea urchin injuries have been reported from the Mediterranean Sea. If their numbers increase, their presence along coastlines popular with tourists will be a health concern.

Scyphozoa Goette, 1887
Rhizostomatidae Cuvier, 1799
Rhopilema nomadica Galil, 1990 V

The nomadic jellyfish *Rhopilema nomadica* (Fig. 1.3) is a native of the tropical Indian Ocean, known from Mozambique and the Red Sea (Stiasny, 1938 as *R. hispidum*; Berggren, 1994). The first record of *R. nomadica* in the Mediterranean Sea was noted in the mid-1970s off Israel (Galil *et al.*, 1990). The species has since spread throughout the Levant and was recently reported from Pantelleria, Sardinia and the Aegadian archipelago off Sicily, Italy, Malta and Tunisia (Deidun *et al.*, 2011; Daly Yahia *et al.*, 2013; Crocetta *et al.*, 2015; Balistreri *et al.*, 2017). In the south-eastern Levant it forms huge swarms, 100 km long, each summer since the early 1980s, though small clusters occur throughout the year.

The venom apparatus of *R. nomadica* consists of nematocyst-laden tentacles. These cell organoids comprise a capsule containing a tightly coiled and pleated tubule armed with spines and packed with venom. When stimulated, the coiled tubule everts, penetrating the epidermis of human skin, and venom is injected from the capsule through the tubule into

Fig. 1.3. *Rhopilema nomadica* Galil, 1990. (S. Rothman.)

the victim's tissues (Halstead, 1988; Rifkin *et al.*, 1996). Three categories of nematocysts were identified in *R. nomadica* (Avian *et al.*, 1995).

In summer 2009 alone, 815 hospitalizations due to *R. nomadica* envenomations were recorded along the south-eastern coast of Turkey (Öztürk and İşinibilir, 2010). The annual swarms off the Israeli coast resulted in envenomation victims suffering adverse effects that have lasted weeks and even months after the event (Benmeir *et al.*, 1990; Galil *et al.*, 1990; Menahem and Shvartzman, 1995; Yoffe and Baruchin, 2004; Sendovski *et al.*, 2005). Benmeir *et al.* (1990) report that, in the summer of 1987, 30 patients, mainly children, were treated in their emergency ward alone. They describe a case where 2 days after envenomation (mistakenly attributed to *Aurelia aura* [sic.]) the pains subsided, after 14 days the swelling and urticaria had disappeared, but the arm and chest preserved marked discoloration for 3 months. Silfen *et al.* (2003) report that on the second day following envenomation the victim suffered raised, demarcated, very itchy skin eruptions, which lasted 5 weeks. Severe systemic manifestation may occur: a female surfer who had suffered five envenomations over the previous 3 weeks was stung in the arm. Within minutes she experienced shortness of breath, severe peri-orbital swelling and facial oedema, and marked erythema with papulovesicular eruptions and itching on her arm. Her anaphylactic reaction was aborted by timely intervention (Friedel *et al.*, 2016; Glatstein *et al.*, 2017).

The countries affected by the nomadic jellyfish swarming are major tourist destinations in the Mediterranean Sea. Jellyfish envenomations pose a threat to the countries' investments in marine recreational tourism (Öztürk and İşinibilir, 2010). A socioeconomic survey in Israel, carried out in summer 2013, captured the impacts of a swarm of *R. nomadica* on seaside recreation. It was estimated that a swarm reduces the number of seaside visits by 3–10.5%, with an annual monetary loss of €1.8–6.2 million. An additional 41% of the respondents stated that their recreational activities were affected (Ghermandi *et al.*, 2015).

Hydrozoa Owen, 1843
Aglaopheniidae Marktanner-Turneretscher, 1890
Macrorhynchia philippina Kirchenpauer, 1872 V

The feathery stinging hydroid *Macrorhynchia philippina* (Fig. 1.4) is distributed circumglobally in tropical and subtropical regions (Rees and Vervoort, 1987) and is common in the Red Sea (Vervoort, 1993). It occurs on rocky reefs and on artificial hard substrates. The species has been known in the south-eastern Mediterranean since the 1990s (Bitar and Bitar-Kouli, 1995; Morri *et al.*, 2009) and has since expanded northwards to the Turkish coast (Çinar *et al.*, 2006).

A brush with its feathery nematocyst-laden branches may cause a mild stinging sensation, but a more extensive contact results in a burning sensation (Vine, 1986). Victims generally develop pinpoint lesions, blotchy red rash, blisters and raised itchy weals, which may last up to 10 days before fading. Systemic effects are rare (Rifkin *et al.*, 1993, as *Lytocarpus philippinus*).

Fig. 1.4. *Macrorhynchia philippina* Kirchenpauer, 1872. (S. Rothman.)

An increase in the abundance of *M. philippina* along the Levant coastline may hurt recreational activities such as swimming and snorkelling (Çinar *et al.*, 2006).

1.2 Discussion and Conclusion

Alien species that are a concern to human health have recently gained notoriety, but broad assessments, spanning over a taxonomic or geographic range, have been scarce (Mazza *et al.*, 2014; Schindler *et al.*, 2015). Unsurprisingly, the information is even more limited for marine alien species (Ojaveer *et al.*, 2015). Among the alien species recorded in the Mediterranean Sea, nine are noted as human health hazards – six fish, one sea urchin, scyphozoan jellyfish and hydrozoan. Most have either been post-millennial records (e.g. *P. lineatus*, *L. sceleratus*, *D. setosum*, *S. verrucosa*) or have greatly increased their spread in the last decade (e.g. *P. miles*, *R. nomadica*). All originate in the Indian Ocean or the Indo-west Pacific Ocean and are considered to have entered the Mediterranean through the Suez Canal. Four species are confined to the Levant and Tunisia, and even for those recorded further west and north (e.g. *L. sceleratus*, *S. luridus*, *S. rivulatus*, *R. nomadica*, *P. miles*), the largest populations occur in the Levant. With rising temperature, it is likely these thermophilic species will expand their range.

Though published records attest to the increasing spread and abundance of marine alien species of human health concern, only fragmentary information is available concerning the spatial and temporal trends of their impacts. In fact, even for common species with acute symptoms such as *L. sceleratus* and *R. nomadica*, incidents are poorly documented. Öztürk and İşinibilir (2010) reported that in summer 2009 *R. nomadica* envenomations caused 815 hospitalizations along the south-eastern coast of Turkey, but no data are available for other years and other locations. Half a dozen publications by Israeli physicians deal mostly with single cases (see above). A similar pattern emerges from the records of pufferfish poisoning: Bentur *et al.* (2008) reported 13 victims of TTX poisoning hospitalized between 2005 and 2008 in Israel, but most publications deal with symptoms, and their authors admit in passing that other cases remained undocumented (Ben Souissi *et al.*, 2014). Erroneous identifications occur due to miscommunication between physicians and marine scientists (Benmeir *et al.*, 1990; Menahem and Shvartzman, 1995). The lack of nation-wide, if not region-wide, quantitative data on medically-treated health impacts is worrying, as ignorance of the extent and severity of these health hazards and their treatment may lead on one hand to medical errors (Beköz *et al.*, 2013) and on the other prejudice risk analyses undertaken by management.

The littoral countries are called on to prepare themselves for these new health hazards.

Regulations should be legislated, implemented and strictly enforced to prevent marketing and consumption of *L. sceleratus*, coupled with education and training of medical staff, tourism industry personnel, marine recreational industry personnel, as well as the general public. This article provides ample examples of risks to native members of the public tempted by easily accessible toxic fish species, especially those who depend on subsistence fisheries as a source of protein and income. Medical and healthcare staff should be familiarized with clinical syndromes and trained in treatment of marine health hazards. Early diagnosis and supportive management help ensure recovery. As incidents involving large numbers of patients may be expected to become more frequent with changing environmental conditions, improving emergency healthcare systems should be made a public health priority (Glatstein *et al.*, 2017).

Public health authorities in littoral countries should be encouraged to initiate and sponsor their national marine health hazards database – a clinical database sourced from hospital registry data and first-aid facilities including injury locations, dates, species identities, symptoms, treatments and outcomes. The information may help healthcare authorities to assess and manage trends of marine health hazards, bioinvasion ecologists to link trends in alien species presence and abundance to impacts, economists to assess their costs and environmental authorities to make accurate and detailed risk assessments.

References

Ahasan, H.A., Mamun, A.A., Karim, S.R., Bakar, M.A., Gazi, E.A. and Bala, C.S. (2004) Paralytic complications of puffer fish (tetrodotoxin) poisoning. *Singapore Medical Journal* 45(2), 73–74.

Akyol, O., Unal, V., Ceyhan, T. and Bilecenoglu, M. (2005) First confirmed record of *Lagocephalus sceleratus* (Gmelin, 1789) in the Mediterranean. *Journal of Fish Biology* 66, 1183–1186.

Ali, M., Saad, A. and Soliman, A. (2015) Expansion confirmation of the Indo-Pacific catfish, *Plotosus lineatus* (Thunberg, 1787), (Siluriformes: Plotosidae) into Syrian marine waters. *American Journal of Biology and Life Sciences* 3(1), 7–11.

Avian, M., Spanier, E. and Galil, B. (1995) Nematocysts of *Rhopilema nomadica* (Scyphozoa: Rhizostomeae), an immigrant jellyfish in the eastern Mediterranean. *Journal of Morphology* 224, 221–231.

Awada, A., Chalhoub, V., Awada, L. and Yazbeck, P. (2010) Coma profond aréactif reversible apres intoxication par des abats d'un poisson méditerraneen [Deep non-reactive reversible coma after a Mediterranean neurotoxic fish poisoning]. *Revue neurologique* 166, 337–340.

Azzurro, E., Stancanelli, B., Di Martino, V. and Bariche, M (2017) Range expansion of the common lionfish *Pterois miles* (Bennett, 1828) in the Mediterranean Sea: an unwanted new guest for Italian waters. *BioInvasions Records* 6(2), 95–98.

Balistreri, P., Spiga, A., Deidun, A., Gueroun, S. and Daly Yahia, M.N. (2017) Further spread of the venomous jellyfish *Rhopilema nomadica* Galil, Spannier & Ferguson, 1990 (Rhizostomeae, Rhizostomatidae) in the western Mediterranean. *BioInvasions Records* 6(1), 19–24.

Bariche, M., Torres, M. and Azzurro, E. (2013) The presence of the invasive lionfish *Pterois miles* in the Mediterranean Sea. *Mediterranean Marine Science* 14(2), 292–294.

Beköz, A.B., Beköz, S., Yilmaz, E., Tüzün, S. and Beköz, Ü. (2013) Consequences of the increasing prevalence of the poisonous *Lagocephalus sceleratus* in southern Turkey. *Emergency Medicine Journal* 30, 954–955.

Benmeir, P., Rosenberg, L., Sagi, A., Vardi, D. and Eldad, A. (1990) Jellyfish envenomation: a summer epidemic. *Burns* 16, 471–472.

Ben Souissi, J., Rifi, M., Ghanem, R., Ghozzi, L., Boughedir, W. and Azzurro E. (2014) *Lagocephalus sceleratus* (Gmelin, 1789) expands through the African coasts towards the Western Mediterranean Sea: a call for awareness. *Management of Biological Invasions* 5(4), 357–362.

Ben-Tuvia, A. (1964) Two siganid fishes of Red Sea origin in the eastern Mediterranean. *Bulletin of the Sea Fisheries Research Station, Haifa* 37, 1–9.

Bentur, Y., Ashkar, J., Lurie, Y., Levy, Y., Azzam, Z.S., Litmanovich, M., Golik, M., Gurevych, B., Golani, D. and Eisenman, A. (2008) Lessepsian migration and tetrodotoxin poisoning due to *Lagocephalus sceleratus* in the eastern Mediterranean. *Toxicon* 52(8), 964–968.

Berggren, M. (1994) *Periclimenes nomadophila* and *Tuleariocaris sarea*, two new species of Pontoniine shrimps (Decapoda: Pontoniinae), from Inhaca island, Moçambique. *Journal of Crustacean Biology* 14(4), 782–802.

Bilecenoğlu, M. (2012) First sighting of the Red Sea originated stonefish (*Synanceia verrucosa*) from Turkey. *Journal of Black Sea/Mediterranean Environment* 18(1), 76–82.

Bilecenoğlu, M., Kaya, M. and Akalin, S. (2006) Range expansion of silverstripe blaasop, *Lagocephalus sceleratus* (Gmelin, 1789) to the northern Aegean Sea. *Aquatic Invasions* 1, 289–291.

Bitar, G. and Bitar-Kouli, S. (1995) Impact de la pollution sur la répartition des peuplements de substrat dur à Beyrouth (Liban – Méditerranée Orientale). *Rapports et Procès-Verbaux des Réunions, Vol XIV (Nouvelle série) Commission internationale pour l'exploration scientifique de la mer Méditerranée* 34, 19.

Boltachev, A.R., Karpova, E.P., Gubanov, V.V. and Kirin, M.P. (2014) The finding of *Lagocephalus sceleratus* (Gmelin, 1789) (Osteichthyes, Tetraodontidae) in the Black Sea, Sevastopol Bay, Crimea. *Journal of Marine Ecology* 4(13), 14 (in Ukrainian).

Chamandi, S.C., Kallab, K., Mattar, H. and Nader, E. (2009) Human poisoning after ingestion of puffer fish caught from Mediterranean Sea – a case report. *Middle East Journal of Anaesthiology* 20(2), 285–288.

Chopra, S.A. (1967) A case of fatal puffer-fish poisoning in a Zanzibari fisherman. *East African Medical Journal* 44(12), 493–496.

Chowdhury, F.R., Nazmul Ahasan, H.A.M., Al Mamun, A., Mamunur Rashid, A.K.M. and Al Mahboob, A. (2007) Puffer fish (Tetrodotoxin) poisoning: an analysis and outcome of six cases. *Tropical Doctor* 37, 263–264.

Çinar, M.E., Bilecenglu, M., Öztürk, B. and Can, A. (2006) New records of alien species on the Levantine coast of Turkey. *Aquatic Invasions* 1, 84–90.

Coppard, S.E. and Campbell, A.C. (2006) Taxonomic significance of test morphology in the echinoid genera *Diadema* Gray, 1825 and *Echinothrix* Peters, 1853 (Echinodermata). *Zoosystema* 28(1), 93–112.

Crocetta, F., Agius, D., Balistreri, P., Bariche, M., Bayhan, Y.K. *et al.* (2015) New Mediterranean biodiversity records (October 2015). *Mediterranean Marine Science* 16(3), 682–702.

Dailianis, T., Akyol, O., Babali, N., Bariche, M., Crocetta, F., Çakir, S., Ciriaco, M., Corsini-Foka, A., Deidun, R., El Zrelli, D. *et al.* (2016) New Mediterranean biodiversity records (July 2016). *Mediterranean Marine Science* 17(2), 608–626.

Daly Yahia, M.N., Kéfi-Daly Yahia, O., Maïte Gueroun, S.K., Aissi, M., Deidun, A. Fuentes, V. and Piraino, S. (2013) The invasive tropical scyphozoan *Rhopilema nomadica* Galil, 1990 reaches the Tunisian coast of the Mediterranean Sea. *BioInvasions Records* 2(4), 319–323.

Daniel, B., Piro, S., Charbonnel, E., Francour, P. and Letourneur, Y. (2009) Lessepsian rabbitfish *Siganus luridus* reached the French Mediterranean coasts. *Cybium* 33(2), 163–164.

Deidun, A., Arrigo, S. and Piraino, S. (2011) The westernmost record of *Rhopilema nomadica* (Galil, 1990) in the Mediterranean – off the Maltese Islands. *Aquatic Invasions* 6(S1), S99–S103.

Diaz, J.H. (2015) Marine Scorpaenidae envenomation in travelers: epidemiology, management and prevention. *Journal of Travel Medicine* 22(4), 251–258.

Doğdu, S.A., Uyan, A., Uygur, N., Gürlek, M., Ergüden, D. and Turan, C. (2016) First record of the Indo-Pacific striped eel catfish, *Plotosus lineatus* (Thunberg, 1787) from Turkish marine waters. *Natural and Engineering Sciences* 1(2), 25–32.

Edelist, D., Spanier, E. and Golani, D. (2011) Evidence for the occurrence of the indo-Pacific stonefish, *Synancea verrucosa* (Actinopterygii: Scorpaeniformes: Synanceiidae), in the Mediterranean Sea. *Acta Ichthyologica et Piscatoria* 41(2), 129–131.

Eisenman, A., Rusetski, V., Sharivker, D., Yona, Z. and Golani, D. (2008) An odd pilgrim in the holyland. *American Journal of Emergency Medicine* 26(3), 383.e3–383.e6.

El-Sayed, M., Yacout, G.A., El-Samra, M., Ali, A. and Kotb, S.M. (2003) Toxicity of the Red Sea pufferfish *Pleuranacanthus sceleratus* 'El-Karad'. *Eco-toxicology and Environmental Safety* 56, 367–372.

Farrag, M.M.S., El Haweet, A.A.K., Akel, A. and Moustafa, M.A. (2015) Stock status of pufferfish *Lagocephalus sceleratus* (Gmelin, 1789) along the Egyptian coast, eastern Mediterranean Sea. *American Journal of Life Sciences* 3, 81–91.

Farrag, M.M.S., El Haweet, A.A.K., Akel, A. and Moustafa, M.A. (2016) Occurrence of puffer fishes (Tetraodontidae) in the eastern Mediterranean, Egyptian coast – filling in the gap. *BioInvasions Records* 5(1), 47–54.

Friedel, N., Scolnik, D., Adir, D. and Glatstein, M. (2016) Severe anaphylactic reaction to Mediterranean jellyfish (*Ropilhema nomadica*) envenomation: Case report. *Toxicology Reports* 3, 427–429.

Froese, R. and Pauly, D. (eds) (2016) FishBase. Available at: http://www.fishbase.org (accessed 19 December 2017).

Galil, B.S., Spanier, E. and Ferguson, W.W. (1990) The Scyphomedusae of the Mediterranean coast of Israel, including two Lessepsian migrants new to the Mediterranean. *Zoologische Mededelingen* 64, 95–105.

Galil, B.S., Marchini, A. and Occhipinti-Ambrogi, A. (2018) East is east and west is west? Management of marine bioinvasions in the Mediterranean Sea. *Estuarine, Coastal and Shelf Science* 201, 7–16 doi:10.1016/j.ecss.2015.12.021.

Galil, B.S., Marchini, A., Occhipinti-Ambrogi, A. and Ojaveer, H. (2017) The enlargement of the Suez Canal – Erythraean introductions and management challenges. *Management of Biological Invasions* 8, 141–152.

George, C.J. and Athanassiou, V. (1967) A two year study of the fishes appearing in the seine fishery of St George Bay, Lebanon. *Annali del Museo Civico di Storia Naturale di Genova* 79, 32–44.

Ghermandi, A., Galil, B., Gowdy, J. and Nunes, P.A.L.D. (2015) Jellyfish outbreak impacts on recreation in the Mediterranean Sea: welfare estimates from a socio-economic pilot survey in Israel. *Ecosystem Services* 11, 140–147.

Glatstein, M., Adir, D., Galil, B., Scolnik, D., Rimon, A., Pivko-Levy, R. and Hoyte, C. (2017) Pediatric enven-omation by *Rhopilema nomadica* jellyfish in the Mediterranean Sea. *European Journal of Emergency Medicine* doi: 10.1097/MEJ.0000000000000479.

Golani, D. (2002) The Indo-Pacific striped eel catfish, *Plotosus lineatus* (Thunberg, 1787), (Osteichtyes: Siluriformes) a new record from the Mediterranean. *Scientia Marina* 66 (3), 321–323.

Golani, D. and Sonin, O. (1992) New records of the Red Sea fishes, *Pterois miles* (Scorpaenidae) and *Pteragogus pelycus* (Labridae) from the eastern Mediterranean Sea. *Japanese Journal of Ichthyology* 39(2), 167–169.

Goren, M. and Dor, M. (1994) *An Updated Checklist of the Fishes of the Red Sea – CLOFRES II.* The Israel Academy of Sciences and Humanities, Jerusalem and Interuniversity Institute for Marine Sciences, Eilat.

Gweta, S., Spanier, E. and Bentur, Y. (2008) Venomous fish injuries along the Israeli Mediterranean coast: scope and characterization. *The Israel Medical Association Journal* 10(11), 783–788.

Halim, Y. and Rizkalla, S. (2011) Aliens in Egyptian Mediterranean waters. A check-list of Erythrean fish with new records. *Mediterranean Marine Science* 12(2), 479–490.

Halstead, B.W. (1988) *Poisonous and Venomous Marine Animals of the World.* The Darwin Press, Princeton, New Jersey, USA.

Herzberg, A. (1973) Toxicity of *Siganus luridus* (Rüppell) on the Mediterranean coast of Israel. *Aquaculture* 2, 89–91.

Iglésias, S.P. and Frotté, L. (2015) Alien marine fishes in Cyprus: update and new records. *Aquatic Invasions* 10(4), 425–438.

Islam, Q.T., Razzak, M.A., Islam, M.A., Bari, M.I., Basher, A. Chowdhury, F.R., Sayeduzzaman, A.B., Ahasan, H.A., Faiz, M.A., Arakawa, O., *et al.* (2011) Puffer fish poisoning in Bangladesh: clinical and toxicological results from large outbreaks in 2008. *Transactions of the Royal Society of Tropical Medicine and Hygiene* 105(2), 74–80.

Johnston, M.W. and Purkis, S.J. (2014) Are lionfish set for a Mediterranean invasion? Modelling explains why this is unlikely to occur. *Marine Pollution Bulletin* 88, 138–147.

Kan, S.K., Chan, M.K. and David, P. (1987) Nine fatal cases of puffer fish poisoning in Sabah, Malaysia. *Medical Journal of Malaysia* 42, 199–200.

Kanchanapongkul, J. (2001) Puffer fish poisoning: clinical features and management experience in 25 cases. *Journal of the Medical Association of Thailand* 84, 385–389.

Kara, M.H., Ben Lamine, E. and Francour, P. (2015) Range expansion of an invasive pufferfish, *Lagocephalus sceleratus* (Actinopterygii: Tetraodontiformes: Tetraodontidae), to the south-western Mediterranean. *Acta Ichthyologica et Piscatoria* 45(1), 103–108.

Katikou, P., Georgantelis, D., Sinouris, N., Petsi, A. and Fotaras, T. (2009) First report on toxicity assessment of the Lessepsian migrant pufferfish *Lagocephalus sceleratus* (Gmelin, 1789) from European waters (Aegean Sea, Greece). *Toxicon* 54(1), 50–55.

Katsanevakis, S., Acar, Ü., Ammar, I., Balci, B.A., Bekas, P., Belmonte, M., Chintiroglou, C.C., Consoli, P., Dimiza, M., Fryganiotis, K., *et al.* (2014) New Mediterranean biodiversity records (October, 2014). *Mediterranean Marine Science* 15(3), 675–695.

Kheifets, J., Rozhavsky, B., Girsh Solomonovich, Z., Rodman, M. and Soroksky, A. (2012) Severe Tetrodotoxin poisoning after consumption of *Lagocephalus sceleratus* (pufferfish, fugu) fished in Mediterranean Sea, treated with Cholinesterase inhibitor. *Case Reports in Critical Care* 2012, 782507.

Kletou, D., Hall-Spencer, J.M. and Kleitou, P. (2016) A lionfish (*Pterois miles*) invasion has begun in the Mediterranean Sea. *Marine Biodiversity Records* 9, 46.

Ktari, F. and Ktari, M.H. (1974) Présence dans le golfe de Gabes de *Siganus luridus* (Rüppell, 1829) et de *Siganus rivulatus* (Forsskal, 1775) (Poissons, siganides) parasites par *Pseudohaliotrematoides polymorphus*. *Bulletin de l'Institut National Scientifique et Technique d'Oceanographie et de Peche de Salammbo* 3, 95–98.

Ktari-Chakroun, F. and Bahloul, M. (1971) Capture de *Siganus luridus* (Rüppell) dans le golfe de Tunis. *Bulletin de l'Institut National Scientifique et Technique d'Oceanographie et de Peche de Salammbo* 2, 49–52.

Lee, J.Y.L., Teoh, L.C. and Leo, S.P.M. (2004) Stonefish envenomations of the hand – a local marine hazard: a series of 8 cases and review of the literature. *Annals Academy of Medicine* 33, 515–520.

Lessios, H.A., Kessing, B.D. and Pearse, J.S. (2001) Population structure and speciation in tropical seas: global phylogeography of the sea urchin *Diadema*. *Evolution* 55(5), 955–975.

Lion, R.M. (2004) Stonefish poisoning. *Wilderness and Environmental Medicine* 15, 284–288.

Liram, N., Gomori, M. and Perouansky, M. (2000) Sea urchin puncture resulting in PIP joint synovial arthritis: case report and MRI study. *Journal of Travel Medicine* 7(1), 43–45.

Louis-Francois, C., Mathoulin, C., Halbwachs, C., Grivois, J.-P., Bricaire, F. and Caumes, E. (2003) Complications cutanées des envenimations par poisson-pierre chez 6 voyageurs au retour de la région maritime indo-pacifique. *Bulletin de la Société de pathologie exotique* 96(5), 415–419.

Mazza, G., Tricarico, E., Genovesi, P. and Gherardi, F. (2014) Biological invaders are threats to human health: an overview. *Ethology Ecology & Evolution* 26(2–3), 112–129.

Menahem, S. and Shvartzman, P. (1995) Recurrent dermatitis from jellyfish envenomation. *Canadian Family Physician* 40, 2116–2118.

Morri, C., Puce, S., Bianchi, C.N., Bitar, G., Zibrowius, H. and Bavestrello, G. (2009) Hydroids (Cnidaria: Hydrozoa) from the Levant Sea (mainly Lebanon), with emphasis on alien species. *Journal of the Marine Biological Association of the United Kingdom* 89, 49–62.

Nader, M.R. and Indary, S.E. (2011) First record of *Diadema setosum* (Leske, 1778) (Echinodermata, Echinoidea, Diadematidae) from Lebanon, Eastern Mediterranean. *Aquatic Invasions* 6, S23.

Nader, M., Indary, S. and Boustany, L. (2012) FAO EastMed The pufferfish *Lagocephalus sceleratus* (Gmelin, 1789) in the eastern Mediterranean. GCP/INT/041/EC – GRE – ITATD – 10.

Ngo, A.S., Ong, J. and Ponampalam R. (2009) Stonefish envenomation presenting to a Singapore hospital. *Annals of Emergency Medicine* 54(3), S113.

Ojaveer, H., Galil, B.S., Campbell, M.L., Carlton, J.T., Canning-Clode, J., Cook, E.J., Davidson, A.D., Hewitt, C.L., Jelmert, A., Marchini, A. *et al.* (2015) Classification of non-indigenous species based on their impacts: considerations for application in marine management. *PLoS Biology* 13(4), e1002130.

Oray, I.K., Sınay, E., Saadet Karakulak, F. and Yıldız, T. (2015) An expected marine alien fish caught at the coast of Northern Cyprus: *Pterois miles* (Bennett, 1828). *Journal of Applied Ichthyology* 31, 733–735.

Öztürk, B. and İşinibilir, M. (2010) An alien jellyfish *Rhopilema nomadica* and its impacts to the Eastern Mediterranean part of Turkey. *Journal of Black Sea/Mediterranean Environment* 16(2), 149–156.

Prentice, O., Fernandez, W.G., Luyber, T.J., McMonicle, T.L. and Simmons, M.D. (2008) Stonefish envenomation. *American Journal of Emergency Medicine* 26(8), 972.e1–972.e2.

Raikhlin-Eisenkraft, B. and Bentur, Y. (2002) Rabbitfish ('Aras'): an unusual source of ciguatera poisoning. *The Israel Medical Association Journal* 4(1), 28–30.

Rees, W.J. and Vervoort, W. (1987) Hydroids from the John Murray Expedition to the Indian Ocean, with revisory notes on *Hydrodendron, Abietinella, Cryptolaria* and *Zygophylax* (Cnidaria: Hydrozoa). *Zoologische verhandelingen* 237, 1–209.

Rifkin, J.F., Fenner, P.J. and Williamson, J.A.H. (1993) First aid treatment of the sting from the hydroid *Lytocarpus philippinus*: the structure of, and *in vitro* discharge experiments with its nematocysts. *Wilderness & Environmental Medicine* 4(3), 252–260.

Rifkin, J.F., Williamson, J.A. and Fenner, P.J. (1996) Anthozoans, Hydrozoans and Scyphozoans. In: Williamson, J.A., Fenner, P.J., Burnett, J.W. and Rifkin, J.F. (eds) *Venomous and Poisonous Marine Animals: a Medical and Biological Handbook*. University of New South Wales Press, Sydney, Australia, pp. 180–235.

Sabrah, M.M., El-Ganainy, A.A. and Zaky, M.A (2006) Biology and toxicity of the pufferfish *Lagocephalus sceleratus* (Gmelin, 1789) from the Gulf of Suez. *Egyptian Journal of Aquatic Research* 32, 283–297.

Schindler, S., Staska, B., Adam, M., Rabitsch, W. and Essl, F. (2015) Alien species and public health impacts in Europe: a literature review. *NeoBiota* 27, 1–23.

Schofield, P. (2009) Geographic extent and chronology of the invasion of nonnative lionfish (*Pterois volitans* [Linnaeus 1758] and *P. miles* [Bennett 1828]) in the Western North Atlantic and Caribbean Sea. *Aquatic Invasions* 4, 473–479.

Sendovski, U., Goffman, M. and Goldshlak, L. (2005) Severe delayed cutaneous reaction due to Mediterranean jellyfish (*Rhopilema nomadica*) envenomation. *Contact Dermatitis* 52, 282–283.

Shiomi, K., Takamiya, M., Yamanaka, H., Kikuchi, T. and Konno, K. (1986) Hemolytic, lethal and edema-forming activities of the skin secretion from the oriental catfish (*Plotosus lineatus*). *Toxicon* 24(10), 1015–1018.

Shiomi, K., Takamiya, M., Yamanaka, H., Kikuchi, T. and Suzuki, Y. (1988) Toxins in the skin secretion of the Oriental catfish (*Plotosus lineatus*): immunological properties and immunocytochemical identification of producing cells. *Toxicon* 26(4), 353–361.

Silfen, R., Vilan, A., Wohl, I. and Leviav, A. (2003) Mediterranean jellyfish (*Rhopilema nomadica*) sting. *Burns* 29, 868–870.

Smith, J.L.B. (1951) A case of poisoning by the stonefish, *Synanceja verrucosa*. *Copeia* 207–210.

Smith, J.L.B. (1957) Two rapid fatalities from stonefish stabs. *Copeia* 249.

Steinitz, W. (1927) Beiträge zur Kenntnis der Küstenfauna Palästinas. I. *Pubblicazioni della Stazione Zoologica di Napoli* 8, 331–353.

Steinitz, H. (1959) Observations on *Pterois volitans* (L.) and its venom. *Copeia* 2, 158–160.

Stiasny, G. (1938) Die Scyphomedusen des Roten Meeres. Verhandelingen der Koninklijke. *Nederlandsche Akademie van wetenschappen te Natuurkunde* 37(2), 1–35.

Temraz, T.A. and Ben Souissi, J. (2013) First record of striped eel catfish *Plotosus lineatus* (Thunberg, 1787) from Egyptian waters of the Mediterranean. *CIESM* 2013, 604.

Toyoshima, T. (1918) Serological study of toxin of the fish *Plotosus anguillaris* lacepede. *Journal of the Japanese Protozoology Society* 6, 45–270.

Tsiamis, K., Aydogan, Ö., Bailly, N., Balistreri, P., Bariche, M., Carden-Noad, S., Corsini-Foka, M., Crocetta, F., Davidov, B., Dimitriadis, C. *et al.* (2015) New Mediterranean biodiversity records (July 2015). *Mediterranean Marine Science* 16(2), 472–488.

Turan, C. and Ozturk, B. (2015) First record of the lionfish *Pterois miles* (Bennett 1828) from the Aegean Sea. *Journal of Black Sea/Mediterranean Environment* 21(3), 334–338.

Turan, C., Erguden, D. and Uygur, N. (2011) On the occurrence of *Diadema setosum* (Leske, 1778) in Antakya Bay, Eastern Mediterranean Sea. *Journal of Black Sea/Mediterranean Environment* 17(1), 78–82.

Turan, C., Erguden, D., Gurlek, M., Yaglioglu, D., Uyan, A. and Uygur, N. (2014) First record of the Indo-Pacific lionfish *Pterois miles* (Bennett, 1828) (Osteichthyes: Scorpaenidae) for the Turkish marine waters. *Journal of Black Sea/Mediterranean Environment* 20(2), 158–163.

Vervoort, W. (1993) Report on hydroids (Hydrozoa, Cnidaria) in the collection of the Zoological Museum, University of Tel-Aviv, Israel. *Zoologische Mededelingen* 67(40), 537–565.

Vine, P. (1986) *Red Sea Safety Guide to Dangerous Marine Animals*. Immel Publishing, London.

Yoffe, B. and Baruchin, A.M. (2004) Mediterranean jellyfish (*Rhopilema nomadica*) sting. *Burns* 30, 503–504.

Yokes, B. and Galil, B.S. (2006) The first record of the needle-spined urchin *Diadema setosum* (Leske, 1778) (Echinodermata: Echinoidea: Diadematidae) from the Mediterranean Sea. *Aquatic Invasions* 1(3), 188–190.

2 Invasive Alien Plant Impacts on Human Health and Well-being

Lorenzo Lazzaro[1]*, Franz Essl[2], Antonella Lugliè[3], Bachisio Mario Padedda[3], Petr Pyšek[4,5] and Giuseppe Brundu[3]

[1]University of Florence, Italy; [2]University of Vienna, Austria; [3]University of Sassari, Italy; [4]The Czech Academy of Sciences, Institute of Botany, Průhonice, Czech Republic; and [5]Charles University, Prague, Czech Republic

Abstract

In this chapter we review, based on information available in scientific literature and reports, the most common negative direct impacts of invasive alien plants and Cyanobacteria on human health and well-being. Poisonous or toxic plants, i.e. plants containing toxic compounds, may impact human health generally after the ingestion of part of the plant (see *Nicotiana glauca*) or of some product derived from toxic plants (see *Senecio inaequidens* poisoned products). Allergenic plants are among the most studied cases of impacts of alien plants, particularly concerning the role of allergenic pollen. Many invasive species are well-known allergenic species (e.g. *Ambrosia artemisiifolia*). Some species may cause contact dermatitis, due to chemical compounds capable of causing adverse cutaneous reactions (e.g. *Heracleum mantegazzianum* and *Parthenium hysterophorus*). A significant cause of impact is linked to plants armed with thorns, spines and prickles. Injuries caused by these plants, and their consequences, represent a common clinical issue, sometimes leading also to dermatitis, infection and chronic wounds. Finally we include a review of human impacts related to Cyanobacteria and toxic algal blooms, highlighting the impacts related to cyanotoxins, which seriously impede water uses and environment, causing hazards to human health and sanitary alarm, with acute and chronic effects.

2.1 Introduction

Some invasive alien plants impact negatively on human well-being and health, both directly and indirectly. Direct negative impacts are often similar in both the native and invaded range, while indirect impacts are likely to cause more harm in the latter because they represent a novel element in existing ecological networks (e.g. Callaway *et al.*, 2012; Sun *et al.*, 2013). Invasive plants may disrupt established ecological conditions such as provision of resources for the native fauna, competitive or mutualistic relationships between species and plant community composition and vegetation structure as well as landscape, water and air quality (Pyšek and Richardson, 2010; Jones, 2017 and reference cited therein). The United

* E-mail: lorenzo.lazzaro@unifi.it

Nations Environment Program (2016) has stated that impacts on the environment by invasive alien species are 'a huge threat to human well-being' and the World Health Organization (2015) has warned that the ecological impacts of invasive alien species are an increasing threat to how people live and interact in their communities (Jones, 2017). Human health ultimately depends on biodiversity and the state of natural systems. Biological invasions can have wide-ranging impacts on human health that may be further exacerbated under climate change (Haines, 2016).

The Millennium Ecosystem Assessment (2005) illustrates the multiple links among biodiversity, ecosystem services and human well-being, and their importance. Indeed, through the benefits they provide, ecosystems are essential for safeguarding multiple dimensions of human well-being, such as the provision of goods and services that are needed for human survival and health (Cruz-Garcia et al., 2017). Recently, it has been shown that contact with natural environments such as parks and gardens fosters recovery from mental fatigue, enhances the ability to cope with and recover from stress, illness and injury, improves concentration and productivity in children, and is related to overall happiness and perceived well-being benefits (Doherty et al., 2014 and reference cited therein).

As briefly remarked above, invasive alien plants have also indirect impacts on ecosystem services, finally resulting in a loss of ecosystem services and interfering with human well-being (Vilà and Hulme, 2017). Among indirect negative impacts, one of the most common is certainly represented by alien aquatic plants offering shelter or favourable habitat for native or non-native pathogens or for their vectors.

Ecological consequences of lake invasion, for example by *Eichhornia crassipes* or *Pistia stratiotes*, including possible harmful algal blooms, oxygen depletion and fish kills, negatively impact people in terms of reduced ecosystem services and increased direct and indirect economic costs (Wilson and Carpenter, 1999). Algal blooms and fish kills create obvious impairments along coastlines that can detract from recreational activities, such as swimming. Low water clarity has been shown to have a substantial negative effect on housing values (Ara, 2007; Roy et al., 2010 and reference cited therein). Invasion by *Eichhornia crassipes* can provide suitable habitats for mosquitoes and snail species such as *Biomphalaria sudanica* and *B. choanomphala*, vectors of *Schistosoma mansoni*, thus favouring the spread of schistosomiasis in Africa (Mack et al., 2000; Plummer, 2005), as well as of malaria (Mazza et al., 2014).

Moving to direct impacts, although the impacts of alien species on human health have recently gained increasing attention in medical research and in invasion ecology, a recent literature review for Europe (Schindler et al., 2015) has shown how little literature is available on this issue. With regard to plant species, very few articles can be found directly linking alien plants to measurable impacts on health. In their intense literature search, Schindler et al. (2015) retrieved only 31 articles (four of which were literature reviews), mostly dealing with few well-known species such as common ragweed (*Ambrosia artemisiifolia*) and giant hogweed (*Heracleum mantegazzianum*). These data highlight the substantial gaps in the literature on human health impacts of alien species and how difficult it may be to make a systematic review on these topics.

Despite this, in this chapter we review, based on information available in scientific literature and reports, the most common negative direct impacts of invasive alien plants on human health and well-being. We also take into account possible impacts exerted by widespread species whose effects on human health are known at least in the native range. We include vascular and non-vascular terrestrial and aquatic alien plant species, as well as Cyanobacteria (Table 2.1). Negative impacts of plants on humans, including socio-economic aspects, are more thoroughly studied in terrestrial habitats, while those in freshwater ecosystems resulting from new introductions of microorganisms (Cyanobacteria, Algae) are comparatively poorly studied. This is probably linked to the difficulty in assessing the alien status of microorganisms in new habitats and in applying a standard terminology relating to

Table 2.1. Alien plant species with human health impacts mentioned in the chapter (see text for details and references). Range of known impacts: N = Native range; A = Alien range; A/N = both Alien and Native range. Nomenclature follows The Plant List online database (http://www.theplantlist.org/; Accessed October 2017).

Species	Family	Native range	Alien range	Type of impact of human health	Range of known impacts
Ambrosia artemisiifolia L.	Compositae	North and Central America	Africa; Asia; Australia; Europe	Allergenic	A/N
Ambrosia trifida L.	Compositae	North America;	Africa; Europe	Allergenic	A/N
Artemisia annua L.	Compositae	Asia	Europe; North America	Allergenic	A/N
Austrocylindropuntia subulata (Muehlenpf.) Backeb.	Cactaceae	South America	Africa; Australia; Europe	Thorns; spines	A/N
Conium maculatum L.	Apiaceae	Europe; Asia; Africa	Africa; America; Asia; Oceania	Poisonous species	A/N
Cupressus arizonica Greene	Cupressaceae	North America	Europe	Allergenic	A/N
Cupressus lusitanica Mill.	Cupressaceae	Central America	Europe	Allergenic	A/N
Cupressus macrocarpa Hartw.	Cupressaceae	North America	Europe	Allergenic	A/N
Cupressus sempervirens L.	Cupressaceae	Eastern Mediterranean region	Africa; Asia; Australia; Europe	Allergenic	A/N
Datura stramonium L.	Solanaceae	Central and South America	Africa; Asia; Europe; North America; Oceania	Poisonous species	A/N
Datura wrightii Regel	Solanaceae	North America	Asia; Southern Europe	Poisonous species	A/N
Datura ferox L.	Solanaceae	North America	Africa; Asia; Central and South America; Europe; Oceania	Poisonous species	A/N
Echium plantagineum L.	Boraginaceae	Mediterranean basin	Africa; America; Oceania	Poisonous species	A/N
Echium vulgare L.	Boraginaceae	Europe; Asia; China	Australia; North America	Poisonous species	A/N
Falcataria moluccana (Miq.) Barneby and J.W. Grimes	Leguminosae	Asia	Africa; America; Asia; Oceania	Thorns; spines	A/N
Heracleum mantegazzianum Sommier and Levier	Apiaceae	Eurasia (Caucasus)	Australia; Europe; North America	Dermatitis	A/N
Heracleum persicum Desf. ex Fisch., C.A. Mey. and Avé-Lall.	Apiaceae	Asia	Europe	Dermatitis	A/N
Heracleum sosnowskyi Manden.	Apiaceae	Eurasia (Caucasus)	Europe	Dermatitis	A/N
Iva xanthiifolia Nutt.	Compositae	North America	Asia; Europe; North America	Allergenic	A/N
Nerium oleander L.	Apocynaceae	Eurasia; Africa (Mediterranean basin)	Africa; Asia; Oceania; America	Poisonous species	A/N

continued

Table 2.1. *continued.*

Species	Family	Native range	Alien range	Type of impact of human health	Range of known impacts
Nicotiana glauca Graham	Solanaceae	South America	Africa; Asia; Australia; Europe; Central and North America	Poisonous species	A
Olea europaea L.	Oleaceae	Asia; Europe	Australia	Allergenic	A/N
Opuntia ficus-indica (L.) Mill.	Cactaceae	Central and South America	Africa; Asia; Europe; North America; Oceania	Thorns; spines	A/N
Opuntia monacantha Haw.	Cactaceae	South America	Asia; Africa; America; Europe; Oceania	Thorns; spines	A/N
Opuntia phaeacantha Engelm.	Cactaceae	Central and North America	Africa; Europe	Thorns; spines	A/N
Opuntia stricta (Haw.) Haw.	Cactaceae	America	Asia; Africa; Europe; Oceania	Thorns; spines	A/N
Parietaria judaica L.	Urticaceae	Mediterranean basin; Europe; Asia	Australia; North America	Allergenic	A/N
Parietaria officinalis L.	Urticaceae	Central-south Europe	Asia; Northern Europe	Allergenic	A/N
Parthenium hysterophorus L.	Compositae	America	Africa; Asia; Australia; Europe	Dermatitis	A/N
Phoenix canariensis Chabaud	Arecaceae	Canary Islands	Cosmopolitan in sub-tropical and temperate areas	Cutting leaves	A/N
Phoenix dactylifera L.	Arecaceae	Asia; Africa	America; Australia; Europe	Allergenic	A/N
Rhus typhina L.	Anacardiaceae	North America	Asia; Europe	Poisonous species	A/N
Ricinus communis L.	Euphorbiaceae	Tropical Asia; Tropical Africa	Africa; America; Europe	Poisonous species	A/N
Senecio inaequidens DC.	Compositae	Africa	Africa; Asia; Europe	Poisonous species	A/N
Senecio vulgaris L.	Compositae	Europe; Asia; Africa	America; Oceania	Poisonous species	A/N
Toxicodendron radicans (L.) Kuntze	Anacardiaceae	North America	Asia	Dermatitis	N
Zantedeschia aethiopica (L.) Spreng.	Araceae	Africa	Australia; Mediterranean basin; North America	Poisonous species	A/N

biological invasions and their impacts across different taxa (Wilk-Woźniak *et al.*, 2016; Bacher *et al.*, 2017). We do not consider alien ectomycorrhizal fungi, although they are increasingly recognized as invasive species. Invasive ectomycorrhizal fungi can be toxic to humans, may compete with native, edible or otherwise valuable fungi, facilitate the co-invasion of trees, and cause major changes in soil ecosystems, but also have positive effects, enabling plantation forestry and, in some cases, becoming a valuable food source (Dickie *et al.*, 2016).

2.2 Poisonous or Toxic Plants

While there are subtle differences among the terms toxic/toxicant (causing injury to a living organism resulting from chemical interaction), toxin (a poison of natural origin, or biotoxin) and poison (a chemical that may harm or kill an organism; see Flanagan *et al.*, 1995), they are commonly used interchangeably in the literature or by some organizations (e.g. by the IUPAC glossary of terms used in toxicology, IUPAC, 2007). These terms are considered as synonyms in the present review, because most of the literature sources considered in this work also did not distinguish between them.

Plants contain numerous compounds, among which those that are beneficial to humans are categorized as 'medicinal' and those that are harmful are termed as 'poisonous'. In addition, the concept of dose response, originally formulated by Paracelsus as 'solely the dose determines that a thing is not a poison' should not be overlooked (Borzelleca, 2000). In addition, the effects of a given plant species might vary strongly due to both the high variability of non-cultivated plants and the high variability of the potential target organism and individual metabolism (Mezzasalma *et al.*, 2017).

Products derived from plants, such as honey, milk and meat, can contain a number of chemical compounds that, depending on their concentration and application, can also be considered medicinal or poisonous (Jansen *et al.*, 2012; Burrows and Tyrl, 2013).

The World Health Organization (WHO) identifies four toxicity classes based on the median lethal dose (LD_{50}) determination in rats: class Ia, extremely hazardous ($LD_{50} \leq 5$ mg/kg bodyweight); class Ib, highly hazardous (5–50 mg/kg bodyweight); class II, moderately hazardous (50–500 mg/kg bodyweight); class III, slightly hazardous (>500 mg/kg bodyweight). Plants that fall into classes Ia and Ib are considered highly poisonous, those assigned to class II as poisonous and the remaining plants belonging to class III are the least poisonous (Wink and van Wyk, 2008). As remarked by Mezzasalma *et al.* (2017), a comprehensive list of worldwide poisonous plants is not available; however, some national or local institutions provide dedicated classifications. For example, the UK Horticultural Trades Association (HTA) proposed the adoption of a code of practice with a labelling system for ornamental species to inform consumers about their poisoning risk.

There are very well-known serious problems caused by the extreme toxicity of some plants, such as those possessing type 2 ribosome-inactivating proteins, most notably castor bean (*Ricinus communis* L.) and many species in the genus *Adenia* (Burrows and Tyrl, 2013 and references cited therein). Considerable information on the mechanisms of intoxication is emerging because of the interest in effects of plant toxicants as models for various human disease problems, such as Huntington's disease, amyotrophic lateral sclerosis (ALS), Alzheimer's disease and Parkinson's disease (Burrows and Tyrl, 2013 and references cited therein).

Clearly, negative impacts on human health of poisonous plants occur independently of their biogeographical status (alien vs native). However, poisonous or toxic alien plants might provide more harm in the invaded range because of ignorance of the negative effects, misidentification or confusion with other local non-toxic species or the formation of very dense and large stands (i.e. increasing the risk of exposure for humans). For example, Furer *et al.* (2011) reported two cases of rare human poisoning in one family in Israel following ingestion of cooked leaves from the tobacco tree plant (*Nicotiana glauca*, Fig. 2.1), after the young leaves of this alien species were mistaken for spinach. Anabasine ($C_{10}H_{14}N_2$), the toxic compound produced by *N. glauca*, is a pyridine alkaloid similar in both structure and effects to nicotine, but appearing to be more potent in humans (Furer *et al.*, 2011).

Many plants of the *Ericaceae* family, *Rhododendron*, *Pieris*, *Agarista* and *Kalmia*, contain diterpene grayanotoxins. Consumption of grayanotoxin-containing leaves, flowers or secondary products such as honey may result in intoxication specifically characterized by

Fig. 2.1. (a) and (c) *Senecio inaequidens* in its alien range (Gran Sasso-Monti della Laga National Park, Central Italy). (b) *Nicotiana glauca* and (d) *Datura stramonium* at Capraia island (Tuscan Archipelago National Park, Central Italy). (Lorenzo Lazzaro.)

dizziness, hypotension and atrial-ventricular block (Jansen *et al.*, 2012). Grayanotoxin-containing honey, called 'mad honey', can cause dramatic effects when ingested, as was recorded by the Greek warrior-writer Xenophon in 401 BC in his *Anabasis* (Jansen *et al.*, 2012).

Particular concern emerged recently in relation to potential negative effects on human health of pyrrolizidine alkaloids (PAs). These compounds have been reported as responsible for various poisoning cases in humans, including deaths, with the liver and lung as the main organs affected; moreover, concern is in particular associated with their genotoxicity and carcinogenicity (Jank and Rath, 2017). PAs are secondary metabolites naturally produced by many plants (approximately 6000 plant species, representing almost 2% of the world's flowering plants) typically as a defence against herbivory (EFSA, 2011). To date, about 660 different PAs have been described, mainly recorded in the plant families *Boraginaceae* (all genera), *Asteraceae* (tribes *Senecioneae* and *Eupatorieae*) and *Leguminosae* (genus *Crotalaria*) (EFSA, 2011). These molecules can cause different kinds of toxic effects reacting with DNA, amino acids and proteins. Acute poisoning can cause high mortality, whereas a sub-acute or chronic onset may lead to liver cirrhosis. In different animal models PAs induced DNA mutations (Frei *et al.*, 1992), teratological forms (i.e. congenital malformations or anomalies; Roeder, 1995) and cancer (Chen *et al.*, 2010). As a result the EFSA Panel of Contaminants in the Food Chain (CONTAM panel) (EFSA, 2011) concluded that 1,2-unsaturated PAs may act as carcinogens in humans.

The main risks associated with PAs linked to human health may arise from the use of plants in herbal 'remedies', herbal teas and folk medicines (Coulombe, 2003), and for honey production (Edgar *et al.*, 2002). This is particularly true for alien plants, whose toxic potential may be unknown in the invaded range, where these plants can be confused with similar native plants or used in food production. Many plants that contain PAs are well-known invasives. Several species of *Senecio* have been linked to human fatalities in incidents of bread poisoning, where seed or other plant parts were incorporated into bread for human consumption. Groundsel (*Senecio vulgaris*), native to Europe, Asia and northern Africa, is nowadays widespread in 169 regions on all continents except Antarctica (Pyšek *et al.*, 2017), in many of them invasive and infesting crops. Another well-known invasive alien species containing PAs is South African ragwort (*Senecio inaequidens*, Fig. 2.1), a herbaceous perennial plant that has spread rapidly in North and Central Europe following its accidental introduction from its native range of South Africa as a contaminant in wool (Scherber *et al.*, 2003). *Senecio inaequidens* is considered one of most invasive species in Europe (Lambdon *et al.*, 2008; Vacchiano *et al.*, 2013), rapidly spreading along roads and railways and into semi-natural ecosystems. As it has still not occupied all its potential niche in the invaded range (currently occurring mainly in sites close to anthropized areas), it is likely that it will spread into open grasslands and pastures in the near future leading to a high risk for cattle and human health (Vacchiano *et al.*, 2013).

Among most *Boraginaceae*, where all genera have representatives containing PAs, there are some well-known invasive plants. *Echium plantagineum* and *E. vulgare* were first introduced from their native range in Europe to Australia in the early 1800s. *Echium plantagineum* is now highly invasive, whereas *E. vulgare* has a limited distribution. In several studies, the presence of pyrrolizidine alkaloids in honey from these species has been well documented (Culvenor *et al.*, 1981; Edgar *et al.*, 2002; Beales *et al.*, 2004). In general, according to Edgar *et al.* (2002), as many pyrrolizidine alkaloid-containing plants are shown to represent a significant source of honey worldwide, contaminated honey may be a potential threat to health, especially for infants and foetuses, and further investigation is warranted.

A substantial number of alien plant species contain other toxic compounds that can harm humans who come into contact with plants, such as several species of *Datura* (mainly *D. stramonium* (Fig 2.1), *D. wrightii* and *D. ferox*), *Conium maculatum*, *Zanthedeschia aethiopica* and *Nerium oleander* (Pimentel, 2011; Burrows and Tyrl, 2013 and references cited therein). Many ornamental plants can be easily confused with other similar species that can have very different poisonous properties. A good example is rhus tree (*Rhus typhina*) which is very similar to Chinese pistachio (*Pistacia chinensis*). The rhus tree causes contact dermatitis and the smoke from burning the wood is toxic, but the similar-looking pistachio is safe.

2.3 Allergenic Plants

Allergenic pollen can be of prime importance in health problems caused by alien plants (Potter and Cadman, 1996; Belmonte and Vilà, 2004). Furthermore, allergenic airborne pollen can interact with other components of environmental global change, synergistically, increasing the risks to public health (Belmonte and Vilà, 2004). In Europe, pollen allergies have a great clinical impact (D'Amato *et al.*, 2007), with a plethora of allergic reactions including hay fever, asthma, allergic rhino-conjunctivitis and eczema. The antigens of allergenic pollen grains are rapidly released when allergen-carrying pollen comes into contact with the oral, nasal or eye mucosa, thereby inducing hay-fever symptoms in sensitized patients (D'Amato *et al.*, 2007). An important proportion of these allergies is caused by invasive plants.

Ragweed (*Ambrosia artemisiifolia*) is probably the best-studied example of an invasive plant producing highly allergenic pollen, causing symptoms in late summer and autumn (typically from August to October in the northern hemisphere) and reportedly inducing asthma about twice as often as other pollen types (Essl *et al.*, 2015 and references therein).

Ambrosia artemisiifolia is considered the main cause of hay fever and allergic rhinitis both in its native range in North America (Ziska *et al.*, 2011; Zhao *et al.*, 2016), and introduced range in Europe and temperate Asia (Smith *et al.*, 2013; Essl *et al.*, 2015). It produces pollen in enormous amounts; one individual plant may produce up to 1 billion pollen grains (Fumanal *et al.*, 2007) that may be transported over long distances (Mandrioli *et al.*, 1998).

The clinical relevance of ragweed effects on health has dramatically increased in Europe in recent decades (Burbach *et al.*, 2009). In Austria, ragweed pollen sensitization rates have increased from 8.5% to 17.5% (Hemmer *et al.*, 2011), and since 2007 ragweed pollen became the second most frequent cause of respiratory allergy around Milan in northern Italy (Asero, 2007). According to a recent European study by Bousquet *et al.* (2009), the sensitivity to ragweed pollen exceeds 66% of the sampled population (3034 patients). The importance of allergenic pollen may rise due to the presence of gas pollutants in the atmosphere (Ziska *et al.*, 2009) or as a consequence of climate change. Air pollution, in particular NOx or particulate matter, can influence the morphology of pollen and increase its allergenicity (Ring *et al.*, 2001). Climate change was reported to favour the spread of allergic alien plants (Plank *et al.*, 2016; Skálová *et al.*, 2017) and prolong their flowering period, increasing the production of pollen; for ragweed, climate warming has been shown to substantially increase pollen production and resulting pollen loads (Ziska *et al.*, 2011).

Plank *et al.* (2016) analysed potential socio-economic effects under climate change of three invasive alien plant species known to cause substantial harm to human health, especially in Central and Eastern Europe: giant ragweed (*Ambrosia trifida*), annual wormwood (*Artemisia annua*) and burweed marshelder (*Iva xanthiifolia*). In China, the pollen of *A. annua* is considered one of the most virulent allergens in autumnal hay fever (Liu *et al.*, 2010). The above study indicates that under different scenarios of climate change huge economic costs could be incurred unless effective action is taken to limit further spread of these species; early and coordinated response yields substantial net benefits under all scenarios (Plank *et al.*, 2016).

Other alien plants with a high allergenic potential include two European *Parietaria* species (Urticaceae), *Parietaria judaica* and *P. officinalis*, that are naturalized in many regions of the world. They are an important cause of allergic reactions both in their native (D'Amato *et al.*, 2007) and alien range, in the USA (Kaufman, 1990) and Australia (Pimentel, 2011).

The cypress family (*Cupressaceae*) also contains many widespread ornamental and forest species producing highly allergenic pollen that is a source of increasing pollinosis in western Mediterranean countries (France, Italy, Spain). These species are also responsible for winter pollinosis in a period of the year when no other allergenic plants are flowering (D'Amato *et al.*, 2007). The most common species are *Cupressus sempervirens* (native to the eastern Mediterranean), *C. arizonica*, *C. macrocarpa* and *C. lusitanica* (native to North America). Many other species have been identified as major sources of pollen responsible for allergenic reactions, some widely planted, such as grasses (especially *Lolium* spp.), olives (*Olea europea*), poplars (*Populus* spp.) (Pimentel, 2011) and date palms (*Phoenix dactylifera*) (Mistrello *et al.*, 2008).

2.4 Dermatitis

Plants are of relevance to dermatology both for their adverse and beneficial effects on skin (Otang *et al.*, 2014). Cutaneous adverse effects induced by plants include: irritant contact dermatitis due to mechanical injury or by irritant chemicals in the plant sap, phytophotodermatitis resulting from skin contamination by plants containing furocoumarins and subsequent exposure to UV light and immediate (type-I) or delayed hypersensitivity contact reactions mediated by the immune system in individuals sensitized to plants or plant products, e.g. poison ivy (*Toxicodendron radicans*) poisoning (Otang *et al.*, 2014). Touching or handling certain plant species can be a serious health hazard. Hence, numerous occupations such as food handlers, caterers, gardeners, farmers, agricultural workers, florists, nursery workers, landscapers, forestry workers and loggers are exposed to the risk of developing adverse cutaneous reactions to plants and woods (Otang *et al.*, 2014). Giant hogweed (*Heracleum mantegazzianum*) is an invasive alien plant that was introduced from the Caucasus to Europe and North America in the early 19th century (Jahodová *et al.*, 2007; Pyšek *et al.*, 2007; Pfurtscheller and Trop, 2014) and spread rapidly across these continents (Pyšek *et al.*, 2008). Its impressive appearance, with a height of up to four metres, and its large leaves make it an attractive plant, also for kids to play with. Furocoumarin derivatives (psoralens and methoxypsoralens), which are in essentially all parts of the plant, are responsible for its harmful properties. After contact with unprotected skin and exposure to sunlight with its long-wave UV radiation, these psoralens are activated and induce inflammation and cell membrane damage by binding to RNA and nuclear DNA (Pfurtscheller and Trop, 2014). This phototoxic reaction leads to erythema and dermatitis that peak after up to 48 hours and are intensified by prolonged exposure to sunlight and moist conditions, such as swimming or bathing after contact. Inhalation of traces of hogweed can also induce obstructive pulmonary symptoms, and eye contact can result in blindness (Pfurtscheller and Trop, 2014). Intense contact with the plant and prolonged exposure to sunlight thereafter can result in a major burn-like wound.

Jahodová *et al.* (2007) demonstrated that within the genus *Heracleum* there are two further closely related species invading Europe, i.e. *Heracleum sosnowskyi*, spreading rapidly in the Baltic States and Russia after its introduction as a fodder plant, and *H. persicum*, which invaded Scandinavia after its introduction as an ornamental plant. All these species contain the photosensitizing furanocoumarins responsible for the toxic reaction after contact with human skin and in combination with ultraviolet radiation (EPPO, 2009).

Parthenium weed (*Parthenium hysterophorus*), native to Central America, spread rapidly in many tropical and subtropical regions (Shrestha *et al.*, 2015). Humans with continued exposure to *P. hysterophorus* can develop allergic eczematous contact dermatitis. Patients with severe dermatitis suffer fatigue and weight loss and severely affected people can die. The pollen of the plant is also allergenic. Cross-sensitivity with other plants, particularly other members of the *Asteraceae* family, may occur, causing allergic reactions to plants to which humans were previously not sensitive (EPPO, 2014).

2.5 Thorns, Spines, Silicates and Other Physical Harms

Thorns, spines and prickles, here collectively termed thorns, are a common anti-herbivory defence in thousands of plant species, especially in arid regions (Grubb, 1992; Halpern *et al.*, 2007). Thorny plants such as cacti, *Agave*, *Euphorbia* and *Aloe* have colourful spines with red, orange, yellow, black or brown colours that may deter herbivores. Thorns may cause infections in animals and humans that may become much more dangerous and painful than the mechanical wounding itself (Halpern *et al.*, 2007 and references cited therein). Among cacti, the genus *Opuntia* account for several species being widely introduced and disseminated around the world, with many invasive (*O. phaeacantha*, *O. ficus-indica*, *O. monacantha* and *O. stricta*, to name but a few) (Novoa *et al.*, 2015). *Opuntia* spp. are well known for their spines or glochidia, which can easily puncture the skin with only minor pressure (i.e. bumping or touching the cactus, but even transported by the wind). Cactus spine injuries represent a serious clinical issue, linked to dermatitis, infection, chronic wounds and granulomas (Lindsey and Lindsey, 1988), with cases reported for both the native (Dieter *et al.*, 2017) and the alien range (Doctoroff *et al.*, 2000). Moreover *Opuntia ficus-indica* thorns may induce arthritis, i.e. typically a monoarticular inflammatory reaction to penetrating plant thorn material. The clinical spectrum is variable and may include chronic monoarthritis, bursitis, tenosynovitis or soft tissue cysts, depending on the site of thorn lodgement (Miller *et al.*, 2000). Many other opuntioid cacti covered in spines (for example, *Austrocylindropuntia subulata* or other *Cylindropuntia* species) are quite popular in cultivation and often widely diffused as alien species (Fig. 2.2).

Many palms commonly used in tropical or subtropical gardens have sharp, annoying spines, thorns, teeth or simply very sharp leaves. These palms are therefore potentially unsafe when planted in public gardens and recreational areas and can be risky to prune or plant. In particular, bodily injuries due to plant thorns are extremely common and frequently involve children (Sugarman *et al.*, 1977). Limbs and joints, especially the hands and feet, are prone to these injuries. Phoenix date palm (*Phoenix canariensis*) develops sharp spines along its fronds which become brittle on drying. These thorns easily puncture and penetrate soft tissue leaving less than a 1–2 mm bloodless puncture site that may harbour a 20–30 mm spike at a variable distance from the site of entry. When the nature of the initial injury is not recognized, a typical inflammatory response can develop. Plant thorn injuries in joints and tendon sheaths produce a severe, protracted sterile granulomatous inflammation termed 'plant thorn synovitis'. When the injury is adjacent to bone, the inflammatory response can lead to 'pseudotumour' formation (Adams *et al.*, 2000; Lozano-Moreno, 2005). Pathogenic bacteria like *Clostridium perfringens*, the causative agent of the life-threatening gas gangrene, and others, were isolated and identified from date palm by Halpern *et al.* (2007). As a result, Adams *et al.* (2000) suggest caution against the planting of *Phoenix* date palms in positions where young children have ready access, such as schools, playgrounds and gardens.

Fig. 2.2.   Some cacti in their alien range: (a) *Austrocylindropuntia subulata*, (b) *Opuntia phaeacantha* (Tuscan Archipelago National Park, Central Italy) and (c) *Opuntia stricta* (Florence, Central Italy). (Lorenzo Lazzaro.)

The large ornamental tree batai wood (*Falcataria moluccana*) has become invasive in forests and in the landscapes of many Pacific islands (Hughes *et al.*, 2011). A fast-growing nitrogen-fixing species, it transforms invaded ecosystems by dramatically increasing nutrient inputs, suppressing native species and facilitating invasion by other weeds. Individuals rapidly reach heights of 35 m, and their massive limbs break easily in storms or with age, creating significant hazards in residential areas and across infrastructure corridors such as roads and power lines. Their management is extremely costly for landowners, utility companies and local governments, since removal of hazardous trees can cost several thousand dollars each (Hughes *et al.*, 2011). Coconut palms are an integral part of life in many areas, given the widespread dependence of subsistence agriculture. Injuries related to the coconut palm are thus inevitable, such as brain injury, which might occur secondary to a falling coconut fruit (Mulford *et al.*, 2001). Parents and young children must be warned of the dangers of playing beneath coconut trees.

2.6 Cyanobacteria and Toxic Algal Blooms

Cyanobacteria, a group of photo-oxygenic microorganisms, are common inhabitants of aquatic (fresh, brackish, marine) and terrestrial environments throughout the world. They are the most competitive and ubiquitous organisms among phytoplankton (Paerl and Otten, 2013), which is the most important primary producer in aquatic ecosystems. Phytoplankton has a fundamental role in the food web and biogeochemical cycles in aquatic environments (Kaiser and Williams, 2011), and its natural ability to produce large proliferations (blooms) is considered a benefit for the secondary productions in natural environments and aquaculture. Nevertheless, these events can also cause negative impacts on humans and biodiversity. This type of bloom is known as harmful algal blooms (HABs). HABs are a complex phenomenon caused by a limited number of taxa, considering the total number of phytoplankton species (Smayda, 1997; Zingone and Oksfeldt Enevoldsen, 2000). They have been known since ancient times, but have increased strongly in recent decades (Carmichael, 2001; Anderson, 2009), a trend attributed to eutrophication and climate change, among the most important drivers, and assumed to become even more pronounced in the future (Heisler *et al.*, 2008; Paerl and Huisman, 2009; O'Neil *et al.*, 2012).

The dispersal of Cyanobacteria to new areas can be due to human activities, migrating animals or winds (Sukenik *et al.*, 2015), and their establishment and growth are favoured by their competitiveness, depending on environmental conditions (e.g. nutrients availability, high temperatures, solar radiation intensity, stability of water column and pH) (Paerl and Huisman, 2008; Salmaso *et al.*, 2012; Paerl and Otten, 2013) and specific biological traits (e.g. regulation of water column position by mean of gas vacuoles, ability to overcome adverse periods forming resting cells and/or to directly fix N_2).

The geographical expansion of harmful species of Cyanobacteria was recently discussed in the context of biological invasion of bloom-forming species of *Nostocales*, from tropical to temperate freshwater environments (Paerl and Paul, 2012; Sukenik *et al.*, 2015). The invasion of aquatic free-living microorganisms, including Cyanobacteria and other autotrophic and heterotrophic representatives, is rather cryptic and difficult to detect (Sukenik *et al.*, 2015). The methodological difficulties in the assessment of the total microbial diversity may have been at the origin of the extremely low number of proven invaders among phytoplankton species (Wyatt and Carlton, 2002). Furthermore, information on prior and current composition of the microbial communities, which is required to determine if a given microorganism is an invader, is often missing (Sukenik *et al.*, 2015). In addition, the presence of resting stages in the life cycles of several Cyanobacteria, similar to other microalgae (e.g. dinoflagellates, diatoms and raphidophytes) which alternate phases of vegetative planktonic and benthic life stages, highlights the continuum among habitats (pelagic and benthic) and the necessity of an integrated water/sediment approach in their study (Satta *et al.*, 2014; Corriero *et al.*, 2016).

Among all the different issues linked to Cyanobacteria, such as altered water quality, the most serious is the ability of ~40 species in 10 genera (Žegura *et al.*, 2011) to produce a great variety of toxic compounds (cyanotoxins) (Messineo *et al.*, 2009). These cyanotoxins seriously impede water uses and environment, causing hazards to human health and sanitary alarm (Hudnell, 2008), with acute and chronic effects (Codd *et al.*, 2005; Metcalf and Codd, 2009, 2012). It is estimated that more than 50% of CyanoHABs are toxic, causing economic losses in addition to the health threats for ecosystems and humans (Codd *et al.*, 1999, 2005; Vasconcelos, 2006).

Cyanotoxins fall into three groups on the basis of their chemical structure (cyclic peptides, alkaloids and lipopolysaccharides) and can be classified into categories that reflect their biological effects on the systems and organs that they affect most strongly (hepatotoxins, neurotoxins, cytotoxins and dermatotoxins; Codd *et al.*, 2005). Cyanotoxin poisoning in humans is mainly caused by three toxic groups: microcystins (MCs), cylindrospermopsins (CYN) and anatoxin-a (ANA-a) (Messineo *et al.*, 2009) and occurs through exposure to contaminated drinking water supplies (Annadotter *et al.*, 2001; Falconer, 2005), recreational waters (Chorus and Bartram, 1999; Behm, 2003) and medical dialysis (Azevedo *et al.*, 2002).

Cylindrospermopsis raciborskii and *Chrysosporum (Aphanizomenon) ovalisporum* (Fig. 2.3) are considered alien species in many water bodies of subtropical and temperate areas (Sukenik *et al.*, 2015). They produce CYN that are alkaloids with multiple organ targets (Codd *et al.*, 2005). In addition, *C. raciborskii* also produces paralytic shellfish poisoning toxins (saxitoxins). *C. raciborskii* is a very abundant pantropical species that expanded rapidly to Europe in the last century and is considered an invasive alien species of the temperate zone (Padisák, 1997; Wilk-Woźniak *et al.*, 2016). It represents a significant health risk for human populations when present in drinking water reservoirs (Kaštovský *et al.*, 2010). The first report of human health impact was in Palm Island, Australia (Bourke *et al.*, 1983). A severe outbreak of hepatoenteritis, requiring hospital treatment of more than 140 people, was attributed to this species, which was found in the lake that supplied drinking water.

2.7 Conclusions

Some invasive alien plants negatively impact human well-being and health, both directly and indirectly. According to the more recent and agreed definitions, health 'is a state of complete physical, mental and social well-being and not merely the absence of disease or infirmity' (World Health Organization, 2015). Health status has important social, economic, behavioural and environmental determinants and wide-ranging impacts. Typically, health has been viewed largely in a human-only context. However, there is increasing recognition of the broader health concept that encompasses other species, our ecosystems and the integral ecological underpinnings of many drivers or protectors of health risks (World Health Organization, 2015).

In the present review we focus on the most common direct negative impacts from invasive alien plants and Cyanobacteria. The former group, in particular, have quite a large array of negative impacts across many families and genera. As our research was based on the available literature there might be many other non-reported high-risk species.

Although invasive alien plants are primarily defined as those producing negative impacts on biodiversity, e.g. in the context of the Convention on Biological Diversity, there is a growing concern about the additional direct negative impacts they can have on human health as well as those that can affect ecosystem services and consequently human well-being. For

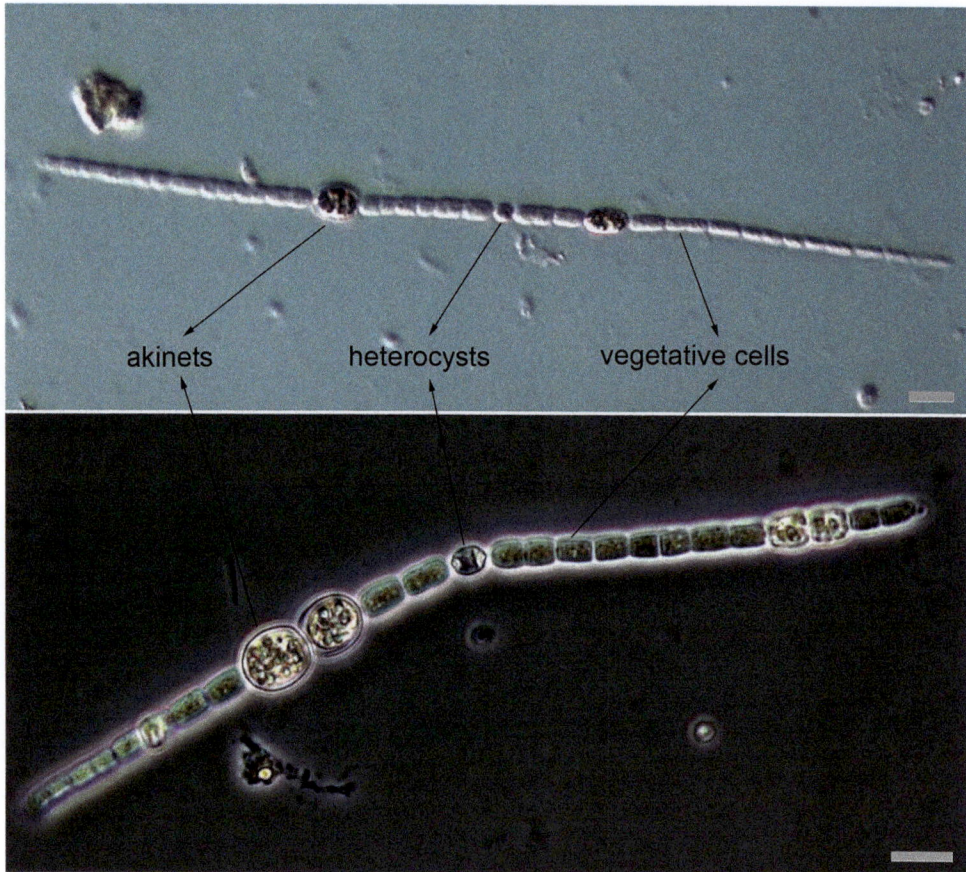

Fig. 2.3. Trichomes of *Chrysosporum* (*Aphanizomenon*) *ovalisporum* showing vegetative cells, heterocysts and akinets (Lake Temo, Sardinia, Italy; 14 October 2004). Scale bars indicate 10 µm. (Antonella Lugliè and Paola Casiddu.)

example, the European Regulation no. 1143/2014 (European Commission, 2014) stresses that invasive alien species can also have a significant adverse impact on human health and the economy. This is in fact considered an aggravating factor when selecting invasive alien species to be included in the List of Invasive Alien Species of Union Concern.

Therefore, the risk of possible negative impacts on human well-being and health must always be taken into account during pre-entry risk assessment schemes, before deciding to introduce a new alien species. It is worth mentioning that historically humans have always used, and introduced, alien plants for many purposes, such as a food source or medicinal plants. It follows that alien plants may exert a plethora of effects on human well-being ranging from positive to negative ones, and both cases have to be carefully evaluated and balanced.

Acknowledgement

We would like to thank Phil Hulme for his insightful comments on the manuscript, which greatly improved the final version of this chapter.

References

Adams, C.D., Timms, F.J. and Hanlon, M. (2000) Phoenix date palm injuries: a review of injuries from the Phoenix date palm treated at the Starship Children's Hospital. *The Australian and New Zealand Journal of Surgery* 70, 355–357.

Anderson, D.M., (2009) Approaches to monitoring, control and management of harmful algal blooms (HABs). *Ocean and Coastal Management* 52, 342–347.

Annadotter, H., Cronberg, G., Lawton, L.A., Hansson, H.B., Gothe, U. and Skulberg, O.M. (2001) An extensive outbreak of gastroenteritis associated with the toxic cyanobacterium *Planktothrix aghardii* (Oscillatoriales, Cyanophyceae) in Scania, South Sweden. In: Chorus, I. (ed.) *Cyanotoxins – Occurrence, Causes, Consequences*. Springer, Berlin, pp. 200–208.

Ara, S. (2007) The influence of water quality on the demand for residential development around Lake Erie. Dissertation. The Ohio State University, Columbus, Ohio.

Asero, R. (2007) The changing pattern of ragweed allergy in the area of Milan, Italy. *Allergy* 62(9), 1097–1099.

Azevedo, S.M.F.O., Carmichael, W.W., Jochimsen, E.M., Rinehart, K.L., Lau, S., Shaw, G.R. and Eaglesham, G.K. (2002) Human intoxication by microcystins during renal dialysis treatment in Caruaru-Brazil. *Toxicology* 181–182, 441–446.

Bacher, S., Blackburn, T.M., Essl F., Genovesi, P., Heikkilä, J., Jeschke, J.M., Jones, G., Keller, R., Kenis, M., Kueffer, C. *et al.* (2017) Socio-economic impact classification of alien taxa (SEICAT). *Methods in Ecology and Evolution* doi: 10.1111/2041-210X.12844.

Beales, K.A., Betteridge, K., Colegate, S.M. and Edgar, J.A. (2004) Solidphase extraction and LCMS analysis of pyrrolizidine alkaloids in honeys. *Journal of Agricultural and Food Chemistry* 52, 6664–6672.

Behm, D. (2003) Coroner cites algae in teen's death. *Milwaukee Journal Sentinel*, 5 September.

Belmonte, J. and Vilà, M. (2004) Atmospheric invasion of non-native pollen in the Mediterranean region. *American Journal of Botany* 91(8), 1243–1250.

Borzelleca, J.F. (2000) Paracelsus: herald of modern toxicology. *Toxicological Sciences* 53(1), 2–4.

Bourke, A.T.C., Hawes, R.B., Neilson, A. and Stallman, N.D. (1983) An outbreak of hepato-enteritis (the Palm Island mystery disease) possibly caused by algal intoxication. *Toxicon* 21, 45–48.

Bousquet, P.J., Burbach, G., Heinzerling, L.M., Edenharter, G., Bachert, C., Bindslev-Jensen, C., Bonini, S., Bousquet-Rouanet, L., Demoly, P., Bresciani, M. *et al.* (2009) GA2LEN skin test study III: minimum battery of test inhalent allergens needed in epidemiological studies in patients. *Allergy* 64(11), 1656–1662.

Burbach, G.J., Heinzerling, L.M., Röhnelt, C., Bergmann, K.C., Behrendt, H. and Zuberbier, T. (2009) Ragweed sensitization in Europe – GA2LEN study suggests increasing prevalence. *Allergy* 64(4), 664–665.

Burrows, G.E. and Tyrl, R.J. (2013) *Toxic Plants of North America*, 2nd edn. Wiley-Blackwell, Oxford, UK.

Callaway, R.M., Schaffner, U., Thelen, G.C., Khamraev, A., Juginisov, T. and Maron, J.L. (2012) Impact of *Acroptilon repens* on co-occurring native plants is greater in the invader's non-native range. *Biological Invasions* 14, 1143–1155.

Carmichael, W.W. (2001) Health effects of toxin-producing cyanobacteria: 'The CyanoHABs'. *Human and Ecological Risk Assessment: An International Journal* 7, 1393–1407.

Chen, T., Mei, N. and Fu, P.P. (2010) Genotoxicity of pyrrolizidine alkaloids. *Journal of Applied Toxicology* 30(3), 183–196.

Chorus, I. and Bartram, J. (1999) *Toxic Cyanobacteria in Water*. World Health Organization, E & FN Spon, London.

Codd, G., Bell, S., Kaya, K., Ward, C., Beattie, K. and Metcalf, J. (1999) Cyanobacterial toxins, exposure routes and human health. *European Journal of Phycology* 34, 405–415.

Codd, G.A., Morrison, L.F. and Metcalf, J.S. (2005) Cyanobacterial toxins: risk management for health protection. *Toxicology and Applied Pharmacology* 203, 264–272.

Corriero, G., Pierri, C., Accoroni, S., Alabiso, G., Bavestrello, G., Barbone, E., Bastianini, M., Bazzoni, A.M., Bernardi Aubry, F., Boero, F. *et al.* (2016) Ecosystem vulnerability to alien and invasive species: a case study on marine habitats along the Italian coast. *Aquatic Conservation: Marine and Freshwater Ecosystems* 26, 392–409.

Coulombe, R.A. (2003) Pyrrolizidine alkaloids in foods. *Advances in Food and Nutrition Research* 45, 61–99.

Cruz-Garcia, G.S., Sachet, E., Blundo-Canto, G., Vanegas, M. and Quintero, M. (2017) To what extent have the links between ecosystem services and human well-being been researched in Africa, Asia, and Latin America? *Ecosystem Services* 25, 201–212.

Culvenor, C.C.J., Edgar, J.A. and Smith, L.W. (1981) Pyrrolizidine alkaloids in honey from *Echium plantagineum* L. *Journal of Agricultural and Food Chemistry* 29, 958–960.

D'Amato, G., Cecchi, L., Bonini, S., Nunes, C., Annesi-Maesano, I., Behrendt, H., Liccardi, G., Popov, T. and Van Cauwenberge, P. (2007) Allergenic pollen and pollen allergy in Europe. *Allergy* 62(9), 976–990.

Dickie, I.A., Nuñez, M.A., Pringle, A., Lebel, T., Tourtellot, S.G. and Johnston, P.R. (2016) Towards management of invasive ectomycorrhizal fungi. *Biological Invasions* 18(12), 3383–3395.

Dieter, R.A. Jr, Whitehouse, L.R. and Gulliver, R. (2017) Cactus spine wounds: a case report and short review of the literature. *Wounds: a Compendium of Clinical Research and Practice* 29(2), E18.

Doctoroff, A., Vidimos, A.T. and Taylor, J.S. (2000) Cactus skin injuries. *Cutis* 65(5), 290–292.

Doherty, S.T., Lemieux, C.J. and Canally, C. (2014) Tracking human activity and well-being in natural environments using wearable sensors and experience sampling. *Social Science and Medicine* 106, 83–92.

Edgar, J.A., Roeder, E. and Molyneux, R.J. (2002) Honey from plants containing pyrrolizidine alkaloids: a potential threat to health. *Journal of Agricultural and Food Chemistry* 50, 2719–2730.

EFSA (2011) EFSA Panel on Contaminants in the Food Chain (CONTAM); Scientific opinion on pyrrolizidine alkaloids in food and feed. *EFSA Journal* 9(11), 2406.

EPPO (2009) *Heracleum mantegazzianum, Heracleum sosnowskyi* and *Heracleum persicum*. *EPPO Bulletin* 39, 489–499.

EPPO (2014) *Parthenium hysterophorus* L. Asteraceae – Parthenium weed. Data sheets on invasive alien plants. *Bulletin OEPP/EPPO Bulletin* 44(3), 474–478.

Essl, F., Biró, K., Brandes, D., Broennimann, O., Bullock, J.M., Chapman, D.S., Chauvel, B., Dullinger, S., Fumanal, B., Guisan, A. *et al.* (2015) Biological flora of the British Isles: *Ambrosia artemisiifolia*. *Journal of Ecology* 103, 1069–1098.

European Commission (2014) Regulation (EU) No 1143/2014 of the European Parliament and of the Council of 22 October 2014 on the prevention and management of the introduction and spread of invasive alien species. *Official Journal of the European Union* 57, 35–55.

Falconer, I.R. (2005) *Cyanobacterial Toxins of Drinking Water Supplies: Cylindrospermopsins and Microcystins*. CRC Press, Boca Raton, Florida, USA.

Flanagan, R.J., Braithwaite, R.A., Brown, S.S., Widdop, B. and de Wolff, F.A. (1995) *Basic Analytical Toxicology*. World Health Organization, Geneva, Switzerland.

Frei, H., Lüthy, J., Brauchli, J., Zweifel, U., Würgler, F.E. and Schlatter, C. (1992) Structure/activity relationships of the genotoxic potencies of sixteen pyrrolizidine alkaloids assayed for the induction of somatic mutation and recombination in wing cells of *Drosophila melanogaster*. *Chemico-Biological Interactions* 83(1), 1–22.

Fumanal, B., Chauvel, B. and Bretagnolle, F. (2007) Estimation of pollen and seed production of common ragweed in France. *Annals of Agricultural and Environmental Medicine* 14, 233–236.

Furer, V., Hersch, M., Silvetzki, N., Breuer, G.S. and Zevin, S. (2011) *Nicotiana glauca* (tree tobacco) intoxication: two cases in one family. *Journal of Medical Toxicology* 7, 47–51.

Grubb, P.J. (1992) A positive distrust in simplicity – lessons from plant defences and from competition among plants and among animals. *Journal of Ecology* 80, 585–610.

Haines, A. (2016) Addressing challenges to human health in the Anthropocene epoch: an overview of the findings of the Rockefeller/Lancet Commission on Planetary Health. *Public Health Reviews* 37, 14.

Halpern, M., Raats, D. and Lev-Yadun, S. (2007) Plant biological warfare: thorns inject pathogenic bacteria into herbivores. *Environmental Microbiology* 9(3), 584–592.

Heisler, J., Glibert, P., Burkholder, J., Anderson, D., Cochlan, W., Dennison, W., Dortch, Q., Gobler, C., Heil, C., Humphries, E. *et al.* (2008) Eutrophication and harmful algal blooms: a scientific consensus. *Harmful Algae* 8, 3–13.

Hemmer, W., Schauer, U., Trinca, A., Neumann, C. and Jarisch, R. (2011) Ragweed pollen allergy in Austria: a retrospective analysis of sensitization rates from 1997 to 2007. *Journal of Allergy and Clinical Immunology* 127(2), AB170.

Hudnell, H.K. (2008) *Cyanobacterial Harmful Algal Blooms: State of the Science and Research Needs. Advances in Experimental Medicine and Biology*. Springer, New York.

Hughes, R.F., Johnson, M.T. and Uowolo, A. (2011) The invasive alien tree *Falcataria moluccana*: its impacts and management. *XIII International Symposium on Biological Control of Weeds – 2011*, pp. 218–223.

IUPAC (2007) IUPAC glossary of terms used in toxicology, 2nd edition – IUPAC recommendations 2007. *Pure and Applied Chemistry* 79(7), 1153–1344.

Jahodová, Š., Trybush, S., Pyšek, P., Wade, M. and Karp, A. (2007) Invasive species of *Heracleum* in Europe: an insight into genetic relationships and invasion history. *Diversity and Distributions* 13, 99–114.

Jank, B. and Rath, J. (2017) The risk of pyrrolizidine alkaloids in human food and animal feed. *Trends in Plant Science* 22(3), 191–193.

Jansen, S.A., Kleerekooper, I., Hofman, Z.L.M., Kappen, I.F.P.M., Stary-Weinzinger, A. and van der Heyden, M.A.G. (2012) Grayanotoxin poisoning: 'mad honey disease' and beyond. *Cardiovascular Toxicology* 12, 208–215.

Jones, B.A. (2017) Invasive species impacts on human well-being using the life satisfaction index. *Ecological Economics* 134, 250–257.

Kaiser, M. and Williams, P. (2011) *Marine Ecology: Processes, Systems, and Impacts.* Oxford University Press, Oxford, UK.

Kaštovský, J., Hauer, T., Mareš, J., Krautová, M., Bešta, T., Komárek, J., Desortová, B., Heteša, J., Hindáková, A., Houk, V. *et al.* (2010) A review of the alien and expansive species of freshwater cyanobacteria and algae in the Czech Republic. *Biological Invasions* 12, 3599–3625.

Kaufman, H.S. (1990) *Parietaria* an unrecognized cause of respiratory allergy in the United States. *Annals of Allergy* 64, 293–296.

Lambdon, P.W., Pyšek, P., Basnou, C., Hejda, M., Arianoutsou, M., Essl, F., Jarošík, V., Pergl, J., Winter, M., Anastasiu, P. *et al.* (2008) Alien flora of Europe: species diversity, temporal trends, geographical patterns and research needs. *Preslia* 80, 101–149.

Lindsey, D. and Lindsey, W.E. (1988) Cactus spine injuries. *The American Journal of Emergency Medicine* 6(4), 362–369.

Liu, A.H., Jaramillo, R., Sicherer, S.H., Wood, R.A., Bock, S.A., Burks, A.W., Massing, M., Cohn, R.D. and Zeldin, D.C. (2010) National prevalence and risk factors for food allergy and relationship to asthma: results from the National Health and Nutrition Examination Survey 2005–2006. *Journal of Allergy and Clinical Immunology* 126(4), 798–806.

Lozano-Moreno, F.J. (2005) Late complications after a palm thorn injury to the hand. *European Journal of Orthopaedic Surgery & Traumatology* 15, 329–332.

Mack, R.N., Simberloff, D., Lonsdale, W.M., Evans, H., Clout, M. and Bazzaz, F.A. (2000) Biotic invasions: causes, epidemiology, global consequences, and control. *Ecological Applications* 10(3), 689–710.

Mandrioli, P., Di Cecco, M. and Andina, G. (1998) Ragweed pollen: the aeroallergen is spreading in Italy. *Aerobiologia* 14, 13–20.

Mazza, G., Tricarico, E., Genovesi, P. and Gherardi, F. (2014) Biological invaders are threats to human health: an overview. *Ethology Ecology and Evolution* 26(2–3), 112–129.

Messineo, V., Bogialli, S., Melchiorre, S., Sechi, N., Lugliè, A., Casiddu, P., Mariani, M.A., Padedda, B.M., Corcia, A. Di, Mazza, R. *et al.* (2009) Cyanobacterial toxins in Italian freshwaters. *Limnologica – Ecology and Management of Inland Waters* 39, 95–106.

Metcalf, J.S. and Codd, G.A. (2009) Cyanobacteria, neurotoxins and water resources: are there implications for human neurodegenerative disease? *Amyotrophic Lateral Sclerosis* 10, 74–78.

Metcalf, J.S. and Codd, G.A. (2012) Cyanotoxins. In: Whitton, B.A. (ed.), *Ecology of Cyanobacteria II: Their Diversity in Space and Time.* Springer-Verlag, Dordrecht, the Netherlands, pp. 651–675.

Mezzasalma, V., Ganopoulos, I., Galimberti, A., Cornara, L., Ferri, E. and Labra, M. (2017) Poisonous or non-poisonous plants? DNA-based tools and applications for accurate identification. *International Journal of Legal Medicine* 131, 1–19.

Millennium Ecosystem Assessment (2005) *Ecosystems and Human Well-being: Biodiversity Synthesis.* World Resources Institute, Washington, DC.

Miller, E.B., Gilad, A. and Schattner, A. (2000) Cactus thorn arthritis: case report and review of the literature. *Clinical Rheumatology* 19, 490–491.

Mistrello, G., Harfi, H., Roncarolo, D., Kwaasi, A., Zanoni, D., Falagiani, P. and Panzani, R. (2008) Date palm pollen allergoid: characterization of its chemical-physical and immunological properties. *International Archives of Allergy and Immunology* 145(3), 224–230.

Mulford, J.S., Oberli, H. and Tovosia, S. (2001) Coconut palm-related injuries in the Pacific islands. *ANZ Journal of Surgery* 71, 32–34.

Novoa, A., Le Roux, J.J., Robertson, M.P., Wilson, J.R. and Richardson, D.M. (2015) Introduced and invasive cactus species: a global review. *AoB Plants* 7.

O'Neil, J.M., Davis, T.W., Burford, M.A. and Gobler, C.J. (2012) The rise of harmful cyanobacteria blooms: the potential roles of eutrophication and climate change. *Harmful Algae* 14, 313–334.

Otang, W.M., Grierson, D.S. and Afolayan, A.J. (2014) A survey of plants responsible for causing irritant contact dermatitis in the Amathole district, Eastern Cape, South Africa. *Journal of Ethnopharmacology* 157, 274–284.

Padisák, J. (1997) *Cylindrospermopsis raciborskii* (Woloszynska) Seenayya et Subba Raju, an expanding, highly adaptive cyanobacterium: worldwide distribution and review of its ecology. *Archiv für Hydrobiologie* 107, 563–593.

Paerl, H.W. and Huisman, J. (2008) Climate. Blooms like it hot. *Science* 320, 57–58.

Paerl, H.W. and Huisman, J. (2009) Climate change: a catalyst for global expansion of harmful cyanobacterial blooms. *Environmental Microbiology Reports* 1, 27–37.

Paerl, H.W. and Otten, T.G. (2013) Harmful cyanobacterial blooms: causes, consequences, and controls. *Microbial Ecology* 65, 995–1010.

Paerl, H.W. and Paul, V.J. (2012) Climate change: links to global expansion of harmful cyanobacteria. *Water Research* 46, 1349–1363.

Pfurtscheller, K. and Trop, M. (2014) Phototoxic plant burns: report of a case and review of topical wound treatment in children. *Pediatric Dermatology* 31(6), e156–e159.

Pimentel, D. (2011) *Biological Invasions: Economic and Environmental Costs of Alien Plant, Animal, and Microbe Species.* CRC Press, Boca Raton, Florida, USA.

Plank, L., Zak, D., Getzner, M., Follak, S., Essl, F., Dullinger, S., Kleinbauer, I., Moser, D. and Gattringer, A. (2016) Benefits and costs of controlling three allergenic alien species under climate change and dispersal scenarios in Central Europe. *Environmental Science and Policy* 56, 9–21.

Plummer, M.L. (2005) Impact of invasive water hyacinth (*Eichhornia crassipes*) on snail hosts of schistosomiasis in Lake Victoria, East Africa. *EcoHealth* 2(1), 81–86.

Potter, P.C. and Cadman, A. (1996) Pollen allergy in South Africa. *Clinical and Experimental Allergy* 26, 1347–1354.

Pyšek, P. and Richardson, D.M. (2010) Invasive species, environmental change and management, and health. *Annual Review of Environment and Resources* 35, 25–55.

Pyšek, P., Cock, M.J.W., Nentwig, W. and Ravn, H.P. (2007) Ecology and management of giant hogweed (*Heracleum mantegazzianum*). CAB International, Wallingford, UK.

Pyšek, P., Jarošík, V., Müllerová, J., Pergl, J. and Wild, J. (2008) Comparing the rate of invasion by *Heracleum mantegazzianum* at the continental, regional and local scale. *Diversity and Distributions* 14, 355–363.

Pyšek, P., Pergl, J., Essl, F., Lenzner, B., Dawson, W., Kreft, H., Weigelt, P., Winter, M., Kartesz, J., Nishino, M *et al.* (2017) Naturalized alien flora of the world: species diversity, taxonomic and phylogenetic patterns, geographic distribution and global hotspots of plant invasion. *Preslia* 89, 203–274.

Ring, J., Eberlein-Koenig, B. and Behrendt, H. (2001) Environmental pollution and allergy. *Annals of Allergy, Asthma and Immunology* 87(6), 2–6.

Roeder, E. (1995) Medicinal plants in Europe containing pyrrolizidine alkaloids. *Pharmazie* 50, 83–98.

Roy, E.D., Martin, J.F., Irwin, E.G., Conroy, J.D. and Culver, D.A. (2010) Transient social–ecological stability: the effects of invasive species and ecosystem restoration on nutrient management compromise in Lake Erie. *Ecology and Society* 15(1), 20.

Salmaso, N., Buzzi, F., Garibaldi, L., Morabito, G. and Simona, M. (2012) Effects of nutrient availability and temperature on phytoplankton development: a case study from large lakes south of the Alps. *Aquatic Sciences* 74, 555–570.

Satta, C.T., Anglès, S., Garcés, E., Sechi, N., Pulina, S., Padedda, B.M., Stacca, D. and Luglié, A. (2014) Dinoflagellate cyst assemblages in surface sediments from three shallow Mediterranean lagoons (Sardinia, North Western Mediterranean Sea). *Estuaries and Coasts* 37(3), 646–663.

Scherber, C., Crawley, M.J. and Porembski, S. (2003) The effects of herbivory and competition on the invasive alien plant *Senecio inaequidens* (Asteraceae). *Diversity and Distributions* 9, 415–426.

Schindler, S., Staska, B., Adam, M., Rabitsch, W. and Essl, F. (2015) Alien species and public health impacts in Europe: a literature review. *NeoBiota* 27, 1.

Shrestha, B.B., Shabbir, A. and Adkins, S.W. (2015) *Parthenium hysterophorus* in Nepal: a review of its weed status and possibilities for management. *Weed Research* 55, 132–144.

Skálová, H., Guo, W.Y., Wild, J. and Pyšek, P. (2017) *Ambrosia artemisiifolia* in the Czech Republic: history of invasion, current distribution and prediction of future spread. *Preslia* 89, 1–16.

Smayda, T.J. (1997) Harmful algal blooms: their ecophysiology and general relevance to phytoplankton blooms in the sea. *Limnology and Oceanography* 42, 1137–1153.

Smith, M., Cecchi, L., Skjøth, C.A., Karrer, G. and Šikoparija, B. (2013) Common ragweed: a threat to environmental health in Europe. *Environment International* 61, 115–126.

Sugarman, M., Stobie, D.G., Quismorio, F.P., Terry, R. and Hanson, V. (1977) Plant thorn synovitis. *Arthritis and Rheumatism* 20, 1125–1128.

Sukenik, A., Quesada, A. and Salmaso, N. (2015) Global expansion of toxic and non-toxic cyanobacteria: effect on ecosystem functioning. *Biodiversity and Conservation* 24, 889–908.

Sun, Y., Collins, A.R., Schaffner, U. and Müller-Schärer, H. (2013) Dissecting impact of plant invaders: do invaders behave differently in the new range? *Ecology* 94, 2124–2130.

United Nations Environment Program (2016) Invasive species – a huge threat to human well-being. Available at: https://www.unenvironment.org/news-and-stories/story/invasive-species-huge-threat-human-well-being (accessed 19 December 2017).

Vacchiano, G., Barni, E., Lonati, M., Masante, D., Curtaz, A., Tutino, S. and Siniscalco, C. (2013) Monitoring and modeling the invasion of the fast spreading alien *Senecio inaequidens* DC in an alpine region. *Plant Biosystems – An International Journal Dealing with all Aspects of Plant Biology* 147(4), 1139–1147.

Vasconcelos, V. (2006) Eutrophication, toxic cyanobacteria and cyanotoxins: when ecosystems cry for help. *Limnetica* 25, 425–432.

Vilà, M. and Hulme, P.E. (2017) *Impact of Biological Invasions on Ecosystem Services*. Springer International Publishing, Switzerland.

Wilk-Woźniak, E., Solarz, W., Najberek, K. and Pociecha, A. (2016) Alien cyanobacteria: an unsolved part of the 'expansion and evolution' jigsaw puzzle? *Hydrobiologia* 764, 65–79.

Wilson, M.A. and Carpenter, S.R. (1999) Economic valuation of freshwater ecosystem services in the United States: 1991–1997. *Ecological Applications* 9(3), 772–783.

Wink, M. and van Wyk, B.E. (2008) *Mind-altering and Poisonous Plants of the World*. Timber Press, Portland, USA.

World Health Organization (2015) *Connecting Global Priorities: Biodiversity and Human Health*. World Health Organization and Secretariat of the Convention on Biological Diversity.

Wyatt, T. and Carlton, J.T. (2002) Phytoplankton introductions in European coastal waters: why are so few invasions reported? Alien marine organisms introduced by ships in the Mediterranean and Black seas. *CIESM Workshop Monographs* 20, 41–46.

Žegura, B., Štraser, A. and Filipič, M. (2011) Genotoxicity and potential carcinogenicity of cyanobacterial toxins – a review. *Mutation Research/Reviews in Mutation Research* 727(1), 16–41.

Zhao, F., Elkelish, A., Durner, J., Lindermayr, C., Winkler, J.B., Ruëff, F., Behrendt, H., Traidl-Hoffmann, C., Holzinger, A., Kofler, W. *et al.* (2016) Common ragweed (*Ambrosia artemisiifolia* L.): allergenicity and molecular characterization of pollen after plant exposure to elevated NO_2. *Plant, Cell and Environment* 39(1), 147–164.

Zingone, A. and Oksfeldt Enevoldsen, H. (2000) The diversity of harmful algal blooms: a challenge for science and management. *Ocean & Coastal Management* 43, 725–748.

Ziska, L.H., Epstein, P.R. and Schlesinger, W.H. (2009) Rising CO_2, climate change, and public health: exploring the links to plant biology. *Environmental Health Perspectives* 117(2), 155–158.

Ziska, L., Knowlton, K., Rogers, C., Dalan, D., Tierney, N., Elder, M.A., Filley, W., Shropshire, J., Ford, L.B., Hedberg, C., Fleetwood, P., Hovanky, K.T., Kavanaugh, T., Fulford, G., Vrtis, R.F., Patz, J.A., Portnoy, J., Coates, F., Bielory, L. and Frenz, D. (2011) Recent warming by latitude associated with increased length of ragweed pollen season in central North America. *Proceedings of the National Academy of Sciences* 108, 4248–4251.

3

Human Health Impact by Alien Spiders and Scorpions

Wolfgang Nentwig*

University of Bern, Switzerland

Abstract

Only a few alien spider species are of medical importance to humans because they are large and/or aggressive enough to inject venom and/or possess peculiar venom components. Spiders of serious medical concern include several *Latrodectus* (Theridiidae) and *Loxosceles* species (Sicariidae), numerously and globally introduced into many new areas. *Loxosceles* species are the only spiders that can cause severe local effects (necrosis) and both mentioned genera contain the only alien spider species that can provoke severe systemic effects. However, fatal issues are very rare. For all other spider species large enough to penetrate the human skin and introduced somewhere as alien species, 42 genera or species from 16 families are considered in detail. The venoms of these spiders cause only modest local (redness at the site of the bite, itching and swelling) and systemic (headache or nausea) effects and can be considered as harmless. In contrast to spiders, information on health issues of alien scorpions is rare and restricted to three species so far. One of them is potentially dangerous to humans, two species are not.

3.1 Introduction

This chapter concentrates on the health issues caused by alien arachnids, i.e. venomous spiders and scorpions in the invaded area. Mites (ticks and dust mites) are excluded here because they are the topic of Chapter 4 (this volume). Among other arachnids, besides spiders, scorpions also have some health impact, but because of the small number of known alien species it is assumed that they are of minor health importance in invaded areas so far. In contrast, spiders are common, highly abundant and species-rich. Furthermore, they easily disperse, actively or passively, around the world. Therefore, they represent, at least theoretically, a considerable health issue for humans.

After insects, spiders constitute the most diverse terrestrial invertebrate group, especially when considering the polyphyly of the classical Acari (Pepato and Klimov, 2015). Spiders occur globally and they represent one of the most important predacious guilds in most terrestrial ecosystems. Most ecosystems usually support 100 or more spider species, leading to 5000–10,000 known species per continent and currently more than 46,000 described species globally (WSC, 2016).

* E-mail: wolfgang.nentwig@iee.unibe.ch

Spiders can cause health issues to humans in various manners. Large spiders such as many mygalomorph groups ('tarantulas') may be defensive or behave aggressively and bite a potential aggressor. Given the structure of their powerful chelicerae and the force these animals can exert, such a bite can cause considerable pain and wounds and potentially follow-up infections. In addition, South American Theraphosidae possess urticating hairs on their opisthosoma, which they can throw towards a potential aggressor, by using fast hind leg movements. These hairs cause skin irritation, especially in the eyes, nose and mouth. Most spiders, however, use their fangs to bite and inject venom, usually in an act of defence, because humans definitely do not belong to their normal prey range. Finally, spiders provoke fear by their mere appearance (arachnophobia). Scorpions, in contrast to spiders, only sting humans with their dorsal sting. Spiders and scorpions are not known to transfer pathogens or parasites to humans.

Spiders and scorpions seem to be dispersed by human trade quite easily (Kobelt and Nentwig, 2008), but their establishment rate is different, depending on the species. In both taxa, a few species can cause serious health issues to humans, including lethal outcomes. Therefore, the situation of alien and invasive spiders and scorpions has to be considered seriously.

3.2 Spiders as Alien and Invasive Species

3.2.1 Pathways

Due to the lack of quantitative data, it is not possible to estimate how many spiders are transported between the main biogeographical regions of the world. When considering general reviews on imported alien arthropods, the impression arises that spiders may constitute a relevant proportion (Kobelt and Nentwig, 2008; Rabitsch, 2010; Roques *et al.*, 2010). Following the pathway classification scheme of Hulme *et al.* (2008), spiders are mainly introduced as contaminants of traded commodities (primarily with potted plants and fruits) and as stowaways in transport containers and in packaging material. According to the review by Nentwig (2015) for Europe, all instars of spiders including egg masses can be found on leaves or stems of potted plants. Traded fruits (mainly bananas and grapes) also occasionally contain spiders, either between the fruits or in the folds of the boxes. Changes in global transportation technologies during the last 50 years has led to the decreasing importance of the fruit pathway (the artificial atmosphere during fruit shipment now largely impedes the survival of living animals; Hallman, 2007), whereas the potted plant pathway (increasing potted plant trade) and the container pathway (increased trade volume) have become more and more important. A very small but potentially quite important pathway concerns the import of used sports cars from the USA to Europe, which have contained black widow spiders (*Latrodectus mactans*) in high numbers, also with cocoons (Van Keer, 2010). Pathways such as intentional release and accidental escape from captivity, in contrast to many other animal groups, do not play a relevant role for alien spiders. It can be assumed that the increasing pet and internet trade will also affect the spread of alien spiders, but there are no good data for this so far.

3.2.2 Establishment

Despite the frequent appearance of alien spiders in new habitats, their establishment rate seems to be very low, with remarkable differences between the major pathways. Following a recent review for Europe (Nentwig, 2015), the used car pathway did not result in the establishment of any spider populations, and spiders following the fruit shipment pathway succeeded in establishing in only 5% of occasions. The transport container and packaging material pathway resulted in the establishment of a viable population in 47% of cases. The most successful pathway involved the potted plants, with an establishment rate of 65%. For obvious reasons, the following text concerns only established species.

3.2.3 Spiders as venomous animals

Typically, spiders possess one pair of venom glands in the basal segment of the chelicerae (still realized in mygalomorph spiders such as Theraphosidae, also called 'tarantulas'), which later expanded into the prosoma (araneomorph spiders). This prosomal location allows larger venom glands even in small spiders, i.e. with relatively small chelicerae. Only a few spider taxa (e.g. Uloboridae) reduced the venom glands completely. The venom glands are surrounded by a layer of muscle cells, allowing precise release of small venom quantities into the venom duct. In araneomorph spiders, the openings are situated on the concave side of the cheliceral fangs, close to the tip. To catch and envenomate prey, the fangs are inserted into a prey item and then can functionally best be compared with a syringe. Spiders are optimized to prey on arthropods, while vertebrates are rare prey items, and large animals cannot be overwhelmed (Nentwig and Wissel, 1986). In the case of humans, spiders do not have a normal reflex to bite and their chelicerae must have a given size and strength to be able to penetrate the relatively thick human skin.

Mostly, spider venom is a very complex mixture of 100 or more compounds. It usually includes highly active neurotoxins (small cysteine-knotted mini-proteins and large proteins), membranolytical peptides (also called cytolytical or antimicrobial), enzymes and a variety of small molecular mass compounds (for a major review see Kuhn-Nentwig et al., 2011).

3.2.4 Verified spider bites

Media reports on introduced (and native) spiders, spider bites and associated health issues need to be treated with great caution, especially when superlatives are used. Media like to play the arachnophobia game, claiming that spiders are a very frightening and dangerous arthropod group. This is astonishing, because the opposite is in fact true. In a comparison of different venomous animal groups, spiders are among the least dangerous (Nentwig and Kuhn-Nentwig, 2013). Many media reports are just a hoax or very close to it, devoid of an objective or verifiable background. To scientifically discriminate 'true' spider bites from assumed or supposed spider bites, four criteria have been set up: the spider bite has to be observed, the spider must be caught during the bite, an expert must identify it and the bite must cause symptoms usually associated with a spider bite (Isbister and White, 2004; Vetter and Isbister, 2008). If all criteria are met, it is a 'verified' spider bite. Such careful examination of generally available information reduces the high number of reported bites to a much smaller number of verified bites. According to their analyses, only 2–4% of reported spider bites could reliably be attributed to a spider (Vetter et al., 2009; Suchard, 2011). In a European review on spider bites, Nentwig et al. (2013) found no verified spider bite with a lethal outcome from either native or alien spiders.

3.3 Bites of Alien Spider Species

In a recent global analysis, Dawson *et al.* (2017) analysed the origin and invaded area of 205 alien spider species. Only some species may potentially pose a threat to humans, or are known to be of medical importance. This is because they are large and/or aggressive enough to:

- provoke fear,
- cause wounds which later may potentially lead to infections; and
- inject venom, thus leading to an envenomation.

Whether or not a spider is physically able to bite a human, i.e. to penetrate the human skin, depends primarily on the structure and power of its chelicerae and its behaviour. Data on cheliceral length are not available for most spider species and behaviour is difficult to quantify. The body length of a spider (measured from the beginning of the prosoma to the end of the opisthosoma, without appendages), however, may be a good proxy for this ability and is commonly proposed as a suitable indicator. While Nentwig *et al.* (2013) mention 10 mm as lower limit for harmful spiders, there are a few reports of slightly smaller species occasionally biting humans, and therefore 9 mm is used here as a rule of thumb to separate potentially harmful spiders from harmless species.

Following this framework and considering the known alien spider species, those from Amphinectidae, Anapidae, Cithaeronidae, Clubionidae, Dictynidae, Linyphiidae, Mimetidae, Nesticidae, Ochyroceratidae, Oecobiidae, Oonopidae, Philodromidae, Pholcidae, Prodidomidae, Salticidae, Scytodidae, Selenopidae, Stiphidiidae, Tetragnathidae, Thomisidae, Titanoecidae, Trachelidae, Uloboridae and Zodariidae are not considered here as being potentially dangerous to humans. They are too small, not aggressive and there are no relevant reports on bites either in their native or invaded areas.

In contrast, all alien spider species from Amaurobiidae, Desidae, Dysderidae, Eutichuridae, Lamponidae, Lycosidae, Nephilidae, Oxyopidae, Segestriidae, Sicariidae, Sparassidae and Zoropsidae are here considered to be potentially harmful to humans. They are large enough to bite humans and/or there is published information about bites. A few families comprise small and large spiders and publications refer only to a few species, so a generalization on family level is not justified. This concerns Agelenidae, Araneidae, Corinnidae, Desidae, Gnaphosidae and Theridiidae.

Among the 205 alien spider species considered here (Dawson *et al.*, 2017), 152 are regarded as harmless, following the criteria above, while 53 species are harmful or at least potentially harmful. They are discussed below in detail, listing the families in alphabetical order.

3.3.1 Agelenidae (funnelweb spiders)

Six large agelenids of Palearctic origin have been introduced to other continents: *Eratigena agrestis* (Walckenaer, 1802) and *Eratigena atrica* (C.L. Koch, 1843) to North America, *Tegenaria pagana* C.L. Koch, 1840 to both Americas, *Tegenaria domestica* (Clerck, 1757) to both Americas, Asia and Australia, *Tegenaria parietina* (Fourcroy, 1785) to South America, Africa and Asia, and *Tegenaria ferruginea* (Panzer, 1804) to Venezuela. It should be noticed that *Eratigena* species were split from *Tegenaria* in 2013 (WSC, 2016), thus in the older literature, they are also mentioned as *Tegenaria*. These species usually occur synanthropically, are probably among the most common spiders in and around buildings and can easily be spread with human activities, even world-wide. It has to be assumed that in the near future more large agelenid species will be detected as aliens in other areas.

Some species reach body lengths of up to 20 mm and human contact with these nocturnal spiders occurs frequently. While these agelenids are not aggressive and usually try to escape, they may bite, usually when one tries to catch or remove them by hand. The bite is not very painful and an occasional moderate local swelling disappears within a few hours (Nentwig *et al.*, 2013). In the USA, *Eratigena agrestis* has been called the hobo spider and was seen as a major cause of dermonecrotic injuries. However, an in-depth analysis by Vetter and Isbister (2004) showed that there is no proof for this, indicating that this species is harmless to humans in North America and confirming traditional European opinion on this species.

3.3.2 Amaurobiidae (meshweb weavers)

Only *Amaurobius ferox* (Walckenaer, 1830) from the Palearctic, with a body length of up to 16 mm, has been introduced to North America. This species occurs in caves, basements and humid mural chinks, thus potentially in close vicinity to humans, and a few more ecologically similar species could be expected as aliens. Usually, *A. ferox* is well hidden and interactions with humans are rare. Nevertheless, this species can bite humans when it is accidently squeezed. The bite is as painful as a wasp sting and disappears completely after 12 h. There may be moderate local swelling and redness (Nentwig *et al.*, 2013).

3.3.3 Araneidae (orb web spiders)

Among the species-rich family of Araneidae, only a few species are known as aliens. Among them are species from the genera *Argiope*, *Cyrtophora*, *Neoscona* and *Zygiella*, which, given their size, could potentially bite humans. *Argiope trifasciata* (Forsskål, 1775) from North America has been introduced to southern Europe, Africa, Asia and several Pacific Islands. *Cyrtophora citricola* (Forsskål, 1775), of African origin, has been introduced to Central and South America. *Neoscona crucifera* (Lucas, 1838), from North America, has been introduced to several Pacific islands and Macaronesia. *N. moreli* (Vinson, 1863), from Africa, now also occurs in the Caribbean and in South America. *N. nautica* (L. Koch, 1875), from the Pacific area, has been introduced to both America and Africa. *N. theisi* (Walckenaer, 1841), from Asia, also occurs in the Seychelles, which biogeographically belong to Africa. *Zygiella atrica* (C.L. Koch, 1845) and *Z. x-notata* (Clerck, 1757), both from Europe, have been introduced to North and South America.

These large orb-weaving species typically show a very shy behaviour and escape or hide when one approaches their orb-web or tries to catch them. No bites from species of these genera have been reported so far. However, for other large orb-weavers with comparable ecology, bites have been reported and so it can be concluded that the mentioned alien species react correspondingly and may cause similar symptoms after a bite. Terhivuo (1993) and Nentwig *et al.* (2013) mention the bite of *Araneus* sp., Hansen (1996) cites a bite by *Nuctenea umbratica* (Clerck, 1757) and Nentwig (unpublished) experienced a bite of *Micrathena* sp. In all cases, the bite caused only local pain, lasting a few minutes, and no further complications were mentioned. Isbister and Gray (2002) list bites from araneids as the second most common spider family in a large review of Australian spider bites. The pain from araneid bites was defined as less severe by 83% of victims and as severe by 17% of victims. Pain duration was 5 min on average.

3.3.4 Corinnidae (dark sac spiders)

The South American *Xeropigo tridentiger* (O. Pickard-Cambridge, 1869) has been introduced to the island of St Helena, biogeographically belonging to Africa. No information on bites from this species is available, but according to an Australian study (Isbister and Gray, 2002), corinnids can bite humans in rare cases. In most cases (15 of 16), the bite caused only moderate pain, lasting for a few minutes.

3.3.5 Desidae (intertidal spiders, house spiders)

The two *Badumna* species, *longinqua* (L. Koch, 1867) and *insignis* (L. Koch, 1872), the black house spiders, both from Australia, have been introduced to Asia, both Americas and New Zealand. They are commonly found in urban areas, usually in the corner of windows or under ceilings. When disturbed, the spiders retreat, so contact with humans and bites are infrequent. A bite causes moderate (72% of cases) or severe pain (in 28%), on average lasting 5 min. The bite also leads to redness (68%), swelling (12%) and itchiness (20%) (Isbister and Gray, 2004).

3.3.6 Dysderidae (woodlouse hunters)

Dysdera crocata (C.L. Koch, 1838), the woodlouse spider, is native to the European Mediterranean area and has been introduced to South Africa, Australia, New Zealand, Hawaii and both Americas, so that it has now a nearly cosmopolitan distribution. With its synanthropic lifestyle, it occurs frequently in buildings. *D. crocata* is up to 15 mm long and has unusual large and powerful chelicerae, so bites to humans occur occasionally. This spider is not aggressive, but its numerous movements of the chelicerae may be misinterpreted. The venom causes minor (five cases) to severe (three cases) pain, typically lasting for less than 1 h, followed by local erythema, swelling and itchiness (Vetter and Isbister, 2006).

3.3.7 Eutichuridae

This concerns the genus *Cheiracanthium* (yellow sac spiders), which has been transferred in the last 20 years from Clubionidae to Miturgidae and then to Eutichuridae. It currently contains 210 species (WSC, 2016), two of which are known as alien species. The Neotropical *Cheiracanthium inclusum* (Hentz, 1847) has been introduced to La Reunion (biogeographically Africa), and the Palearctic *C. mildei* (L. Koch, 1864) now occurs in all parts of North America and Argentina. However, due to difficulties with proper species identification, it is probable that more species have been introduced as aliens.

Several *Cheiracanthium* species have an affinity to urban areas and live more or less synanthropically, thus they may enter houses and have been found hiding in clothes. Furthermore, several species have shown spreading tendencies in the last decades, for example *C. punctorium* (Villers, 1789) has spread from southern to northern Europe and become more frequent. This provokes more encounters with humans and also bites, especially when the spiders are squeezed. Medical reports on bites of yellow sac spiders involve at least five species, including *C. inclusum* and *C. mildei*. A typical *Cheiracanthium* bite causes an immediate pain (described as medium in 50% and as severe in 50% of cases) reaching its maximum intensity after 5–20 min and lasting up to 3 h. Usually, a bite causes redness (90%), swelling

(30%) and itchiness (30%), but only rarely systemic effects (headache and nausea in single cases) (Vetter *et al.*, 2006; Nentwig *et al.*, 2013). In contrast to older literature, it has become clear that *Cheiracanthium* bites never cause dermonecrosis.

3.3.8 Gnaphosidae (ground spiders)

Several large gnaphosid species have been introduced into new areas: *Hemicloea rogenhoferi* L. Koch, 1875 and *Intruda signata* Hogg, 1900 from Australia to New Zealand, *Scotophaeus blackwalli* (Thorell, 1871) from Europe to both Americas, *Sosticus loricatus* (L. Koch, 1866) from Asia to Europe, *Trachyzelotes barbatus* (L. Koch, 1866) from Europe to California, *Urozelotes rusticus* (L. Koch, 1872) from Asia to Europe and Macaronesia, and *Zelotes tenuis* (L. Koch, 1866) from Europe to California. Gnaphosids may be rather abundant in the environment, but most are nocturnal and they live a rather cryptic life, thus the probability of a human encounter is low. There is only one report of a bite by *Zelotes* sp. (only local pain, for a few minutes) (Nentwig *et al.*, 2013), indicating that spiders of this family are of very low medical importance for humans.

3.3.9 Lamponidae (white-tailed spiders)

Lampona cylindrata (L. Koch, 1866) and *L. murina* L. Koch, 1873, both native to Australia, have been introduced to New Zealand and adjacent islands. As synanthropic species, they occur frequently in houses and gardens and bites occur relatively often, especially because the spiders tend to hide in clothing or towels. The bite causes mild (73%) to severe (27%) pain, initial redness (83%), local swelling (8%) and itchiness (44%). Systemic effects are rare (9%) and primarily include nausea or headache. Symptoms are generally considered to be mild and disappear (medium duration) after 24 h (Isbister and Gray, 2003a).

3.3.10 Lycosidae (wolf spiders)

The European *Trochosa ruricola* (De Geer, 1778) has been introduced to North America. Common in agricultural landscapes, these medium-sized lycosids (up to 14 mm body length) usually do not interact with humans. However, Terhivuo (1993) mentions the bite of the closely related *T. spinipalpis* (F.O. Pickard-Cambridge, 1895) in Finland (pain like a wasp sting and local epidermal swelling) and Russell (1991) refers to North American lycosid bites ('little more than localised pain'). Two further comprehensive studies from Brazil and Australia confirm that wolf spider bites generally cause mild pain (Ribeiro *et al.*, 1990). Medium pain duration is 10 min and other effects include swelling (20%), redness (67%) and itchiness (13%). Systemic effects are rare (7%) and concern mainly nausea and headache (Isbister and Framenau, 2004).

3.3.11 Nephilidae (giant or golden orb web spiders)

This is a species-poor spider family, common in the tropics and subtropics of the world. Only one species, the large (body length 30 mm) African *Nephilingis cruentata* (Fabricius, 1775), has been introduced to several South American countries. It lives there as a highly synanthropic spider, commonly found in and around human dwellings; however, there are no reports on nephilids biting humans, from anywhere in the world.

3.3.12 Oxyopidae (lynx spiders)

Two African *Peucetia* species, *P. striata* Karsch, 1878 and *P. viridis* (Blackwall, 1858) have been introduced to Atlantic and Caribbean islands. The behaviour of these up to 14 mm long spiders has been described as partly aggressive when females guard their egg sac, but bites do not seem to be frequent. Following descriptions from Bush *et al.* (2000), bites from the related green lynx spider *P. viridans* (Hentz, 1832), native to North and Central America, caused only local pain, itchiness and redness. They also describe one case of a *Peucetia* that spat venom in the face of a person, resulting in eye injury, and moderate conjunctivitis for 3 days. Such spitting behaviour is highly unusual in spiders and the reliability of this report (dating back to 1946) is not clear.

3.3.13 Segestriidae (tube-web spiders)

One species, the European *Segestria florentina* (Rossi, 1790), has been introduced into several South American countries. It can be found in and around buildings, but it is a nocturnal and very secretive spider, thus interactions with humans are rare. There are no records of bites.

3.3.14 Sicariidae (brown spider, violin spiders)

Loxosceles species definitely belong to the few spider taxa of major medical importance to humans. Among the 114 currently known *Loxosceles* species, most of them from the Neotropics, medical reports concern about a dozen species, and five are known as alien species. Three species from South America have been introduced to other continents: *L. gaucho* Gertsch, 1967 to Tunisia; *L. laeta* (Nicolet, 1849) to North America, Finland and Australia; *L. rufipes* (Lucas, 1834) to tropical Africa. The European Mediterranean *L. rufescens* (Dufour, 1820) now has a cosmopolitan distribution (Africa, Asia, Australia, both Americas) and the African *L. amazonica* Gertsch, 1967 is found in Brazil.

Sicariid spiders possess a highly unusual venom insofar as it contains large amounts of an enzyme (sphingomyelinase D) (Binford, 2013) that occasionally may provoke severe local skin necrosis and a deep ulcerating wound (cutaneous loxoscelism), which heals slowly and skin grafting may become necessary. This enzyme is, among spiders, only found in sicariids, thus skin necrosis can, in contrast to many older publications, only be caused by *Loxosceles* (and a few rare relatives) (Vetter, 2015). Another peculiarity of *Loxosceles* is that its bite, for yet unknown reason, does not cause pain. Since sudden pain is usually the typical alarm signal for a human that they have been bitten or stung by a spider or insect, many if not most people do not realize that they had been bitten by a violin spider.

Loxosceles species are highly synanthropic and may reach high densities of up to 2000 spiders per house (Vetter, 2008). However, due to its nocturnal and secretive nature, *Loxosceles* bites are rather uncommon. Four types of human reactions to *Loxosceles* bites can be distinguished (Da Silva *et al.*, 2004; Vetter, 2008, 2015): (i) unperceived bites with little damage and self-healing; (ii) mild reaction with skin redness, itching and slight lesion but self-healing; (iii) necrotic skin lesions, two-thirds of which heal without complication, but severe cases result in large lesions, healing only after months and leaving ugly scars; (iv) systemic or viscerocutaneous reactions. These are rare and characterized by severe intra-vascular haemolysis and disseminated intravascular coagulation, eventually leading to renal failure and death, particularly in children. While reactions (i) and (ii) are probably most com-mon and heal without treatment, reaction (iii) is less frequent and fatal issues following reaction (iv) are extremely rare. Due to frequent misidentification of *Loxosceles* and confu-sion with other reasons, reliable frequency data seem to be unavailable (Vetter, 2015).

3.3.15 Sparassidae (huntsmen spiders, giant crab spiders)

Three sparassid species are known to have been introduced to other continents: *Delena can-cerides* Walckenaer, 1837 and *Isopeda villosa* L. Koch, 1875, both from Australia, have been introduced to New Zealand and *Heteropoda venatoria* (Linnaeus, 1767) of tropical Asian ori-gin to all other continents and many islands. Sparassids are large, nocturnal spiders, usually not interacting with humans, and most species are not aggressive. However, they occur quite frequently inside houses and easily bite when humans try to catch or remove them. Their bite is rather harmless, mostly described as moderate (73%), rarely severe pain (27%), lasting on average 5 min. Systemic effects, such as headache or nausea, are unusual (4%). Swelling occurred in only 16%, itchiness in 14%, redness in 57% of cases (Isbister and Hirst, 2003).

3.3.16 Theridiidae (cob web spiders)

The 26 theridiid species which are currently known as alien somewhere in the world are usu-ally smaller than 9 mm body length, and thus should not be of medical importance to humans. However, two theridiid groups are known to bite humans and to cause medical problems. The genus *Latrodectus*, widow spiders, comprises 35 spider species, of which three have been introduced to and established in other biogeographical regions of the world: the brown widow *L. geometricus* C.L. Koch, 1841 from Africa has invaded temperate and tropical Asia, Australia, both Americas and several islands; the black widow *L. mactans* (Fabricius, 1775) from North America occurs meanwhile also in Sri Lanka; the redback spider *L. hasselti* (Fabricius, 1775), native to South-east Asia, has been introduced to Japan and New Zealand.

Widow spiders occasionally bite humans and the symptoms they cause cover a wide range. Pain is, as usual for most spider bites, the first reaction. In 38% pain is described as moderate, in 62% as severe. Typical for *Latrodectus* bites, however, is that the pain lasts rela-tively long (>24 h in 66%), that it may increase in intensity during the first hour (54%), radiate into the proximal limb (38%) and that pain may be so severe that it prevents sleep in the first 24 h (32%). Local effects include redness (74%) and diaphoresis (34%), but only rarely swelling (7%). Systemic effects are more frequent than after most other spider bites (35%) and include nausea, vomiting, headache and lethargy, but lasted on average only 1–2 days (Isbister and Gray, 2003b). Death following widow bites is very rare. There are no confirmed cases for the last decades for this globally, and older records appear to be due to secondary infections and other complications, not to direct venom effects (Nentwig and Kuhn-Nentwig, 2013).

The second theridiid group comprises *Steatoda* and *Parasteatoda* species, sometimes named 'false black widow'. *Parasteatoda tabulata* (Levi, 1980) from tropical Asia can now be found in Europe, temperate Asia and North America; *P. tepidariorum* (C.L. Koch, 1841) from South America now occurs in North America, Europe, Africa, Asia and several islands; *P. tesselata* (Keyserling, 1884), also from South America, occurs in Asia; *Steatoda capensis* Hann, 1990 from South Africa has been introduced to Australia and New Zealand; *S. castanea* (Clerck, 1757) from Europe can be found in Canada; *S. grossa* (C.L. Koch, 1838) from temperate Asia is now found in Europe, Macaronesia, both Americas and many islands; *S. nobilis* (Thorell, 1875) from Macaronesia occurs in Europe and California; *S. triangulosa* (Walckenaer, 1802) from Europe can be found in Macaronesia and North America.

Most of these spiders live synanthropically, and thus occur in and at buildings, sometimes in high densities. These spiders are not aggressive but coming in contact with them, e.g. while cleaning a window, may urge a spider to bite. For *Steatoda* species, pain was described as moderate (74%) or severe (26%). Pain increased in the first hour in 30% of cases and lasted for 6 h on average. Local or regional diaphoresis did not occur. Systemic effects (30% of cases) included nausea, headache and lethargy. *Steatoda* bites also caused prolonged pain, and systemic effects were similar to *Latrodectus*, but were less severe (Vetter and Visscher, 1998; Isbister and Gray, 2003c).

3.3.17 Zoropsidae (false wolf spiders)

The Mediterranean *Zoropsis spinimana* (Dufour, 1820) has been introduced into North America. It occurs synanthropically and bites usually occur when people try to remove a spider by hand. Most bites were described as moderately painful, disappearing after a few minutes, and once it was described as severely painful. Swelling and redness occur but disappear soon after the bite. Systemic effects seem to be rare but in one case nausea was mentioned (Nentwig *et al.*, 2013).

3.4 Scorpions

There is not much information available on non-native scorpions. This is remarkable because it is well-known that scorpions are quite frequently introduced unintentionally as stowaways with transported goods. They seem to be often detected (and removed), and further dispersal seems to be rather limited and the establishment probability in the new environment must be low. Nevertheless, three scorpion species are known as alien species in invaded areas, but there are certainly many others that have not yet been reported.

All scorpions are venomous and many species are of human concern. In their global analysis of health issues due to scorpion stings, Chippaux and Goyffon (2008) conclude (for areas where scorpions occur) that there are 3271 deaths per year, which relates (for the concerned human population) to a death rate of 1.4 per million people. This makes scorpions, after snakes, the second most lethal venomous animal group, and indicates that an established alien population of a venomous scorpion species may cause considerable health problems to humans.

It is unclear if the reported mortality concerns 'verified' scorpion stings. This topic (see discussion above) seems to be absent from the scorpion literature, but it may be possible that reported fatal issues are a consequence of insufficient medical infrastructure, as frequently encountered in the third world where most scorpion stings occur. The discrepancy between the ease of unintentionally introduced scorpions and the detection of viable alien populations, finally, may point to an insufficient knowledge of scorpion spread and distribution over the last few decades.

3.4.1 Buthidae

Centruroides gracilis (Latreille, 1804)

This scorpion (up to 10 cm long) is native to Central America and northern South America. It has been introduced to Caribbean islands (Cuba, Jamaica), the USA (Florida), Africa (Cameroon, Gabon) and the Canary Islands (Fet and Lowe, 2000). This species has some synanthropic affinities, since it can be found in the walls of houses and under rubbish piles in yards. It is sometimes kept as a pet, thus transport and introduction seem to be facilitated by the pet trade. *Centruroides gracilis* can reproduce sexually and via parthenogenesis, which facilitates establishment in a new environment.

This species is venomous, but much less toxic than others of its genus. Its venom is neurotoxic and cardiotoxic, causing the release of catecholamines. Local effects from the sting include pain, redness, itching and swelling. Systemic effects are not always present and may include cardiac effects (such as arrhythmia, pulmonary oedema, tachycardia or bradycardia, and hyper- or hypotension), nausea, vomiting, sweating, diarrhoea, shock and convulsions, occasionally leading to coma and potentially death (CTS, 2016).

Isometrus maculatus (DeGeer, 1778)

This Asian species now has a pantropical distribution, due to introductions to both Americas, Africa, Australia and many islands. The only European population reported, in Spain, seems to be extinct (Fet, 2010). The global spread of this species has been facilitated by human activity and it is often found near human habitation (CABI, 2016). Local effects of the sting of this non-deadly scorpion species consist of severe pain and swelling. Systemic effects are rare (nausea). There seem to be no reported fatalities due to stings of this scorpion (Yates, 2016).

3.4.2 Euscorpiidae

Euscorpius flavicaudis (De Geer, 1778)

The native range of *E. flavicaudis* extends through north-west Africa and southern Europe, but it has been introduced to South America (Brazil, Uruguay). The species is common in and around human habitations, which certainly facilitates spread to other areas. It is also kept as a pet and offered by commercial suppliers. The few medical data available suggest that it is a harmless species, which rarely will use its stinger. A sting seems not to be very painful, and other symptoms have not been reported.

3.5 Conclusions

Among spiders, it is remarkable that the group of very large and often frightening 'tarantulas' (Mygalomorphae) have never been recorded as an alien introduction. This also includes the South American Theraphosidae with urticating hairs, which can potentially cause a serious health issue. Also among those spider species which present the most serious threat to human health, many taxa have (so far) never been introduced elsewhere and/or did not establish. This concerns Australian Hexathelidae species (*Hadronyche* sp., *Atrax* sp.) and South American *Phoneutria* sp. (Ctenidae). Among spiders of serious medical concern, only several *Latrodectus* (Theridiidae) and *Loxosceles* species (Sicariidae) have been numerously and globally introduced into many new areas. The considerable number of other large spider species that have been introduced into new areas may occasionally bite humans, but their bites are not considered to be medically important (Table 3.1).

Spider bites cause medical problems on several levels. First, pain is noticed. It is induced by histamine, which may be a constituent of all spider venoms, but at various concentrations. This may explain the varying intensity and duration of pain caused by different species. Physiologically, for vertebrates, pain after a spider bite is considered to be only a side-effect of the proposed main venom-enhancing function of histamine (Kuhn-Nentwig et al., 2011). Second, local effects occur. This primarily concerns redness at the site of the bite, itching and swelling, all together harmless reactions that usually disappear after a short time. Only sicariid spiders are able to induce dermatonecrosis as a local reaction, because their venom contains sphingomyelinase D (Binford, 2013). Third, the various neurotoxins in a spider's venom are probably the reason for systemic reactions. Overall, in most cases they seem to cause only weak effects to humans (such as headache or nausea), probably because of the low venom quantity injected.

Remarkable exceptions from the last statement are *Loxosceles* and *Latrodectus* species, where very potent venom components may, in rare cases, lead to severe systemic reactions. The venom of *Latrodectus* is special because it contains the very large latrotoxins, peptides of 110–140 kDa molecular mass. They provoke a massive transmitter release from nerve endings of vertebrates and form ion channels through cell membranes, leading to a rapid paralysis (Grishin, 1998). As explained above, *Loxosceles* venom may occasionally cause severe intravascular haemolysis and disseminated intravascular coagulation that may lead to renal failure. For both spider genera, fatal issues are very rare and the number of such cases may be in the range of one or a few cases per year, worldwide. While for the vast majority of people bitten by these two spider groups, the quantity of venom injected is too low to cause serious consequences, however, the situation may be more severe with children, older people, persons already weakened or with a deficiency of their immune system.

Acute allergic reactions to hymenopteran bites are the best characterized allergic reactions of arthropods and quite regularly lead to fatal issues. In contrast, comparable allergic reactions to spiders are extremely rare. The only report in the medical literature concerns a huntsman spider (Sparassidae) that, after crawling over both arms of a man, caused a rash, quickly covering the arms and trunk, followed by bradycardia and hypotension. Immediate medical treatment led to recovery within a few hours (Isbister, 2002). In this context it may be interesting that there is evidence for the pathogenesis of loxoscelism at least partly due to a hypersensitivity reaction (Isbister, 2002).

Table 3.1. Summary of the medical effects of the bites of alien spider species. Only species of potential medical concern to humans are considered, i.e. species ≥9 mm body length and species that are known to cause medical problems. NA = not applicable.

Family (genus)	Species globally introduced	Pain	Pain duration	Local effects	Systemic effects	Fatal issues
Sicariidae (*Loxosceles*)	5	No	NA	Harmless to severe (necrosis)	May be considerable	Very rare
Theridiidae (mainly *Latrodectus*)	11	Yes	Often >24 h, increasing	Harmless	May be considerable	Very rare (not in last decades)
All other spider families	42	Yes	Most species <1 h, some species a few hours	Harmless	Harmless	No

At the psychological level, irrespective of the species origin, spiders may cause serious phobic reactions in some people, which also needs to be considered as a health issue. Such fear is fuelled by sensation-seeking media with reports on newly introduced spiders (e.g. from overseas, with banana shipment, detected in a school class), always referred to as 'the deadliest species', but mostly, this is incorrect. Nevertheless, in principle, introduced spiders may cause or increase the fear of them, but no useful data or publications on this issue are available.

It should also be mentioned that currently our knowledge on pathways, establishment probability and distribution pattern in the invaded habitats of many spider and scorpion species is insufficient. An increased investigation of this topic will probably show that there are even more alien species which have already colonized new habitats, increased in density and range. Finally, it has to be underlined that with more world trade, the number and frequency of alien species will also increase. This will enhance the chances for the establishment of spider and scorpion populations in new habitats, which will lead to more contact with humans and more bites or stings. Additionally, with increasing climate change, it will be easier for these species to establish in new habitats because the prevailing warmer and drier climate meets the demands of many spider and scorpion species.

Acknowledgements

Many thanks for their valuable discussion contributions and comments go to Matt E. Braunwalder, Myles Menz and Wolfgang Rabitsch.

References

Binford, G. (2013) The evolution of a toxic enzyme in sicariid spiders. In: Nentwig, W. (ed.) *Spider Ecophysiology.* Springer, Heidelberg, Germany, pp. 229–240.

Bush, S.P., Giem, P. and Vetter, R.S. (2000) Green lynx spider (*Peucetia viridans*) envenomation. *American Journal of Emergency Medicine* 18, 64–66.

CABI (2016) Invasive species compendium. Datasheet *Isometrus maculatus* (lesser brown scorpion). Available at: http://www.cabi.org/isc/datasheet/78223 (accessed 10 October 2016).

Chippaux, J.-P. and Goyffon, M. (2008) Epidemiology of scorpionism: a global appraisal. *Acta Tropica* 107, 71–79.

CTS (2016) *Clinical Toxinology Resources*. The University of Adelaide. Available at: http://www.toxinology.com (accessed 10 October 2016).

Da Silva, P.H., Da Silveira, R.B., Appel, M.H., Mangili, O.C., Gremski, W. and Veiga, S.S. (2004) Brown spiders and loxoscelism. *Toxicon* 44, 693–709.

Dawson, W., Moser, D., van Kleunen, M., Kreft, H., Pergl, J., Pyšek, P., Weigelt, P., Winter, M., Lenzner, B., Blackburn, T.M., *et al.* (2017) Global hotspots of alien species across taxonomic groups. *Nature Ecology & Evolution* 1 doi: 10.1038/s41559-017-0186.

Fet, V. (2010) Scorpions of Europe. *Acta Zoologica Bulgarica* 62, 3–12.

Fet, V. and Lowe, G. (2000) Family Buthidae C.L. Koch, 1837. In: Fet, V., Sissom, W.D., Lowe, G. and Braunwalder, M.E. (eds) *Catalog of the Scorpions of the World (1758–1998).* The New York Entomological Society, New York, pp. 54–286.

Grishin, E.V. (1998) Black widow spider toxins: the present and the future. *Toxicon* 36, 1693–1701.

Hallman, G.J. (2007) Phytosanitary measures to prevent the introduction of invasive species. In: Nentwig, W. (ed.) *Biological Invasions.* Springer, Berlin, pp. 367–384.

Hansen, H. (1996) L'importanza medica di alcuni ragni viventi negli ambienti urbani di Venezia. *Bollettino Museo Civico di Storia Naturale di Venezia* 45, 21–32.

Hulme, P.E., Bacher, S., Kenis, M., Klotz, S., Kühn, I., Minchin, D., Nentwig, W., Olenin, S., Panov, V., Pergl, J. *et al.* (2008) Grasping at the routes of biological invasions: a framework for integrating pathways into policy. *Journal of Applied Ecology* 45, 403–414.

Isbister, G.K. (2002) Acute allergic reaction following contact with a spider. *Toxicon* 40, 1495–1497.

Isbister, G.K. and Framenau, V. (2004) Australian wolf spider bites (Lycosidae): clinical effects and influence of species on bite circumstances. *Clinical Toxicology* 42, 153–161.

Isbister, G.K. and Gray, M.R. (2002) A prospective study of 750 definite spider bites, with expert spider identification. *Queensland Journal of Medicine* 95, 723–731.

Isbister, G.K. and Gray, M.R. (2003a) White-tail spider bite: a prospective study of 130 definite bites by *Lampona* species. *Medical Journal of Australia* 179, 199–202.

Isbister, G.K. and Gray, M.R. (2003b) Latrodectism: a prospective cohort study of bites by formally identified redback spiders. *Medical Journal of Australia* 179, 88–91.

Isbister, G.K. and Gray, M.R. (2003c) Effects of envenoming by comb-footed spiders of the genera *Steatoda* and *Achaearanea* (family Theridiidae: Araneae) in Australia. *Journal of Toxicology and Clinical Toxicology* 41, 809–819.

Isbister, G.K. and Gray, M.R. (2004) Black house spiders are unlikely culprits in necrotic arachnidism: a prospective study. *Internal Medicine Journal* 34, 287–289.

Isbister, G.K. and Hirst, D. (2003) A prospective study of definite bites by spiders of the family Sparassidae (huntsmen spiders) with identification to species level. *Toxicon* 42, 163–171.

Isbister, G.K. and White, J. (2004) Clinical consequences of spider bites: recent advances in our understanding. *Toxicon* 43, 477–492.

Kobelt, M. and Nentwig, W. (2008) Alien spider introductions to Europe supported by global trade. *Diversity and Distributions* 14, 273–280.

Kuhn-Nentwig, L., Stöcklin, R. and Nentwig, W. (2011) Venom composition and strategies in spiders: is everything possible? *Advances in Insect Physiology* 40, 1–86.

Nentwig, W. (2015) Introduction, establishment rate, pathways and impact of spiders alien to Europe. *Biological Invasions* 17, 2757–2778.

Nentwig, W. and Kuhn-Nentwig, L. (2013) Spider venoms potentially lethal to humans. In: Nentwig, W. (ed.) *Spider Ecophysiology*. Springer, Berlin, pp. 253–264.

Nentwig, W. and Wissel, C. (1986) A comparison of prey length among spiders. *Oecologia* 68, 595–600.

Nentwig, W., Gnädinger, M., Fuchs, J. and Ceschi, A. (2013) A two year study of verified spider bites in Switzerland and a review of the European spider bite literature. *Toxicon* 73, 104–110.

Pepato, A.R. and Klimov, P.B. (2015) Origin and higher-level diversification of acariform mites – evidence from nuclear ribosomal genes, extensive taxon sampling, and secondary structure alignment. *BMC Evolutionary Biology* 15, 178.

Rabitsch, W. (2010) Pathways and vectors of alien arthropods in Europe. *BioRisk* 4, 27–43.

Ribeiro, L.A., Jorge, M.T., Piesco, R.V. and Nishioka, S.A. (1990) Wolf spider bites in Sao Paulo, Brazil: a clinical and epidemiological study of 515 cases. *Toxicon* 28, 715–717.

Roques, A., Kenis, M., Lees, D., Lopez-Vaamonde, C., Rabitsch, W., Rasplus, J.-Y and Roy, D.B. (2010) Alien terrestrial arthropods of Europe. *BioRisk* 4, 1–1028.

Russell, F.E. (1991) Venomous arthropods. *Veterinary and Human Toxicology* 33, 505–508.

Suchard, J.R. (2011) 'Spider bite' lesions are usually diagnosed as skin and soft-tissue infections. *Journal of Emergency Medicine* 41, 473–481.

Terhivuo, J. (1993) Novelties to the Finnish spider fauna (Araneae) and notes on species having bitten man. *Memoranda Societatis pro Fauna et Flora Fennica* 69, 53–56.

Van Keer, K. (2010) An update on the verified reports of imported spiders (Araneae) from Belgium. *Nieuwsbrief van de Belgische Arachnologische Vereniging* 25, 210–214.

Vetter, R.S. (2008) Spiders of the genus *Loxosceles* (Araneae, Sicariidae): a review of biological, medical and psychological aspects regarding envenomations. *Journal of Arachnology* 36, 150–163.

Vetter, R.S. (2015) *The Brown Recluse Spider*. Cornell University Press, Ithaca, New York.

Vetter, R.S. and Isbister, G.K. (2004) Do hobo spider bites cause dermonecrotic injuries? *Annals of Emergency Medicine* 44, 605–607.

Vetter, R.S. and Isbister, G.K. (2006) Verified bites by the woodlouse spider, *Dysdera crocata*. *Toxicon* 47, 826–829.

Vetter, R.S. and Isbister, G.K. (2008) Medical aspects of spider bites. *Annual Review of Entomology* 53, 409–429.

Vetter, R.S. and Visscher, P.K. (1998) Bites and stings of medically important venomous arthropods. *International Journal of Dermatology* 37, 481–496.

Vetter, R.S., Isbister, G.K., Bush, S.P. and Boutin, L.J. (2006) Verified bites by yellow sac spiders (Genus *Cheiracanthium*) in the United States and Australia: where is the necrosis? *American Journal of Tropical Medicine and Hygiene* 74, 1043–1048.

Vetter, R.S., Hinkle, N.C. and Ames, L.A. (2009) Distribution of the brown recluse spider (Araneae: Sicariidae) in Georgia with comparison of poison center reports of envenomation. *Journal of Medical Entomology* 46, 15–20.

WSC (2016) *World Spider Catalog, version 17.5.* Natural History Museum, Bern, Switzerland. Available at: http://wsc.nmbe.ch (accessed 10 October 2016).

Yates, J.R. (2016) Urban Knowledge Master. *Isometrus maculatus* (De Geer). Lesser Brown Scorpion. Available at: http://www.extento.hawaii.edu/kbase/urban/site/scorp.htm (accessed 10 October 2016).

4

Ticks and Dust Mites: Invasive and Health-affecting Borderline Organisms

Sauro Simoni[1]* and Giulio Grandi[2,3]

[1]CREA-DC Research Council for Agriculture and Economics – Research Centre for Plant Protection and Certification, Florence, Italy; [2]Swedish University of Agricultural Sciences, Uppsala, Sweden; and [3]National Veterinary Institute, Uppsala, Sweden

Abstract

According to the situations and cases presented here, organisms belonging to tick and house dust mite groups can be included within alien invasive pests representing a risk for human health. *Rhipicephalus sanguineus*, the brown dog tick, a vector for many tick-borne pathogens, is able to efficiently colonize households at quite wide ranges in temperature and relative humidity. Cases of the spread of this tick by movements of pets and people are reported. House dust mites and main storage mites – although not strictly alien species – can act not only as allergens' disseminators but even as potential indicators of environmental quality in different settings (newly colonized areas, both geographic and within human dwellings). The impact of alien mites affecting agricultural production has been generally measured by the ecological consequences on biodiversity, but the evaluation of the effects of these species on human health has been mostly neglected. Trading, travelling and modified home conditions can increase the chance for diffusion of ticks and mites affecting human health, i.e. causing higher exposures to vector-borne pathogens and allergens, respectively. A more interdisciplinary approach that takes into account both the dynamics in ecological evaluation and the consequences on human health of these organisms is needed.

4.1 Introduction

Globalization, characterized by an intense exchange of natural resources and goods between distant geographical areas, together with increased movement of people and animals, as well as climate change have led to a progressively higher risk of biological invasion in many nations throughout the planet. The majority of studies, especially those from the past two decades, focus on the impact of the spread of alien pests on ecosystem services (Charles and Dukes, 2007; Gilioli *et al.*, 2014).

When considering the effect of climatic change on the spread of pests, the assessment of invasion rate in relation to weather frequently refers to long-term statistical models of climatic conditions (Wu *et al.*, 2016). However, in order to study and manage the emergence of alien pests more accurately from the phytosanitary point of view, evidence should be

---

* E-mail: sauro.simoni@crea.gov.it

acquired from a more limited time scale (NIEHS, 2010). For example, the modelling of the impact of pests and diseases on agricultural systems may be improved by using data of better quality in terms of temporal proximity to the events (Donatelli *et al.*, 2017).

Many studies address the relationship between ecosystems and invasion by alien species (Venette, 2015); in other cases, governmental authorities have implemented studies focusing on both the assessment of potential effects of alien organisms on the public and on the effects exerted by climate change-mediated indoor environmental alterations on occupant health (Aries and Bluyssen, 2009; NRC, 2010).

Programmes and laws to protect ecosystems, agriculture, forestry and human health against alien species, as well as programmes designed to face the effects of economic globalization (resulting in an ever-increasing risk of invasion, including the invasion/spreading of mites) are being developed at the national level by most countries.

Since the year 2000, more than 100 mite species belonging to the superorders Acariformes (Actinotrichida) and Parasitiformes (Anactinotrichida) have been considered as alien to Europe, 96 of them having originated from other continents and 5 of them being cryptogenic species (see Navajas *et al.*, 2010). Most of the recorded alien species belong to two orders of Acariformes, Prostigmata and Astigmata; the former includes the superfamilies Tetranychoidea and Eriophyoidea which include herbivorous species representing major pests of agricultural crops. More and more studies expanding the knowledge on the role of some mite groups have been performed in the light of their spreading and colonization of new areas: for example, eriophyoid mites have been identified as plant virus vectors (Stenger *et al.*, 2006). The mite *Tetranychus urticae* Koch, 1836, one of the major phytophagous pests of numerous plants, has been involved with increasing frequency in cases of sensitization and allergenic problems in people: 20 years ago, it was diagnosed as an occupational risk (Astarita *et al.*, 1996), was later detected in fruit growers (Kim and Kim, 2002) and recently was found in a population without occupational exposure (Santos *et al.*, 2014).

The aim of this chapter is to identify some situations and cases involving tick and house dust mite groups that mean these organisms should be included as alien invasive pests. In these times characterized by climate change, extreme weather events and global trading, mites belonging to the cited groups and others like ticks and house dust mites may be designated as invasive species representing a risk for human health.

4.2 Ticks as Alien Pests: Biology and Risks of Pathogen Transmission to Humans and Animals

Ticks are blood-feeding obligate parasites that do not have the potential to cover long distances and therefore spread out of an endemic area by themselves: they either rely on the recurrent occupation of burrows and nests by their hosts (nidiculous ticks) or on the random passage of a suitable host nearby (non-nidiculous ticks) to encounter a host and begin or continue their life cycle. Their spread to new geographic areas is therefore related to their passive transport either on their hosts (travelling or imported pets, as well as migratory birds) or by accidental transportation (Sonenshine and Roe, 2014).

Establishment of ticks outside their native range has been demonstrated only for a few species, since these parasites need the presence of a suitable host (during the parasitic phase) and/or the presence of specific environmental conditions (during the off-host phase) to continue and eventually complete their life cycle (Sonenshine and Roe, 2014).

According to a recent review (Uspensky, 2014), several tick species can be considered pests; these ticks include either non-nidiculous species living mostly outside (i.e. *Ixodes* sp., *Dermacentor* sp.) or those well-adapted to urban environments since they show some appropriate biological features (like adaptability to a wide host range, tolerance to long star-vation periods and tolerance to indoor climatic conditions, i.e. ticks of the *Argas reflexus* group – soft ticks infesting pigeons – and *Rhipicephalus sanguineus sensu lato (s.l.)*, commonly known as the brown dog tick). The phenomenon of urbanization has been supporting both groups of ticks, with expansion of urban areas into rural surroundings as well as expansion of green areas in the cities creating favourable conditions to non-nidiculous ticks and expan-sion of the size of human settlements increasing the chances for the development of nidicu-lous ones (Uspensky, 2014).

In order to apply the 'alien pest' definition to ticks, one should distinguish between: (i) ticks that are introduced into non-endemic geographical areas but that do not have the potential to establish themselves outside their endemic range and (ii) ticks like those in (i) that are able to complete their life cycle and therefore increase their population in the new, non-endemic geographical area where they are introduced – in the case of ticks often indoors, therefore independently from the geographical area. This applies mostly to *R. sanguineus s.l.* (Gray *et al.*, 2013).

Usually the concern regarding alien ticks is limited to colder climates, especially the northern hemisphere, where the variety of tick species is usually smaller compared with that of tropical areas. Continuous monitoring of the situation is needed because the role of exotic ticks, well known for some decades, has increased in importance as a result of both ongoing climate change and the huge increase in pet travel and travel in general. New climatic condi-tions could lead to the establishment of new foci of infestation for alien tick species with consequent larger potential of transmission of tickborne pathogens. A role for climate change has been demonstrated only in terms of expansion of the geographical distribution of non-nidiculous ticks, like *Ixodes ricinus* (Linnaeus, 1758) and *Dermacentor reticulatus* (Fabricius, 1794) and does not seem to be relevant for indoor-infesting ticks (Gray *et al.*, 2013).

4.2.1 *Rhipicephalus sanguineus sensu lato* (the brown dog tick)

Rhipicephalus sanguineus sensu lato (Fig. 4.1a) is a three-host tick that usually infests dogs. The duration of the life-cycle phases, including feeding periods, are influenced by both biotic (host-related) and abiotic (environmental) factors. Larvae can hatch from eggs after 6 days to some weeks, then they find a suitable host and they can feed for 2–6 days until they drop to the ground and moult into nymphs in 5–23 days. Nymphs attach to a host, feed for 3–9 days, drop to the ground and moult into the adult stage. Male and female ticks then mate on the host; fertilized females feed for about 9–14 days and then drop to the ground and lay eggs (up to 5000) in sheltered spots and then die. The whole life cycle can be com-pleted in 63 days in optimal environmental conditions; unfed ticks can survive for 6–19 months (depending on the unfed stage) (Taylor *et al.*, 2007; Dantas-Torres 2010; Gray *et al.*, 2013). The potential of this tick to colonize households efficiently is linked to the quite wide range of relative humidity (35–95%) and temperature (20–35°C) that *R. sanguineus* can tolerate. It seems that the fitness of *R. sanguineus* is increased at higher relative humidity rates, making relatively dry households not optimal for the development of these ticks. Another important biological feature of *R. sanguineus s.l.* is negative geotaxis, i.e. these ticks tend to climb upwards along different surfaces (walls, furniture, curtains) (Dantas-Torres, 2010; Gray *et al.*, 2013).

The importance of *R. sanguineus s.l.* is especially related to its ability to be a vector for many tickborne pathogens, some of them being human pathogens (*Rickettsia conorii, Rickettsia rickettsii*), some others being parasites (*Babesia vogeli, Hepatozoon canis*) or bacte-ria (*Ehrlichia canis*) of primary veterinary interest (Gray *et al.*, 2013).

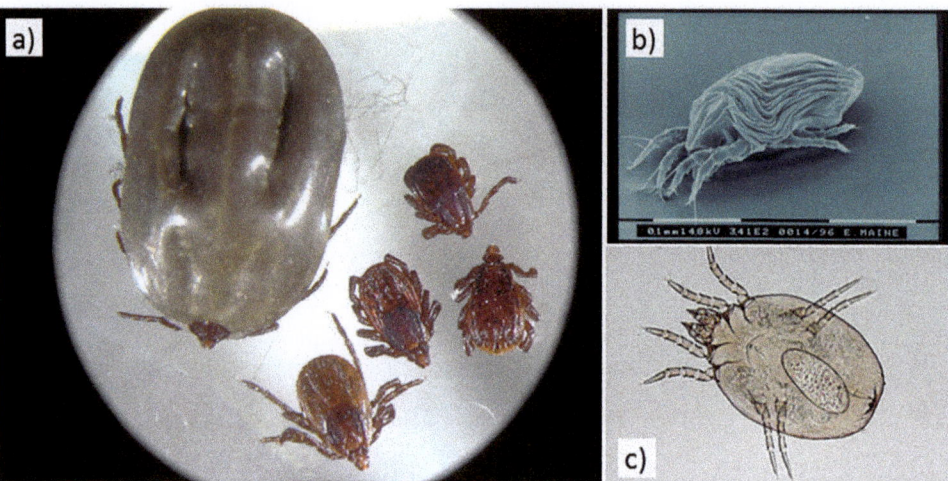

Fig. 4.1. *Rhipicephalus sanguineus*, the brown dog tick, (a) female and immatures; *Euroglyphus maynei*, (b) dorsal view at SEM; (c) female with egg, ventral view. (4.1a: SVA (National Veterinary Institute, Uppsala, Sweden); 4.1b,c (CREA-DC, Florence, Italy).)

4.2.2 Travelling pets, travels and the brown dog tick: an emerging problem?

Several ways of introduction of *R. sanguineus s.l.* in a country have been described or hypothesized, here listed in order of assumed frequency of occurrence: (i) ticks infesting dogs imported from endemic countries; (ii) ticks infesting dogs returning from travels to endemic areas; and (iii) ticks carried with luggage by people returning from travel in endemic countries. Some examples of these patterns will be provided by the case studies described below.

Despite the anecdotal role of *R. sanguineus s.l.* as a pest tick able to infest households even in colder climatic zones, it is difficult to gather a complete picture of the published reports of indoor infestation by this tick because they were often published long ago or in languages other than English. Moreover, many cases of household infestation by this tick are not reported, as in the case of Finland (Uspensky, 2014). Usually detailed and individual case descriptions come from north-western Europe, where the occurrence of household infestation has been considered somehow exceptional and therefore deserving of increased attention. In fact, cases of house infestation or accidental introduction of *R. sanguineus s.l.* have been described in detail, e.g. in the UK, Denmark, Poland, Belgium and Sweden (Winding *et al.*, 1970; Szymiański, 1979; Fox and Sykes, 1985; Sibomana *et al.*, 1986; Forshell, 2012).

In Denmark, after two initial reports of this tick in 1966 and in 1968 after the introduction of infested dogs from African countries, an 'outbreak' of 29 tick infestation events was reported in 1969 (Winding *et al.*, 1970), demonstrating the potential for a rapid spread of the parasite. In other southernmost countries, these events seem to be relatively more common: for example, 22 events of introduction of *R. sanguineus s.l.* (cases of dogs with established tick infestation) have been reported in Germany between January and December 1995. Two of the dogs had been imported from areas where this tick is endemic, 11 dogs had travelled to endemic countries with their owners and came back infested, in the remaining cases dogs had never left the country (n=6) or the cases could not be traced back. Interestingly, 16 abundant and recurrent household infestations were reported and in five households even people were attacked by the ticks (Dongus *et al.*, 1996). *Rhipicephalus sanguineus* presence and events of home tick infestations are not reported in countries like Italy, while major attention is focused on monitoring the incidence of related human diseases, like Botonneuse Fever caused by *R. conorii* (Maroli *et al.*, 1996).

The case of the UK deserves special consideration since in this country the progressive changes in regulation have led to an increase in the introduction of *R. sanguineus s.l.* and, especially in recent years, reports of house infestation. Strict quarantine laws were in force in the UK until 2001, when the Pet Travel Scheme (PETS) was introduced. After 2012, when a new and more relaxed PETS was introduced and the obligatory application of acaricides for pets entering the UK was withdrawn, a significant increase in dog brown tick introduction events was recorded. A programme for the passive surveillance of ticks found in the UK (Tick Surveillance Scheme, TSS) was introduced in 2005 by Public Health England; TSS also receives information on ticks from the Animal and Plant Health Agency. This programme, aimed at monitoring epidemiological features of ticks in general, has been shown to be useful for detecting less common tick species as well as the introduction of alien ticks. One more source of information on the occurrence of ticks in the UK is the University of Bristol's Big Tick Project. Through the aggregated records of these different information sources, at least 40 importation events of *R. sanguineus s.l.* have been detected in the UK since 2012; the impression of the authors is that not all cases are detected (Hansford *et al.*, 2017).

Regarding cases of home infestation, while between 1969 and 2013 only sporadic events had been reported and described in the UK (Best *et al.*, 1969; Fox and Sykes, 1985; the latter is summarized here as case study 1), from 2013 to 2016 at least three cases of house infestation have been reported (Hansford *et al.*, 2017). Before the introduction of PETS, *R. sanguineus s.l.* ticks in the UK were only found in quarantine kennels (Hoyle *et al.*, 2001).

Case study 1: London, UK

In London in July 1983, a dog owner submitted for identification ticks collected from a dog infested despite repeated monthly acaricide treatments. After the ticks were identified as *R. sanguineus s.l.*, the dog owner sprayed with acaricides the rooms in the household where ticks had been seen crawling around. In October 1983, when the infestation was considered over, the dog appeared to be re-infested. *R. sanguineus s.l.* were still present in the house and they were mostly found in the bedroom (mostly in the dog's bed and underneath it) and lounge (in this case in old furniture). Around 200 ticks (of which 66% were larvae, 15% nymphs and 18% adults, equally distributed between males and females) and more than 2300 eggs were present, demonstrating that the whole life cycle was taking place. The infestation was considered definitively extinct in September 1984, after the most heavily infected rooms of the house had been sprayed with acaricides (gamma benzene hexachloride and synergized tetramethrin) and after the infested dog and some other visiting dogs had been properly treated. In this case it is unclear what was the most likely source of infection. The infested dog had boarded in a quarantine establishment one month before the first ticks were observed, but no ticks were observed in the establishment at the time, only later on. That ticks were brought in within the luggage of a relative arriving from southern France shortly before the onset of the problem cannot be excluded (Fox and Sykes, 1985).

Case study 2: Warsaw, Poland

The first report of the establishment of *R. sanguineus s.l.* in a flat in Warsaw, Poland is dated 1977. According to the original description (Szymiański, 1979), some dog owners had a vacation in Puszcza forest (Mazurian region, Poland) which was rich in ticks. Despite regularly removing the ticks, some ticks dropped off during a short stay in their home in Warsaw in the middle of their vacation. Two weeks after coming back from holiday, they found a lot of ticks on their dog, and these ticks were identified as *R. sanguineus s.l.* The ticks were later found infesting furniture, walls and floors, even clothes and bed linen, but they were never found attached to people. At the time of first description, the source of infection was not identified (since the dog had never left Poland), even if it was suspected that the dog had acquired ticks from another dog at the place vacation, which was frequented by many tourists. Additional information (Nowak-Chmura, 2014) identified another dog brought to the same vacation place by a family from Italy was the likely source of infection for the Polish dog. The peculiarity of this case is that no direct import of ticks by means of travelling abroad had been involved, making it difficult to trace the source of infection.

Case study 3: Stockholm, Sweden

The first documented occurrence of *R. sanguineus s.l.* in Sweden dates to 1988, when four cases of infestation were reported from different places in the country (Christensson, 1988). Despite some regulatory changes in Sweden after it became a member of the European Union (e.g. abolition of quarantine for dogs and cats in 1994), introductions of *R. sanguineus* are still regarded as occasional. For example, *R. sanguineus s.l.* were submitted for identification at the National Veterinary Institute twice a year in the period 2007–2012 (Forshell, 2012). Since in Sweden infestation by *R. sanguineus s.l.* does not have to be officially reported, no records are available for its occurrence and therefore it is not possible to identify an increase of the frequency of *R. sanguineus s.l.* importation events.

The only detailed description of the infestation of an apartment in Sweden dates back to 2009, when in late summer a villa in Stockholm was found to be infested by this tick. In this case the owners of a toy poodle dog found many unidentified arthropods on its fur that in the beginning they thought were fleas. Afterwards they submitted the arthropods to the National Veterinary Institute and they were identified as nymphs of *R. sanguineus s.l.* Three other dogs and two cats were living in the same household, a large villa dating from the 1800s with wooden floors and other features typical of an old house. Nearly at the same time as tick identification, one of the owners discovered that the same ticks were present in several places in the house. The treatment proved to be difficult both for the dogs and for the household. The dog received a first treatment with pyriprol, which was not completely effective, and a subsequent treatment with metaflumizon + amitraz, although this is toxic to cats so they had to be removed from the house. The villa was fumigated with pyrethroids several times, with some damage to the house and the furniture because a persistent yellow layer remained after every application. The source of infection was identified as a journey the owners – but not their animals – had taken in Thailand in August, eight months before the owners discovered the infestation. The occupants of the household, as well as the cats, were never attacked by the ticks; the preferred non-canine component of the family, a small girl, was only bitten once. The infestation was brought under control in the summer of 2010 and it was definitively over in November of the same year (Forshell, 2012).

4.2.3 Secondary and potential alien tick species

Other tick species that could be considered as alien invasive species, potentially pests, are those imported on animals, especially reptiles and birds (Nowak-Chmura, 2014).

Ticks introduced on reptiles mainly belong to the genus *Amblyomma* and *Hyalomma*. A complete literature on the phenomenon of invasive tick species introduction is available for the USA (Burridge, 2011). The author reported that 100 species of alien ticks were introduced in the continental USA over the past half-century. Of these, some have been able to circulate between exotic animals (16 species, 12 of which belong to the genus *Amblyomma*), others have been found on animals native to the USA (10 species, 9 of them belonging to genus *Amblyomma*). Regarding the success of the invasive ticks in establishing permanently in the country, only three species have become established in limited foci in the southern USA (*A. dissimile* Koch, 1844 and *A. rotundatum* Koch, 1844 in Florida and *A. cajennense* (Fabricius, 1787) in southern Texas; this species is particularly interesting since it is a mammalian tick). Two more species are suspected of being established in Florida (*A. auricularium* (Conil, 1878) and *A. exornatum* Koch, 1844), while established foci of five species (*Rhipicephalus evertsi* Neumann, 1897, *Dermacentor nitens* Neumann, 1897, *A. marmoreum* Koch, 1844, *A. sparsum* Neumann, 1899, *Aponomma komodense* Oudemans, 1928) have been eradicated (Burridge, 2011).

The importance of the role of alien tick pests is historically well known in the USA. The introduction of *R. (Boophilus) microplus* (Canestrini, 1888) and *R. (B.) annulatus* (Say, 1821) by explorers and colonists in infected cattle threatened the American cattle industry in the first half of 1900s because of a disease – Texas fever – caused by protozoan parasites (*Babesia bigemina*) transmitted by these ticks. Both ticks and the disease were successfully eradicated in the USA in 1960, but they are intermittently reintroduced into the USA from across the Mexican border. The major concerns related to invasive ticks in the USA are the risk of introducing (or reintroducing) in the country diseases like heartwater (a bacterial disease of ruminants caused by *Ehrlichia ruminantium*, transmitted by several *Amblyomma* species, but especially by *A. variegatum* (Fabricius, 1794)), that is endemic in the east Caribbean (despite several programmes undertaken to eradicate it) and the abovementioned Texas fever (Burridge, 2011).

Burridge (2011) underlined the importance of regulations and tried to summarize the complex regulatory system for the movement of animals and animal products to the USA, identifying some incompleteness (like the fact that no general rule is available regarding reptile importation) that could be partly responsible for the introduction of alien ticks in the USA. In some cases reactionary rather than precautionary actions have been taken by US governmental agencies to avoid the introduction of alien ticks. For example, in 2000, after discovering that some tortoise species imported in the USA were often infected by tick vectors of heartwater, importation of these animals was prohibited. In 1989, after imported ostriches were found to be harbouring several species of non-native ticks capable of transmitting various exotic diseases, an interim rule was issued by the USDA prohibiting importation of ratites into the country (Burridge, 2011).

Such an exhaustive summary about invasive ticks is not available for other geographical areas. In Europe, a literature review (period 1979–2014) of records of the events of ticks introduced on reptiles has been published (Mihalca, 2015). According to the author, at least 12 *Amblyomma* species together with *Hyalomma aegyptium* (L.) have been introduced in Europe. The establishment of these ticks in the new environment has been described in only two cases, for *H. aegyptium* (L.) in a colony of *Testudo graeca* L. hosted in a garden with Mediterranean climate and for *A. latum* Koch, 1844 in the terrarium of *Python regius* Shaw, 1802, both in Spain (Brótons and Estrada-Peña, 2004). A summary of alien tick introduction in Poland has also been published (Nowak-Chmura, 2014) and in this review even ticks introduced by migratory birds are named. Out of 14 alien tick species, three (*Ixodes eldaricus* Dzhaparidze, 1950, *I. festai* Tonelli-Rondelli, 1926 and *H. marginatum* Koch, 1844) had been introduced by migratory birds, two on mammals (cattle: *R. rossicus* Yakimov and Kol-Yakimova, 1911; dog: *R. sanguineus*) and the rest on reptiles.

The relevance of ticks imported with reptiles has so far been considered quite small, since usually these ticks belong to species that are not able to establish in colder climates; this is also demonstrated by the fact that these events occurred only in the southern USA and southern Europe.

In order to better understand the risks related to the introduction of alien ticks in previously tick-free geographical areas, improvement is needed in several tools and fields, including predictive models able to identify areas at risk of establishment of introduced ticks; standardization of rules regarding animal trading in order to include both host species (reptiles) and ticks in the controls; and surveillance of the occurrence of new tick species, using the TSS in the UK as a model. Finally, the vectorial capacity of reptile and other less known tick species should be studied to assess the risk that these ticks pose to human and animal health.

Alien mammal species can alter the ecology of native ticks or even cause the spread of non-native tick pathogens introduced along with these species. For example, it has been reported that *Tamias* contribute to a higher density of nymphs infected by *Borrelia* sp. bacteria and harbouring *Borrelia* sp. characterized by a higher diversity (Marsot *et al.*, 2013).

4.3 House Dust Mites

Generally, humans spend most of their life in houses and other indoor environments, cohabiting with arthropods that may act as pests, disease vectors and producers of sensitizing allergens (Madden *et al.*, 2016). Their presence is often detected as result of the onset of problems relevant to human health and well-being; however, relatively little information is available about arthropods (mites *in primis*) living in homes and the factors driving their ecology.

'Domestic mites' refers to a variety of small (<0.4–0.5 mm in size) arthropods found living in close proximity to humans and include house dust mites and storage mites (Spieksma, 1991). The house dust mites commonly found in human homes typically belong to the family Pyroglyphidae (Analgoidea, Astigmata, Acariformes). Other mites can be found in house dust, especially several species from the families Glycyphagidae, Aeroglyphidae, Chortoglyphidae, Echimyopodidae, Tyroglyphidae, Lardoglyphidae and Suidasiidae, as well as species of the families Tarsonemidae and Cheyletidae. Some of these (i.e. Glycyphagidae, Tyroglyphidae) are typically regarded as storage mites because they occur widely in stored products and contribute to their deterioration in quality (Fang and Cui, 2009).

Regarding mites' global frequency and abundance, the main house dust mite species belong to the family Pyroglyphidae and are *Dermatophagoides pteronyssinus* (Trouessart, 1897), *D. farinae* Hughes, 1961 and *Euroglyphus maynei* (Cooreman, 1950) (Colloff, 2009). *Dermatophagoides farinae* is common in continental Europe and North America, but is rare in the UK and Australia. *Blomia tropicalis* (van Bronswijk, Cock and Oshima, 1973) (family Echimyopodidae) has emerged as a particularly important species in the tropics and subtropics. In rural homes in temperate areas, species of *Glycyphagus* and *Lepidoglyphus* (family Glycyphagidae) may be very abundant. Traditionally, the common name 'house dust mite' has been used to include those members of the family Pyroglyphidae that live permanently in house dust. Terms such as 'domestic mites' have been used to define pyroglyphid mites as well as stored products species such as *Lepidoglyphus destructor* (Schrank, 1781).

Taxa of home mites are also found in non-synanthropic habitats and several astigmatid species are parasitically associated with animals for part or all of their life cycles (Colloff, 2009). Pyroglyphids are frequently associated with birds' nests, Acarids live in soil and plant litter and may be associated with insects, birds and small mammals, Glycyphagids live in mammal burrows while species of the superfamily Glycyphagoidea are mainly associated with mammals (Baker *et al.*, 1976; Hughes, 1976). The complexity in relationships and associations, temporary or stable over the mites' life cycle, leads to increased chances of contact with human environments, either through the intermediary role of domestic animals or through their high resistance to changing environmental conditions (Mathison and Pritt, 2014).

The importance of some domestic mites is highly related to the allergen presence they cause (Raulf *et al.*, 2015); some may also be considered responsible for other, non-allergic, diseases in humans, e.g. acariasis. To date, not many reports on human acariasis have been published, but an increase in reports is likely in the coming years (Cui, 2014). Little is known about acariasis: mites can parasitize the human body in various tissues from the gastrointestinal tract to the lungs. These interactions, still considered as sporadic and non-specific, should in the future be approached as a true parasitic relationship.

4.3.1 House dust mites as colonizers of new environments

Due to their cryptogenic origin and current domestic distribution, house dust mites cannot directly fit in the concept of 'alien species': first, they are strictly adapted to indoor environments and second their dynamics are generally determined by human habits that are generally independent of outdoor conditions.

In this chapter, we present the role of house dust mites and main storage mites not only as disseminators of allergens but also as potential indicators of environmental quality in different settings (newly colonized areas, both geographic and within human dwellings). Regarding these aspects, it could be useful to record both the presence of new species in the environment and the onset of new associations among different domestic mites. House dust mites, *D. pteronyssinus*, *D. farinae* and *E. maynei*, frequently can be found together. Doctors and allergologists commonly recommend lowering the relative humidity and moisture in households to reduce allergy problems. Reducing the availability of water to dust mites is a key aspect for decreasing their populations and spread, but it is not known if all the three species are affected in the same way by the increased ratio water loss/water uptake.

4.3.2 Domestic mites associated in indoor detrital ecosystems

Although simplifying the concept of the discrete ecosystem in the trophic relationship among various organisms, van Bronswijk (1972) supported the idea of house dust as a discrete ecosystem where skin scales, microorganisms and mites basically represent food for house dust mites.

The composition in species of the domestic mite fauna has changed over the centuries (Colloff, 2009): acaroid and glycyphagoid mite species were reported from European houses during 17th–19th centuries. The year 1964 can be considered a milestone, when pyroglyphids (*D. pteronyssinus*, *D. farinae* and *E. maynei*) in European houses were mentioned and their involvement in dust allergy considered (Voorhorst *et al.*, 1964). It is highly likely that pyroglyphids were not absent from houses before this period but were unlikely to be detected because of the body size of the mite species; furthermore, the capability of acaroids and glycyphagoids to develop dense populations was limited by conditions in domestic hygiene and food storage (i.e. refrigeration; Colloff, 2009). In the last 40 years of the 20th century, screenings were performed aimed at evaluating their relative abundance and composition in apartments in different cities, for example in Hungary (Halmai, 1984), and in Moscow (Petrova and Zheltikova, 1990) and Glasgow, Scotland (Colloff, 1987).

A lot of research regarding house dust mites was mainly undertaken to assess their clinical relevance because of their allergenicity. The ecology of these mites in human environments and how this was affected by climate changes and human habits were somehow ignored. The presence or absence of a mite species can be due to factors relating to dispersal, behaviour, interaction with other species, as well as physical and chemical factors (i.e. temperature, moisture and oxygen): again, these parameters were evaluated only if they could determine differences in allergen exposure.

In the past 20 years, advances in biomolecular analyses and proteomics have allowed studies to address the ecology of allergens by paying attention to allergen diversity (An *et al.*,

2013). In 2000, Crowther *et al.* (2000) published the first interim report of the project 'A hygrothermal model for predicting house-dust mite response to environmental conditions in dwellings'. This investigated differences in the indoor/outdoor distribution of the common species, focusing on geographic and climatic conditions, and suggesting the ecological mechanisms that could allow new areas to be colonized by domestic mites. Even if home characteristics, at first, seem particularly useful in predicting arthropod diversity in homes (Madden *et al.*, 2016), indoor conditions are not independent of external ones. Information on different species association and new colonized areas may be useful to assess mite species allergen exposure and to identify phenomena of reactivity to cross-exposure.

4.3.3 *Euroglyphus maynei*, a house dust mite species of increasing relevance

Euroglyphus maynei (Fig. 4.1b), described by Cooreman in 1950, is one of the top three pyroglyphid species (with *D. pteronyssinus* and *D. farinae*) in terms of global frequency and abundance (Colloff, 2009). The first reports of this mite were in atypical habitats (i.e. mouldy cottonseed cake). It is able to live and develop in different environments and niches, in association with the other main dust mites and storage mites.

Among house dust mites, *E. maynei* is the third most common species, with a presence of about 30%, and demonstrates no consistent difference between temperate and tropical/subtropical regions, coastal and high rainfall areas, at high latitudes in low temperature and precipitation conditions (Colloff, 2009).

The biological potential of *E. maynei*, in experimental conditions, has lower results than those registered for the other Pyroglyphidae and Glycyphagidae (data reported in Colloff, 2009): at 25°C and 75% relative humidity it needs 39 days to double its population, the other dust mites require 20–25 days and Glycyphagidae 2–8 days. Both the mean fecundity for *E. maynei* (1.3 eggs/day; Fig. 4.1c) and mean oviposition period (11 days) are about half of the values registered for *D. pteronyssinus* and *D. farinae*.

Many dust and storage mite species may develop to desiccation-resistant stages, as a response or feedback to environmental conditions: this can be a homeomorphic protonymph as in *D. farinae*, but not *D. pteronyssinus* or *E. maynei*; or a heteromorphic deutonymph (hypopus) as in *G. domesticus* (De Geer, 1778) (Knulle, 1995), *L. destructor* (Knulle, 1995) and *Acarus siro* L. (Knulle, 1995), but not *Tyrophagus putrescentiae* (Schrank, 1781).

Detailed screenings performed in Scotland, California (Colloff, 2009) and Italy (S. Simoni, Florence, 2017, personal communication) confirmed the high presence of *D. pteronyssinus*, *E. maynei* and *G. domesticus*. *Euroglyphus maynei* is present in about 30–40% of homes.

The females of *E. maynei* are smaller than those of *D. pteronyssinus* and *D. farinae*, but the eggs of the two species are about the same size, implying that *E. maynei* loses significantly more body water per oviposition event.

Typical figures come from the screening in Glasgow apartments: *E. maynei* is present in 32% of homes and is likely to contribute to a significant level of allergen exposure such that it cannot be ignored, while *G. domesticus* is present in 14% of homes but is more abundant than *E. maynei*, suggesting that *E. maynei* tends to live in more houses/areas without developing very high populations.

Despite the general advice provided by medical doctors usually recommending people with allergy or sensitization to house dust mites to lower the relative humidity in rooms, it has to be noted that lower relative humidity may not be particularly detrimental to the survival of *E. maynei* adults (Colloff, 2009). Temperature seems to be the major driver of population growth (Frazier *et al.*, 2006) and the three house dust mites species show different responses to modified environmental conditions: the onset of resistant stages for *D. pteronyssinus* and *D. farinae*, the tendency to move for *E. maynei*.

Dermatophagoides farinae and *D. pteronyssinus* are often found together in houses in North America and Europe, sometimes with *E. maynei* or other species. Where multiple species are present, one is usually dominant, and this has been ascribed to differences in microclimatic requirements. However, laboratory experiments showed that when *E. maynei* was added to single-species cultures, *D. pteronyssinus* and *D. farinae* showed slower growth rates, indicating an inhibiting effect by the former species (S. Simoni, Florence, 2017, personal communication).

There are models that describe how, in suboptimal growth conditions, a greater proportion of the population of *E. maynei* survives longer than *T. putrescentiae* (Colloff, 2009).

Although *E. maynei* is characterized by a lower biological potential, it tends to respond to unsuitable conditions not by staying and resisting them, but by moving and searching for more agreeable conditions, therefore distributing allergens further afield.

Taken together, these considerations, along with the allergological properties identified at the genomic level in this species (Morgan *et al.*, 2017), suggest *E. maynei* is a potential candidate for the increase of environmental allergens and for becoming an alien invasive species in different environments.

4.4 Conclusions

Local, regional and global environmental changes can alter exposure rates not only to chemical and physical factors but also to other stressors like ticks and dust mites that have the potential to produce new or adverse impacts on human health (Smith *et al.*, 2014; Levy and Patz, 2015; LaKind *et al.*, 2016).

The impact of alien mites affecting agricultural production has been generally measured by the ecological consequences and threats to biodiversity and ecosystem functioning, but the evaluation of the effects of these species on human health has been in part neglected (MEA, 2005; Pascal *et al.*, 2010).

Trading, travelling and modified home conditions can increase the diffusion of ticks and mites affecting human health, i.e. causing higher exposures to vector-borne pathogens and allergens, respectively.

A more interdisciplinary approach to the spread of new (alien) species of ticks and house dust mites is needed that evaluates their ecological dynamics and their effects on human health.

Acknowledgements

G. Grandi would like to thank all those who helped in retrieving and translating original references, Edwin Claerebout (University of Ghent, Belgium), Giovanna Olivieri (University of Parma, Italy), Anna Maria Pyziel (Polish Academy of Sciences, Poland), Anna Rydzik (National Veterinary Institute, Uppsala) and Jakob Skov (University of Copenhagen, Denmark).

S. Simoni would like to express gratitude to Marisa Castagnoli, Roberto Nannelli, Marialivia Liguori (acarologists of CREA, Firenze, Italy) for enhancing the knowledge in the topic and to the staff of Anallergo company (Scarperia and San Piero, Firenze, Italy) for supporting continuing research programmes.

References

An, S., Chen, L., Long, C., Liu, X., Xu, X., Lu, X. and Lai, R. (2013) *Dermatophagoides farinae* allergens diversity identification by proteomics. *Molecular & Cellular Proteomics* 12(7), 1818–1828.

Aries, M.B.C. and Bluyssen, P.M. (2009) Climate change consequences for the indoor environment. *Heron* 54(1), 49–70.

Astarita, C., Di Martino, P., Scala, G., Franzese, A. and Sproviero, S. (1996) Contact allergy: another occupational risk to *Tetranychus urticae*. *Journal of Allergy and Clinical Immunology* 98(4), 732–738.

Baker, E.W., Delfinado, M.D. and Abbatiello, M.J. (1976) Terrestrial mites of New York II. Mites in birds' nests (Acarina). *Journal of the New York Entomological Society* 84, 48–66.

Best, J.M.J., Butt, K.M. and Rohrbach, J.A. (1969) Occurrence of *Rhipicephalus sanguineus* in London. *Veterinary Record* 85, 633.

Brótons, N.J. and Estrada Peña, A. (2004) Survival of tick colonies on captive imported reptiles in Spain. In: Seybold, J. and Mutschmann, F. (eds) *Proceedings of the 7th International Symposium on Pathology and Medicine in Reptiles and Amphibians*, Berlin, Germany, pp. 84–89.

Burridge, M.J. (2011) *Non-native and Invasive Ticks. Threats to Human and Animal Health in the United States.* University Press of Florida, Gainesville, Florida, USA.

Charles, H. and Dukes, J. (2007) Impacts of invasive species on ecosystem services. In: Nentwig, W. (ed.) *Biological Invasions.* Springer-Verlag, Berlin, pp. 217–237.

Christensson, D. (1988) Brun hundfästing, *SVAVet* 1, 13.

Colloff, M.J. (1987) Mites from house dust in Glasgow. *Medical and Veterinary Entomology* 1, 163–168.

Colloff, M.J. (2009) *Dust Mites.* Springer, Dordrecht, the Netherlands.

Crowther, D., Horwood, J., Baker, N., Thomson, D., Pretlove, S., Ridley, I. and Oreszczyn, T. (2000) House dust mites and the built environment: a literature review. Available at: http://www.ucl.ac.uk/bartlett-housedustmites/Publications/Publications/review10Oct02.pdf (accessed 19 December 2017).

Cui, Y. (2014) When mites attack: domestic mites are not just allergens. *Parasites & Vectors* 7, 411.

Dantas-Torres, F. (2010) Biology and ecology of the brown dog tick, *Rhipicephalus sanguineus. Parasites & Vectors* 3, 26.

Donatelli, M., Magareyb, R.D., Bregaglio, S., Willocquet, L., Whish, J.P.M. and Savary, S. (2017) Modelling the impacts of pests and diseases on agricultural systems. *Agricultural Systems* 155, 213–224.

Dongus H., Zahler, M. and Gothe, R. (1996) Die Braune Hundezecke, *Rhipicephalus sanguineus* (Ixodidae), in Deutschland: eine epidemiologische Studie und Bekämpfungsmaßnahmen [The brown dog tick, *Rhipicephalus sanguineus* (Ixodidae), in Germany: an epidemiological study]. *Berliner und Münchener tierärztliche Wochenschrift Journal* 109, 245–248.

Fang, W. and Cui, Y.B. (2009) A survey of stored product mites of traditional Chinese medicinal materials. *Pan-Pacific Entomologist* 85, 174–181.

Forshell, U. (2012) Invasion i villa av brun hundfästing [Invasion of a villa by the brown dog tick]. *Svensk Veterinärtidning* 64(6), 15–18.

Fox, M.T. and Sykes, T.J. (1985) Establishment of the tropical dog tick, *Rhipicephalus sanguineus*, in a house in London. *Veterinary Record* 116, 661–662.

Frazier, M.R., Huey, R.B. and Berrigan, D. (2006) Thermodynamics constrains the evolution of insect population growth rates: 'warmer is better'. *American Naturalist* 168, 512–520.

Gilioli, A., Schrader, G., Baker, R.H., Ceglarska, E., Kertész, V.K., Lövei, G., Navajas, M., Rossi, V., Tramontini, S. and van Lenteren, J.C. (2014) Environmental risk assessment for plant pests: a procedure to evaluate their impacts on ecosystem services. *Science of the Total Environment* 468–469, 475–486.

Gray, J., Dantas-Torres, F., Estrada-Peña, A. and Levin, M. (2013) Systematics and ecology of the brown dog tick, *Rhipicepalus sanguineus. Ticks and Tick-borne Diseases* 4, 171–180.

Halmai, Z. (1984) Changes in the composition of house-dust mite fauna in Hungary. *Parasitologia Hungarica* 17, 59–70.

Hansford, K.M., Phipps, L.P., Cull, B., Pietzsch, M.E. and Medlock, J.M. (2017) *Rhipicephalus sanguineus* importation into the UK: surveillance, risk, public health awareness and One Health response. *Veterinary Record* 180, 119.

Hoyle, D.V., Walker, A.R., Craig, P.S. and Woolhouse, M.E.J. (2001) Survey of parasite infections not endemic to the United Kingdom in quarantine animals. *Veterinary Record* 149, 457–458.

Hughes, A.M. (1976) *The Mites of Stored Food and Houses.* Her Majesty's Stationery Office, London.

Kim, Y.K. and Kim, Y.Y. (2002) Spider-mite allergy and asthma in fruit growers. *Current Opinion in Allergy and Clinical Immunology* 2(2), 103–107.

Knulle, W. (1995) Expression of a dispersal trait in a guild of mites colonising transient habitats. *Evolutionary Ecology* 9, 341–353.

LaKind, J.S., Overpeck, J., Breysse, P.N., Backer, L., Richardson, S., Sobus, J., Sapkota, A., Upperman, C.R., Jiang, C., Beard, C.B. *et al.* (2016) Exposure science in an age of rapidly changing climate: challenges and opportunities. *Journal of Exposure Science and Environmental Epidemiology* 26, 529–538.

Levy, B.S. and Patz, J.A. (eds) (2015) *Climate Change and Public Health.* Oxford University Press, Oxford, UK.

Madden, A.A., Barberán, A., Bertone, M.A., Menninger, H.L., Dunn, R.R. and Fierer, N. (2016) The diversity of arthropods in homes across the United States as determined by environmental DNA analyses. *Molecular Ecology* 25, 6214–6224.

Maroli, M., Khoury, C., Frusteri, L. and Manilla, G. (1996) Diffusione della zecca del cane (*Rhipicephalus sanguineus* Latreille, 1806) in Italia: un problema di salute pubblica. [Distribution of dog tick in Italy: a public health problem (in Italian)]. *Annali Istituto Superiore di Sanità* 32, 387–397.

Marsot, M., Chapuis, J.-L., Gasqui, P., Dozières, A., Massèglia, S., Pisanu, B., Ferquel, B. and Vourc'h, G. (2013) Introduced Siberian chipmunk (*Tamias sibiricus barberi*) contribute more to Lyme borreliosis risk than native reservoir rodents. *PloS ONE* 8, e55377.

Mathison, B.A. and Pritt, B.S. (2014) Laboratory identification of arthropod ectoparasites. *Clinical Microbiology Reviews* 27, 48–67.

MEA (Millennium Ecosystem Assessment) (2005) *Ecosystems and Human Well-being: Synthesis.* Island Press, Washington, DC.

Mihalca, A.D. (2015) Ticks imported to Europe with exotic reptiles. *Veterinary Parasitology* 213, 67–71.

Morgan, M.S, Rider, S.D. and Arlian, L.G. (2017) A draft genome of *Euroglyphus maynei* with predicted allergens. *Journal of Allergy and Clinical Immunology* 139(2) Supplement, AB118.

Navajas, M., Migeon, A., Estrada-Peña, A., Mailleux, A.C., Servigne, P. and Petanović, R. (2010) The Acari. In: Roques, A. *et al.* (eds) *Invasive Terrestrial Invertebrates in Europe. BioRisk* (special issue) 4(1), 149–192.

NIEHS (2010) The Interagency Working Group on Climate Change and Health. *A Human Health Perspective on Climate Change. Report Outlining the Research Needs on the Human Health Effects of Climate Change.* Available at: http://www.noaanews.noaa.gov/stories2010/PDFs/HHCC_Final_v5-4.pdf (accessed 5 August 2017).

Nowak-Chmura, M. (2014) A biological/medical review of alien tick species (Acari: Ixodida) accidentally transferred to Poland. *Annals of Parasitology* 60, 49–59.

NRC (National Research Council) (2010) *Informing an Effective Response to Climate Change.* The National Academies Press, Washington, DC.

Pascal, M., Le Guyader, H. and Simberloff, D. (2010) Biological invasions and the conservation of biodiversity. *Revue scientifique et technique (International Office of Epizootics)* 29, 387–403.

Petrova, A.D. and Zheltikova, T.M. (1990) The seasonal dynamics of allergenic mite numbers (Acariformes, Pyroglyphidae) in house dust of three apartments in Moscow. *Biologicheskie Nauki* 10, 37–45.

Raulf, M., Bergmann, K.C., Kull, S., Sander, I., Hilger, Ch., Brüning, T., Jappe, U., Müsken, H., Sperl, A., Vrtala, S., *et al.* (2015) Mites and other indoor allergens – from exposure to sensitization and treatment. *Allergo Journal International* 24(3), 68–80.

Santos, N., Areola, V. and Placido, J.L. (2014) *Tetranychus urticae* allergy in a population without occupational exposure. *European Annals of Allergy and Clinical Immunology* 46(4), 137–141.

Sibomana, G., Geert, S. and De Vries, T. (1986) L'etablissment de *Rhipicephalus sanguineus* (Latreille, 1806) a l'interieur des maisons en Belgique. *Annales des Societes Belges de Medecine Tropicale* 66, 79–81.

Smith, K.R., Woodward, A., Campbell-Lendrum, D.D., Chadee, D., Honda, Y., Liu, Q., Olwoch, J.M., Revich, B. and Sauerborn, R. (2014) Human health: impacts, adaptation, and co-benefits. In: Field, C.B., Barros, V.R. *et al.* (eds) *Climate Change 2014: Impacts, Adaptation, and Vulnerability. Part A: Global and Sectoral Aspects. Contribution of Working Group II to the Fifth Assessment Report of the Intergovernmental Panel on Climate Change.* Cambridge University Press, Cambridge, UK, pp. 709–754.

Sonenshine, D.E. and Roe, M. (2014) *Biology of Ticks*, 2nd edn. Oxford University Press, New York.

Spieksma, F.T.M. (1991) Domestic mites, their role in respiratory allergy. *Clinical and Experimental Allergy* 21, 655–660.

Stenger, D.C., Hein, G.L. and French, R. (2006) Nested deletion analysis of Wheat streak mosaic virus HC-Pro: mapping of domains affecting polyprotein processing and eriophyid mite transmission. *Virology* 350, 465–474.

Szymiański, S. (1979) Przypadek masowego rozwoju kleszcza *Rhipicephalus sanguineus* (Latreille, 1806) w warszawskim mieszkaniu [A case of massive development of the tick *Rhipicephalus sanguineus* (Latreille, 1806) in a Warsaw flat]. *Wiadomości Parazytologiczne* 25, 453–459.

Taylor, M.A., Coop, R.L. and Wall, R. (2007) *Veterinary Parasitology*. Blackwell, Oxford.

Uspensky, I. (2014) Tick pests and vectors (Acari: Ixodoidea) in European towns: introduction, persistence and management. *Ticks and Tick-borne Diseases* 5, 41–47.

van Bronswijk, J.E.M.H. (1972) Food preference of pyroglyphid house-dust mites (Acari). *Netherlands Journal of Zoology* 22, 335–340.

Venette, R.C. (2015) *Pest Risk Modelling and Mapping for Invasive Alien Species*. CAB International, Wallingford, UK.

Voorhorst, R., Spieksma-Boezeman, M.I.A. and Spieksma, F.T.M. (1964) Is a mite the producer of the house dust allergen? *Allergie und Asthma* 10, 329–334.

Winding, O., Willeberg, P. and Haarløv, N. (1970) Husflåten (*Rhipicephalus sanguineus*), en aktuel snylter hos hunden [The brown dog tick (*Rhipicephalus sanguineus*), an ectoparasite of dogs of current interest]. *Nordisk Veterinaermedicin* 22, 48–58.

Wu, X., Lu, Y., Zhou, S., Chen, L. and Xu, B. (2016) Impact of climate change on human infectious diseases: empirical evidence and human adaptation. *Environmental International* 86, 14–23.

5 Bugs, Ants, Wasps, Moths and Other Insect Species

Alain Roques[1]*, Cristina Preda[2], Sylvie Augustin[1] and Marie-Anne Auger-Rozenberg[1]

[1]*INRA Zoologie Forestière, Orléans, France and* [2]*Ovidius University of Constanţa, Romania*

Abstract

A total of 43 insect species non-native to Europe are so far considered to affect human welfare through their biting, urticating and allergenic properties, or by causing domestic nuisances. They involve several orders. In Hymenoptera, species in two families, Formicidae (ants) and Vespidae (wasps and hornets), are known to cause disturbance and health problems. Several moth species (Lepidoptera) have urticating larvae which may induce painful urticarial and allergic reactions. Bugs in five Hemipteran families have direct impacts on health, such as bed bugs and kissing bugs, which are vectors of pathogens, but most are considered to be household pests, causing nuisances to people when invading houses or aggregating on walls. Several non-native species of cockroaches that develop in synanthropic habitats have body parts, saliva or faeces containing powerful indoor allergens, and they can also facilitate mechanical transmission of pathogens to humans. Some species of Siphonaptera (fleas) and Phthiraptera (lice), which are obligate ectoparasitic insects of warm-blooded animals, are of high importance for human health because they cause itches and skin infection, and transmit major diseases such as bubonic plague and murine typhus.

5.1 Introduction

Whatever the continent, insects represent one of the major groups of invading organisms connected with the ever-increasing worldwide trade. This is especially true for Europe (DAISIE, 2009). In this region, the rate of establishment of insect species originating from another continent has nearly doubled in 50 years, increasing from an average of 10.9 species per year for the period 1950–1974 to an estimated 19.6 species per year for 2000–2008 (Roques, 2010). A total of 1418 non-native insect species have established in Europe up to late 2015 (Roques *et al.*, 2016). This recent influx of non-native insects primarily relates to phytophagous species and is probably connected to the increased magnitude in the global movement of live plants for ornamental purposes (Eschen *et al.*, 2015). In contrast, the introduction of parasitic and detrivorous species proportionally decreased during recent decades, in Europe at least (Roques, 2010). Most of the species that directly or indirectly affect human health belong to these two last feeding groups, so their establishment in Europe has followed the same trend (Fig. 5.1). A preliminary list of such species has been

---

* E-mail: alain.roques@inra.fr

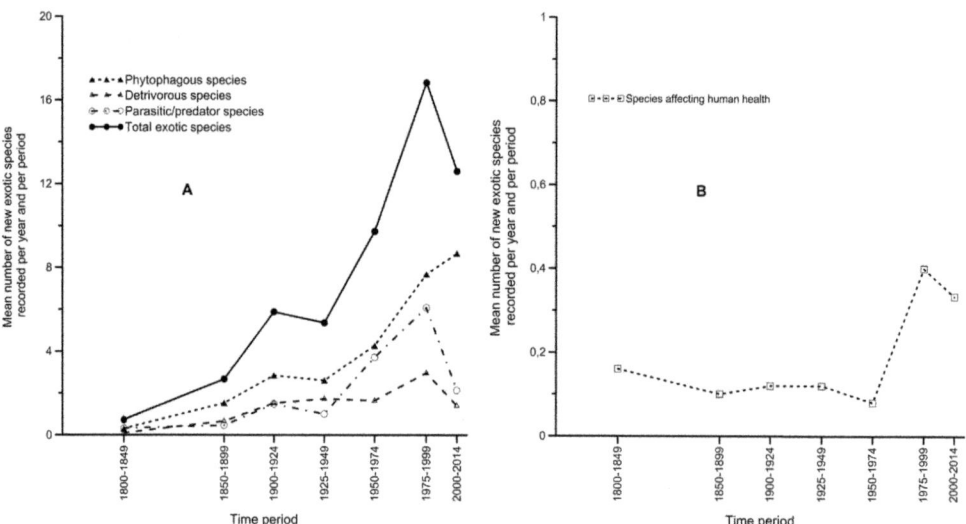

Fig. 5.1.   Temporal changes in the mean number of new records per year of insect species exotic to Europe since 1800. A: Total insects and detail per feeding regime (modified from Roques, 2010 and Roques *et al.*, 2016); B: Insects known to affect human health (Roques, unpublished data extracted from EASIN database, http://easin.jrc.ec.europa.eu)

provided by Kenis and Branco (2010). A further extraction from the DAISIE database (Delivering Alien Species Inventories in Europe; www.europe-aliens.org) updated by the EASIN catalogue (European Alien Species Information Network; http://easin.jrc.ec.europa.eu) resulted in a total of 43 insect species non-native to Europe so far showing biting, urticating and allergenic properties, or causing nuisances to human welfare. However, this number is probably underestimated, especially for the species causing allergies.

The suspected introduction pathway for most of these 43 species is transport as stowaways in various kinds of vehicles and containers, and along with commodities having no direct biological interest for them such as manufactured goods. These species include a number of cockroaches (Rasplus and Roques, 2010), many ants (Rasplus *et al.*, 2010) and the Asian hornet (Rome *et al.*, 2013), some true bugs (e.g. bed bugs and seed bugs; Lesieur, 2014; Lai *et al.*, 2016), and mosquitoes (e.g. travelling with used tyres or 'lucky bamboos'; see Chapter 7, this volume). Another major pathway is as contaminants of animals, where insects are transported as ecto- or endoparasites along with their animal host and then may switch onto humans (e.g. fleas and lice; Kenis and Roques, 2010). However, the development of the ornamental plant trade has also increased the probability of introducing phytophagous species that can affect human health, either directly, such as urticating moths (Gyulai-Garai and Gyulai, 2008) and allergenic bugs (Anderson *et al.*, 2012), or indirectly, as indoor nuisances such as *Leptoglossus* bugs (Wheeler, 1992) and *Harmonia* ladybirds (Cranshaw, 2011). Most of these non-native insects primarily establish near the initial point of entry, i.e. in man-made habitats (Lopez-Vaamonde *et al.*, 2010), especially in cities as household pests favoured by urban heating, thus reinforcing the potential impact on human health. Cranshaw (2011) and Müller *et al.* (2011) published reviews of household pests in the USA and Zurich (Switzerland), respectively.

Non-native insect species can generate a wide range of impacts on human health, but large differences exist between groups. Direct impacts include the transmission of native or non-native diseases through bites by non-native species of mosquitoes and flies (see Chapter 7, this volume) and ants (Roxo *et al.*, 2010), but also through food contamination by cockroaches (Baumholtz *et al.*, 1997). Painful stings are inflicted by non-native hornets, wasps and ants. Dermatitis and other physical symptoms, as well as allergic reactions, are caused by urticating larvae of non-native moths of different families (Battisti *et al.*, 2017) and rarely also by non-native bugs (Anderson *et al.*, 2012). Allergic reactions and asthma are also induced by non-native coleopteran beetles (e.g. carpet beetles in house dust; Cuesta-Herranz *et al.*, 1997), cockroaches (Chapman and Pomes, 2009) and dust lice (Müsken *et al.*, 1998). Finally, an indirect impact on human welfare is also observed with regard to seasonal concentration of numerous individuals of several species of non-native bugs and beetles in houses, especially for overwintering under favourable conditions (Cranshaw, 2011).

5.2 Hymenoptera (Ants, Wasps, Hornets)

Although the Hymenoptera order usually represents a significant part of the non-native faunas of insects (e.g. 18.7% in Europe; Roques, 2010), only species in two families, Formicidae (ants) and Vespidae (wasps and hornets), are known to cause disturbance and human health problems. Rabitsch estimated that 'over 150 ant species are known to have become established outside their native range' (2011, p. 552), while 34 species of Vespidae were considered so by Beggs *et al.* (2011). Besides ecological and economic impacts, a few of these species have serious public health implications, ranging from simple nuisances to life-threatening allergies (Beggs *et al.*, 2011; Mazza *et al.*, 2014). Direct impacts, the most notorious and common, are related to allergens present in the venom of ants, wasps and hornets, which is introduced through human skin by the female stinger developed from its ovipositor. Indirect impacts are mostly attributed to ants as vectors of non-native microorganisms (Schindler *et al.*, 2015).

5.2.1 Formicidae (Ants)

A majority of ants introduced outside of their native habitat are commensal species confined to human-modified habitats (McGlynn, 1999; Holway *et al.*, 2002) and many have few or no impacts on human health. Nevertheless, some non-native ants can pose a serious public health problem and have direct and indirect effects on human health: (i) they sting and bite, injecting venoms similar to those of wasps, which can cause allergic reactions (Srisong *et al.*, 2016), even if they are rarely clinically relevant (Przybilla and Ruëff, 2010); (ii) they carry and transmit pathogenic microorganisms to humans (de Castro *et al.*, 2015), especially in hospital environments (Mello Garcia and Lise, 2013); and (iii) they cause significant nuisance in houses (Cranshaw, 2011).

Anaphylaxis cases resulting from venomous stings by fire ants were reported in the areas invaded by the red imported fire ant, *Solenopsis invicta*, the tropical red fire ant, *S. geminata*, the black imported fire ant, *S. richteri* (Collingwood *et al.*, 1997; Srisong *et al.*, 2016) and the electric ant, *Wasmannia auropunctata* (Mazza *et al.*, 2014). Two non-native species of *Pachycondyla*, the samsun ant, *P. sennaarensis* (Collingwood *et al.*, 1997), and the Chinese needle ant, *P. chinensis* (Srisong *et al.*, 2016), are major causes of ant sting venom allergy in the Asia-Pacific region and the USA, respectively (Nelder *et al.*, 2006).

Non-native ants are also potential vectors of pathogens, being indirectly responsible for the spread of human diseases. In Brazilian hospitals, some species were shown to be mechanical vectors of pathogenic microorganisms (Roxo *et al.*, 2010). Vectoring ants include fire ants (*Solenopsis* spp. and *Wasmannia auropunctata*) and species already referred to as a low threat for human health (*Monomorium pharaonis*, *M. destructor*, *Tapinoma melanocephalum*; Rabitsch, 2011; Bertelsmeier *et al.*, 2016), but also species previously considered as non-hazardous for human health, such as *Linepithema humile*, *Paratrechina longicornis* or *Pheidole megacephala* (dos Santos *et al.*, 2009; de Castro *et al.*, 2015). The identified microorganisms included dangerous pathogens (*Escherichia coli*, *Klebsiella*, *Pseudomonas*, *Staphylococcus* and *Streptococcus*; Mello-Garcia and Lise, 2013) as well as fungi (*Aspergillus*, *Purpureocillium* and *Fusarium*; Aquino *et al.*, 2013) and mycobacteria (Roxo *et al.*, 2010). Cranshaw (2011) also mentions increasing problems of non-native ants as household nuisances in the USA, involving the rover ant, *Brachymyrmex patagonicus*, the odorous house ant, *Tapinoma sessile*, the Argentine ant, *L. humile*, and the Caribbean crazy ant, *Nylanderia pubens*.

5.2.2 Vespidae (Wasps and Hornets)

The bites, stings and allergic reactions to Vespidae venom constitute major hazards (Mazza *et al.*, 2014). The venoms of these insects are various and can even show cross-reactivity (Srisong *et al.*, 2016). According to Beggs *et al.* (2011), the Vespidae species introduced outside their native range are mainly eusocial, and only a few are a threat to humans. They include species in the genera *Vespa* (hornets), *Vespula* ('wasps' in Europe and 'yellow jackets' in America) and *Polistes* (paper wasps).

The Asian black hornet, or yellow-legged hornet, *Vespa velutina nigrithorax*, is the most famous hornet introduced outside its native range. Originating from eastern China, it established in France in the last decade and is presently spreading in Western Europe (Rome *et al.*, 2013). This voracious honeybee predator has the potential to be deadly for allergic people. Although it seems more aggressive in its native range than in France (Monceau *et al.*, 2014), its rapid spread across South Korea, where the species was first recorded in 2003, seems to have a serious effect on people in urban areas (Choi *et al.*, 2012). The congeneric European hornet, *Vespa crabro*, introduced for biological control in the USA during the 19th century and at present widespread on this continent, presents a minimal stinging hazard in its introduced range (Beggs *et al.*, 2011). The situation appears similar with *V. affinis*, an Asian species introduced into North America and New Zealand (Kimsey and Carpenter, 2012).

Two Palearctic wasp species in the genus *Vespula*, *V. vulgaris* and *V. germanica*, have been introduced all over the world, whereas the North American *V. pensylvanica* has been introduced to Hawaii (Rust and Su, 2012). The majority of the deaths caused by stings in the invaded countries are attributable to these wasps (Beggs *et al.*, 2011). At least 14 species of paper wasps, *Polistes* spp., have been reported outside their supposed native range (Beggs *et al.*, 2011). Among them, *Polistes dominula*, a native to Europe, has arrived relatively recently in North America, whereas two Asian species, *P. chinensis* and *P. humilis*, are well-known in New Zealand (Beggs *et al.*, 2011). In their native range, stings from *Polistes* are frequent and painful (Biló *et al.*, 2005), and as for all the Vespidae, they can represent a threat to the invaded human communities according to the degree of allergy and sensitivity to allergens.

5.3 Lepidoptera (Moths)

The best-documented case concerns the pine processionary moth, *Thaumetopoea pityocampa* (Notodontidae), whose larvae carry urticating setae from the third larval instar onwards (Battisti *et al.*, 2017). Originally restricted to the Mediterranean area, this insect has expanded its distribution northwards since the late 1990s in response to climate change. Its gregarious larvae have the peculiar habit during winter of developing in colonies within silky tents. Winter warming is allowing them to survive in areas where they could not establish before. Moreover, recent studies have revealed several human-mediated introductions far beyond the moth's expansion front edge, through the trade of mature pine trees transplanted with soil that includes moth pupae. All the introduced colonies have been found in man-made habitats with recent plantations of mature pine trees, e.g. along highways, on roundabouts, near urban buildings and in recreation parks. The larval colonies have entered highly populated areas and have even survived better within cities because of the heat island effect. Simulations reveal that winter warming is presently making moth establishment possible in a large part of Western Europe (Battisti *et al.*, 2017).

Urticating setae are present on each abdominal tergite of mature larvae, up to 2 million per individual (Petrucco-Toffolo *et al.*, 2014). They can be actively released when the larva is threatened and probably evolved as a defence against natural enemies (Battisti *et al.*, 2011). People can be exposed to setae either through direct contact with larvae or their tents, or through setae spread by the wind (Battisti *et al.*, 2017). The threat to human health is especially high during the procession (hence the name 'processionary') of mature larvae moving down the tree to search for a favourable site on the ground to dig for pupation. The hooked-shape setae penetrate through the skin into peripheral blood vessels but can also enter the eyes or, less frequently, be inhaled into the respiratory system (Battisti *et al.*, 2011). Certain proteins in the setae are considered to be allergens (Rodriguez-Mahillo *et al.*, 2015). Battisti *et al.* (2017) and Moneo *et al.* (2015) reviewed the observed impacts on human health, which include local reactions such as dermatitis, stomatitis, ophthalmia, ptyalism, face oedema and tongue necrosis, but also allergic systemic reactions and in a few cases anaphylactic shock. In outbreak situations, a significant proportion of the human population can show symptoms, especially children (Vega *et al.*, 2011). The introduction of the moth on the Spanish Balearic Islands also resulted in an indirect effect, with people preferring to remove pine trees rather than risk being exposed to the processionary setae (Battisti *et al.*, 2017).

Another urticating processionary species, the oak processionary moth (*Thaumetopoea processionea*), native to continental Europe, has been introduced into the United Kingdom since 2006 as a result of the commercial movement of nursery oaks from the Netherlands infested by moth eggs. Moths have resulted in an outbreak of itchy rash in suburban southwest London (Mindlin *et al.*, 2012).

Non-native moths from other families also possess larvae that may inflict painful urtications. Larvae of slug moths of the Limacodidae family are often covered with protective stinging hairs. The arrival of urticating larvae of saddleback caterpillar moth, *Archaria stimulea*, a polyphagous forest defoliator introduced from North America to Europe in 2002, is posing a human health hazard, especially in gardens in Hungary (Gyulai-Garai and Gyulai, 2008). A related species, the nettle caterpillar, *Darna pallivitta*, a native of South-east Asia, is similarly threatening the island of Hawaii, where this introduced pest is causing defoliation of ornamental nursery stocks (Siderhurst *et al.*, 2007). The urticating larvae of a lymantriid moth introduced to New Zealand and Australia, the painted apple moth (*Teia anartoides*), have been suggested to induce adverse reactions in susceptible people, but no clear evidence yet exists in the field (Derraik, 2008).

5.4 Hemiptera (Bugs)

Hemiptera is a large insect order including scales, aphids, whiteflies, cicadas, planthoppers and true bugs. Only the last group include species directly affecting human health. The true bugs, or Heteroptera, comprise approximately 42,300 described species worldwide, but only a few have been introduced in continents other than the native one (e.g. only 16 non-native species in Europe; Rabitsch, 2010). Most species are phytophagous, many are predatory and some are haematophagous. The major pathway for non-native true bugs is transportation as contaminants with ornamental plants and as stowaways (Rabitsch, 2010). Five families, Cimicidae, Pentatomidae Reduviidae, Tingidae and Thaumastocoroidae, are known to cause human health problems. They include species with a direct impact on health, such as bed bugs and kissing bugs, but most are considered to be household pests, causing nuisances to people when invading houses or aggregating on walls.

Bed bugs in the Cimicidae family, the common bed bug, *Cimex lectularius*, and the tropical bed bug, *C. hemipterus*, are the best documented cases. Both are nocturnal blood-sucking ectoparasites living in proximity with humans in residential dwellings. They can be highly detrimental because their bites provoke dermatological reactions and super-infections can result in important physiological and psychological disorders. *Cimex lectularius* has been known as feeding on humans since 3550 BC at least, and it probably spread with the development of agriculture. In Europe, bed bugs were common until the Second World War because of bad hygiene conditions, but economic progress combined with massive insecticide treatments have led to an almost complete eradication in developed countries, while less-developed countries still experience infestations (Heymann, 2009; Delaunay, 2012). However, since the late 1990s bed bugs have re-invaded, at a world level, countries where they had been considered to have been eradicated (Bencheton *et al.*, 2011), because of the intensification of global travel and trade to and from areas where endemic populations remained (Delaunay, 2012; Lai *et al.*, 2016) and the increase of resistance to insecticides (Romero *et al.*, 2007). Europe, USA, Australia and Canada have recently experienced new, and important, bug infestations in houses and hotels, and *C. lectularius* was detected for the first time in Magallanes province, Chile, in 2004 (Faundez and Carvajal, 2014). Bed bugs can travel short distances via active dispersal to neighbouring apartments (Cooper *et al.*, 2016) and long distances via human-mediated passive dispersal in furniture, luggage or clothing. Bed bugs may also vector diseases, with at least 45 pathogens reported in these insects (e.g. *Bartonella quintana* and *Trypanosoma cruzi*; Lai *et al.*, 2016). However, public health reports have so far failed to produce evidence about the implication of bed bugs in recent disease outbreaks.

The haematophagous triatomine 'kissing' bugs (Reduviidae, subfamily Triatominae) are another important group of non-native species because they are the main vectors of the pathogenic protozoan *Trypanosoma cruzi*, the agent of American trypanosomiasis (Chagas disease) (Carabarin-Lima *et al.*, 2013). This disease, ultimately resulting in cardiomyopathy, gastrointestinal and neural disorders that may lead to death, is endemic from the southern USA to Central and South America. First restricted to rural human populations, it spread to non-endemic regions, and especially to urban centres, with the migration of infected people and the associated passive transport of the bugs at different stages of development (especially eggs), as has been reported in Chile (Neghme *et al.*, 1991). Although most kissing bugs can potentially vector *T. cruzi*, only a few (e.g. *Triatoma infestans*, *T. rubrofasciata*, *Rhodnius prolixus*) are common human commensals, and thus have to be considered as the most threatening ones. *T. rubrofasciata* is spreading throughout the tropical and subtropical regions, becoming a cause of concern for health authorities in cities in Vietnam (Dujardin *et al.*, 2015). Natural transmission of the parasite occurs through bug faeces, skin lesions or mucous membranes. In non-endemic areas, transmission may occur via organ transplantation, blood transfusion, congenital transmission and domestic animals (Dias, 2000). Chagas disease is considered to be one of the most important emerging health problems in non-endemic areas of the USA and several European countries (Gascon *et al.*, 2010).

Several non-native bug species cause a nuisance to people when entering apartments and houses or aggregating on walls or windows in large numbers. Since the 1970s, *Oxycarenus lavaterae* (Lygaeidae), a native of the Mediterranean basin, has increasingly been reported in northern European countries, where it may gather in very large numbers in urban environments (Reynaud, 2000). The conifer seed bug, *Leptoglossus occidentalis* (Coreidae), a native of western North America, was first introduced to eastern North America, and then to Europe and Asia (Lesieur, 2014), where it aggregates during autumn in its thousands on walls and houses for overwintering. It has thus become a nuisance to people (Wheeler, 1992). Similar nuisances have been reported for the brown marmorated stink bug, *Halyomorpha halys* (Pentatomidae), an Asian fruit pest introduced in North America and Europe, which moves during autumn, often in thousands, to human habitats (Müller *et al.*, 2011). In addition, fruit crop workers have complained of a slight allergic reaction to the chemicals released by the bug (Anderson *et al.*, 2012). Two other non-native species have been reported in urban parks as stinging people. The lace bug, *Corythucha ciliata* (Tingidae), a north American pest of plane trees that is at present recorded in most of Europe, Russia and northern Asia, is presumed to cause dermatitis to people in urban parks and playgrounds (Dutto and Bertero, 2013). The Bronze bug, *Thaumastocoris peregrinus* (Thaumastocoridae), an Australian species introduced to Italy and South Africa with Eucalypts, may have a similar impact (Jacobs and Neser, 2005).

5.5 Coleoptera (Beetles)

Beetles are usually the most species-rich group in non-native faunas (e.g. accounting for 25% in Europe; Roques, 2010). Except for ladybirds (Coccinellidae) and a few other species intentionally released for biological control of pests or weeds, most non-native beetles have been introduced accidentally as contaminants of plants, compost, seeds and wood-derived products but also as stowaways. Very few of these species have been shown to affect human health.

The Asian harlequin ladybird, *Harmonia axyridis*, largely introduced as a biological control agent against aphid pests, has spread to countries within which it was not intentionally released (Roy *et al.*, 2016). Besides negative effects on non-target species and wine production, it may affect human health. The large aggregations of adult beetles in houses when migrating for overwintering sites represent a nuisance through staining damage to carpets and furniture. In addition, *H. axyridis* can bite humans (Ramsey and Losey, 2012) and is suspected to cause urticaria and allergic reactions such as rhino-conjunctivitis and asthma (Goetz, 2008). The massive colonization of an intensive care unit in an Austrian hospital led to patients being relocated and the temporary closure of the unit, thus generating considerable financial costs (Roy *et al.*, 2016).

Non-native detrivorous species such as carpet beetles (Dermestidae), whose larvae are typical components of house dust, have been reported to cause allergic reactions, especially for wool workers and museum personnel, who developed occupational rhinoconjunctivitis and asthma. The larvae of *Dermestes maculatus*, a cryptogenic species, can provoke papular urticaria (Rustin and Munro, 1984). Allergens were also identified in carpet beetles of the genus *Attagenus* (Cuesta-Herranz *et al.*, 1997).

An accidentally introduced beetle, the emerald ash borer, *Agrilus planipennis* (Buprestidae) seems to affect human health indirectly by the magnitude of its impact on forest trees. This insect killed over 100 million ash trees in the USA since its first detection in 2002. The loss of trees was considered to increase human mortality through cardiovascular and lower-respiratory illness (Donovan *et al.*, 2013). More specifically, women living in a US county infested with emerald ash borer had an increased risk of cardiovascular disease (Donovan *et al.*, 2015). Indirect effects of ecosystem disruption by forest insects expanding in new areas with climate change have also been observed with the dramatic expansion of the native mountain pine beetle, *Dendroctonus ponderosae* in the USA. Embrey *et al.* (2012) suggested that the devastation of millions of hectares of pine forests could increase the risk of fire and the subsequent potential exposure to smoke could exacerbate respiratory and cardiac problems, bronchitis and asthma.

5.6 Blattodea (Cockroaches)

Roughly 4500 species of cockroaches have been described so far from almost all regions of the world, but the number of human-associated species is rather low (less than 1%; Gullan and Cranston, 2014). Cockroaches are typical stowaways that can easily be transported from one country to another, e.g. as contaminants of commercial goods. Eighteen species in the families Blaberidae, Blattellidae and Blattidae are considered non-native to Europe, most of them developing in synanthropic habitats like houses, buildings, greenhouses and cultivated areas (Rasplus and Roques, 2010). The Madagascar hissing roach, *Gromphadorhina portentosa*, is sold in pet shops, e.g. as food, and is becoming a relatively popular tool in education. Besides *Blattella germanica* (the German cockroach), four widespread non-native species are considered very important from a medical perspective: *Blatta orientalis* (the oriental cockroach), *Periplaneta americana* (the American cockroach), *P. australasiae* (the Australian cockroach) and *Supella longipalpa* (the brown-banded cockroach) (Service, 2012).

The medical importance of cockroaches refers to their potential to induce allergies and asthma because their body parts, saliva or faeces may contain powerful indoor allergens. Thus, cockroaches are one of the main health threats in human dwellings in Europe (Franchi *et al.*, 2006). Ingestion of contaminated food and inhalation of airborne particles represent common means of sensitization to cockroaches, a key factor inducing bronchial asthma, particularly in children living in lower socio-economic standards (Kang, 1990). Inner cities and urban areas are more exposed to allergic diseases caused by cockroaches. Cockroach allergens can persist for long periods of time in the environment, facilitating exposure and thus increasing rates of morbidity and mortality associated with asthma (Chapman and Pomes, 2009). These allergens are found throughout the infested dwellings (e.g. bedrooms), but their highest levels are encountered within kitchens (Chapman and Pomes, 2009).

By indiscriminately visiting various substrates, including food destined for human consumption and food-related facilities, cockroaches can facilitate the mechanical transmission of pathogens that can cause disease in humans, like viruses, bacteria, fungi, protozoa and helminths (Baumholtz *et al.*, 1997). More than 40 bacterial agents and protozoans pathogenic to humans have been isolated from cockroaches (e.g. *Bacillus subtilis*, *Entamoeba histolytica*, *Escherichia coli*, *Klebsiella pneumoniae*, *Mycobacterium leprae*, *Shigella dysentariae*, *Salmonella* spp. including *S. typhi* and *S. typhimurium*; Burns, 2009; Rasplus and Roques, 2010; Service, 2012). Cockroaches also host helminths (e.g. *Schistosoma haematobium*, *Taenia saginata*, *Ascaris lumbricoides*, *Ancylostoma duodenale*; Rasplus and Roques, 2010; Service, 2012) and fungi and moulds, including *Aspergillus*, *Candida*, *Mucor* and *Penicillum*, especially the Madagascar hissing roach (Yoder *et al.*, 2008). The Pacific beetle cockroach, *Diploptera punctata*, and the Speckled-feeder cockroach, *Nauphoeta cinerea*, have also been associated with foodborne pathogens (Blazar *et al.*, 2011). A growing concern exists about the possible transfer by cockroaches of bacterial pathogens and antibiotic-resistant microorganisms between livestock production systems and neighbouring urban dwellings (Ahmad *et al.*, 2011).

5.7 Siphonaptera (Fleas) and Phthiraptera (Lice)

Siphonaptera (fleas) and Phthiraptera (lice) are obligate ectoparasitic insects of warm-blooded animals, including humans. Many of them are of unknown origin because their present cosmopolitan distribution has been shaped by multiple and complex human migrations together with their associated herds and pets (Lounibos, 2002; Kenis and Roques, 2010). Some of these species are of high importance for human health because they cause itches and skin infection, and transmit major diseases such as bubonic plague and murine typhus.

Among the seven non-native fleas reported in Europe, three have rats and one has mice as their main host (Kenis and Roques, 2010), both of which usually live close to humans. Urban heating has allowed the tropical rat flea species *Xenopsylla cheopis* to become synanthropic in most of southern Europe, while *X. brasiliensis* has invaded the Canary Islands and is found sporadically in port areas (Beaucournu and Launay, 1990). A third rat flea species, *Nosopsyllus fasciatus*, is a temperate species from Asia, whereas the mouse flea, *Leptopsylla segnis*, is cryptogenic. Rat fleas can transmit the bubonic plague by carrying the bacteria *Yersinia pestis* (Beaucournu and Launay, 1990). *Xenopsylla cheopis* is also a vector of murine typhus fever caused by the bacteria *Rickettsia typhi* (Beaucournu and Launay, 1990), which can be occasionally transmitted by the mouse flea, *L. segnis* (Sekeyová *et al.*, 2013). Transmission to humans occurs through the flea bite, contaminated skin, conjunctiva or the respiratory route by aerosols of contaminated flea faeces (Kenis and Roques, 2010). Fleas are capable of multiple feeding and thus potentially can transmit bacteria to several hosts (Sekeyová *et al.*, 2013). Some other species affecting human health are probably non-native to Europe but arrived in ancient times (Kenis and Roques, 2010). They include the cat flea, *Ctenocephalides felis felis*, which is the only known vector of *Ricksettia felis*, the spotted fever agent, and occasionally transmits *R. typhi* (Sekeyová *et al.*, 2013), and the human flea (*Pulex irritans*) reported in Europe since the Bronze Age (Kenis and Roques, 2010). Allergic urticaria has also been reported due to multiple bites by pigeon fleas, *Ceratophyllus columbae*, that had massively colonized an apartment (Haag-Wackernagel and Spiewak, 2004).

A total of 31 lice are considered non-natives to Europe, but only a few of them directly affect human health. A rat louse, *Polyplax spinulosa*, originating from Asia, is occasionally responsible for the transmission of *Rickettia typhi* (Sekeyová *et al.*, 2013). In a similar pattern as for fleas, a number of other species are probably non-native to Europe but arrived through phoretic human transport long ago. They likely include the body louse (*P. humanus corporis*), the head louse (*Pediculus capitis*) and the crab louse (*Phtirus pubis*) (Kenis and Roques, 2010). Whereas the two latter species cause itches and skin infections, *P. humanus* is responsible for the transmission of *Rickettia prowazekii*, the agent of epidemic typhus, which has caused millions of deaths since the Middle Ages to nowadays, for example through infected lice carried by troops during wartime (Lounibos, 2002; Sekeyová *et al.*, 2013).

5.8 Psocoptera (Psocids)

Dust lice and book lice regularly aggregate in human dwellings, feeding on microflora and organic debris. They can be found in old papers, books, stored food products and more generally in dust. In Europe, 49 species are considered as non-natives (Schneider, 2010). One of them, *Liposcelis bostrichophila*, has been reported to be associated with allergic reactions in Germany. Specific IgE antibodies could be detected in 34% of the analysed patients living in cities but only in 15% of those living in rural habitats (Müsken *et al.*, 1998). IgE-mediated responses to a related species, *Liposcelis divinatorius*, were also detected in India in 20% of city dwellers with nasobronchial allergy (Patil *et al.*, 2001). Immunoblot analysis revealed 67, 59, 43 and 27 kD proteins as major allergens of Psocoptera (Patil *et al.*, 2001).

5.9 Conclusions

The increase in introductions of new non-native species shows no sign of saturation at a global level, especially in insects (Seebens *et al.*, 2017). Moreover, once introduced, these species tend to spread in the invaded region faster than before, in relation with the dramatic changes in trade volume and trade regulations (Roques *et al.*, 2016). Therefore, it is essential that we find new methods to identify insect species that are likely to be introduced and directly or indirectly affect human health, especially because many of these species may not be reported in their native range as threats to humans because their natural enemies control them. Attention must also be paid to new cross-combinations between native/non-native insects and pathogens that could emerge in the future.

References

Ahmad, A., Ghosh, A., Schal, C. and Zurek, L. (2011) Insects in confined swine operations carry a large antibiotic resistant and potentially virulent enterococcal community. *BMC Microbiology* 11, 1–13.

Anderson, B.E., Miller, J.J. and Adams, D.R. (2012) Irritant contact dermatitis to the brown marmorated stink bug, *Halyomorpha halys*. *Dermatitis* 23, 170–172.

Aquino, R.S.S., Silveira, S.S., Pessoa, W.F.B., Rodrigues, A., Andrioli, J.L. *et al.* (2013) Filamentous fungi vectored by ants (Hymenoptera: Formicidae) in a public hospital in north-eastern Brazil. *Journal of Hospital Infection* 83, 200–204.

Battisti, A., Holm, G., Fagrell, B. and Larsson, S. (2011) Urticating hairs in arthropods: their nature and medical significance. *Annual Review of Entomology* 56, 203–220.

Battisti, A., Larsson, S. and Roques, A. (2017) Processionary moths and associated urtication risk: global-change driven effects. *Annual Review of Entomology* 62, 323–342.

Baumholtz, M.A., Parish, L.C., Witkowski, J.A. and Nutting, W.B. (1997) The medical importance of cockroaches. *International Journal of Dermatology* 36, 90–96.

Beaucournu, J.C. and Launay, H. (1990) *Les puces (Siphonaptera) de France et du Bassin méditerranéen occidental. Faune de France 76.* Fédération Française des Sociétés de Sciences Naturelles, Paris.

Beggs, J., Brockerhoff, E.G., Corley, J., Kenis, M., Masciocchi, M., Muller, F., Rome, Q. and Villemant, C. (2011) Ecological effects and management of invasive alien Vespidae. *BioControl* 56, 505–526.

Bencheton, A.L., Berenger, J.M., Del Giudice, P., Delaunay, P., Pages, F. and Morand J.J. (2011) Resurgence of bedbugs in southern France: a local problem or the tip of the iceberg? *Journal of the European Academy of Dermatology and Venereology* 25, 599–602.

Bertelsmeier, C., Blight, O. and Courchamp, F. (2016) Invasions of ants (Hymenoptera: Formicidae) in light of global climate change. *Myrmecological News* 22, 25–42.

Biló, B.M., Rueff, F., Mosbech, H., Bonifazi, F., Oude-Elberink, J.N.G. *et al.* (2005) Diagnosis of Hymenoptera venom allergy. *Allergy* 60, 1339–1349.

Blazar, J.M., Kurt Lienau, E. and Allard, M.V. (2011) Insects as vectors of foodborne pathogenic bacteria. *Terrestrial Arthropod Reviews* 4, 5–16.

Burns, D.A. (2009) Diseases caused by arthropods and other noxious animals. In: Burns, D.A., Breathnach, S.M., Cox, N.H. and Griffiths, C.E.M. (eds) *Rook's Textbook of Dermatology*, 8th edn. Wiley-Blackwell, Oxford, pp. 11–46.

Carabarin-Lima, A., Gonzalez-Vazquez, M.C., Rodriguez-Morales, O., Baylon-Pacheco, L., Rosales-Encina, J.L. *et al.* (2013) Chagas disease (American trypanosomiasis) in Mexico: an update. *Acta Tropica* 127, 126–135.

Chapman, M.D. and Pomes, A. (2009) Cockroach allergens, environmental exposure, and asthma. In: Kay, A.B., Kaplan, A.P., Bousquet, J. and Holt, P.G. (eds) *Allergy and Allergic Diseases*, 2nd edn. Wiley-Blackwell, Oxford, pp. 1131–1145.

Choi, M.B., Martin, S.J. and Lee, J.W. (2012) Distribution, spread, and impact of the invasive hornet *Vespa velutina* in South Korea. *Journal of Asia-Pacific Entomology* 15, 473–477.

Collingwood, C.A., Tigar, B.J. and Agosti, D. (1997) Introduced ants in the United Arab Emirates. *Journal of Arid Environments* 37, 505–512.

Cooper, R.A., Wang, C.L. and Singh, N. (2016) Evaluation of a model community-wide bed bug management program in affordable housing. *Pest Management Science* 72, 45–56.

Cranshaw, W. (2011) A review of nuisance invader household pests of the United States. *American Entomologist* 57, 165–169.

Cuesta-Herranz, J., de las Heras, M., Sastre, J., Lluch, M., Fernández, M., Lahoz, C. and Alvarez-Cuesta, E. (1997) Asthma caused by Dermestidae (black carpet beetle): a new allergen in house dust. *Journal of Allergy and Clinical Immunology* 99, 147–149.

DAISIE (2009) *Handbook of Alien Species in Europe*. Springer, Berlin.

De Castro, M.M., Prezoto, H.H.S., Fernandes, E.F., Bueno, O.C. and Prezoto, F. (2015) The ant fauna of hospitals: advancements in public health and research priorities in Brazil. *Revista Brasileira de Entomologia* 59, 77–83.

Delaunay, P. (2012) Human travel and traveling bedbugs. *Journal of Travel Medicine* 19, 373–379.

Derraik, J.G.B. (2008) The potential direct impacts on human health resulting from the establishment of the painted apple moth (*Teia anartoides*) in New Zealand. *New Zealand Medical Journal* 121, 35–40.

Dias, J.C. (2000) Epidemiological surveillance of Chagas disease. *Cad Saude Publica* 16 (Suppl. 2), 43–59.

Donovan, G.H., Butry, D.T., Michael, Y.L., Prestemon, J.P., Liebhold, A.W., Gatziolis, D. and Mao, M.Y. (2013) The relationship between trees and human health: evidence from the spread of the emerald ash borer. *American Journal of Preventive Medicine* 44, 139–145.

Donovan, G.H., Michael, Y.L., Gatziolis, D., Prestemon, J.P. and Whitsel, E.A. (2015) Is tree loss associated with cardiovascular-disease risk in the Women's Health Initiative? A natural experiment. *Health and Place* 36, 1–7.

Dos Santos, P.F., Fonseca, A.R. and Sanches, N.M. (2009) Formigas (Hymenoptera: Formicidae) como vetores de bactérias em dois hospitais do município de Divinópolis, Estado de Minas Gerais. *Revista da Sociedade Brasileira de Medicina Tropical* 42, 565–569.

Dujardin, J.P., Thi, K.P., Xuan, L.T., Panzera, F., Pita, S. and Schofield, C.J. (2015) Epidemiological status of kissing-bugs in South East Asia: a preliminary assessment. *Acta Tropica* 151, 142–149.

Dutto, M. and Bertero, M. (2013) Dermatosis caused by *Corythuca ciliata* (Say, 1932) (Heteroptera, Tingidae). Diagnostic and clinical aspects of an unrecognized pseudoparasitosis. *Journal of Preventive Medicine and Hygiene* 54, 57–59.

Embrey, S., Remais, J.V. and Hess, J. (2012) Climate change and ecosystem disruption: the health impacts of the North American Rocky Mountain pine beetle infestation. *American Journal of Public Health* 102, 818–827.

Eschen, R., Roques, A. and Santini, A. (2015) Taxonomic dissimilarity in patterns of interception and establishment of alien arthropods, nematodes and pathogens affecting woody plants in Europe. *Diversity and Distributions* 21, 36–45.

Faundez, E.I. and Carvajal, M.A. (2014) Bed bugs are back and also arriving is the southernmost record of *Cimex lectularius* (Heteroptera: Cimicidae) in South America. *Journal of Medical Entomology* 51, 1073–1076.

Franchi, M., Carrer, P., Kotzias, D., Rameckers, E.M.A.L., Seppänen, O., Bronswijk, J.E.M.H., Viegi, G., Gilder, J.A. and Valovirta, E. (2006) Working towards healthy air in dwellings in Europe. *Allergy* 61, 864–868.

Gascon, J., Bern, C. and Pinazo, M.J. (2010) Chagas disease in Spain, the United States and other non-endemic countries. *Acta Tropica* 115, 22–27.

Goetz, D.W. (2008) *Harmonia axyridis* ladybug invasion and allergy. *Allergy and Asthma Proceedings* 29, 123–129.

Gullan, P.J. and Cranston, P.S. (2014) *The Insects: An Outline of Entomology*, 5th edn. Wiley-Blackwell, Oxford.

Gyulai-Garai, A. and Gyulai, P. (2008) The appearance of the saddleback caterpillar moth (*Archaria* (= *Sibine*, = *Stibine*) *stimulea* (Clemens, 1860) pest species in Hungary (Lepidoptera: Limacodidae). *Növényvédelem* 44, 226–228.

Haag-Wackernagel, D. and Spiewak, R. (2004) Human infestation by pigeon fleas (*Ceratophyllus columbae*) from feral pigeons. *Annals of Agricultural and Environmental Medicine* 11, 343–346.

Heymann, W.R. (2009) Bed bugs: a new morning for the night time pests. *Journal of the American Academy of Dermatology* 60, 482–483.

Holway, D.A., Lach, L., Suarez, A.V., Tsutsui, N.D. and Case, T.J. (2002) The causes and consequences of ant invasions. *Annual Review of Ecology and Systematics* 33, 181–233.

Jacobs, D.H. and Neser, S. (2005) *Thaumastocoris australicus* Kirkaldy (Heteroptera: Thaumastocoridae): a new insect arrival in South Africa, damaging to Eucalyptus trees. *South African Journal of Science* 101, 233–236.

Kang, B.C. (1990) Cockroach allergy. *Clinical Reviews in Allergy* 8, 87–98.

Kenis, M. and Branco, M. (2010) Impact of alien terrestrial arthropods in Europe. *BioRisk* 4, 51–71.

Kenis, M. and Roques, A. (2010) Lice and fleas (Phthiraptera and Siphonaptera). *BioRisk* 4, 833–848.

Kimsey, L.S. and Carpenter, J.M. (2012) The Vespinae of North America (Vespidae, Hymenoptera). *Journal of Hymenoptera Research* 28, 37–65.

Lai, O., Ho, D., Glick, S. and Jagdeo, J. (2016) Bed bugs and possible transmission of human pathogens: a systematic review. *Archives of Dermatological Research* 308, 531–538.

Lesieur, V. (2014) Expansion de la punaise invasive nord-américaine prédatrice des graines de conifères *Leptoglossus occidentalis* (Heteroptera Coreidae). Traçage génétique de l'origine des populations invasives et impact écologique et économique des populations introduites en Europe. PhD thesis, Orléans University, France.

Lopez-Vaamonde, C., Glavendekić, M., Paiva, M.R. *et al.* (2010) Invaded habitats. *BioRisk* 4, 45–50.

Lounibos, L.P. (2002) Invasions by insect vectors of human disease. *Annual Review of Entomology* 47, 233–266.

Mazza, G., Tricarico, E., Genovesi, P. and Gherardi, F. (2014) Biological invaders are threats to human health: an overview. *Ethology Ecology and Evolution* 26, 112–129.

McGlynn, T.P. (1999) The worldwide transfer of ants: geographical distribution and ecological invasions. *Journal of Biogeography* 26, 535–548.

Mello-Garcia, F.R. and Lise, F. (2013) Ants associated with pathogenic microorganisms in Brazilian hospitals: attention to a silent vector. *Acta Scientiarum. Health Sciences* 35, 9–14.

Mindlin, M.J., Le Polain de Waroux, O., Case, S. and Walsh, B. (2012) The arrival of oak processionary moth, a novel cause of itchy dermatitis in the UK: experience, lessons and recommendations. *Public Health* 126, 778–781.

Monceau, K., Bonnard, O. and Thiéry, D. (2014) *Vespa velutina*: a new invasive predator of honeybees in Europe. *Journal of Pest Science* 87, 1–16.

Moneo, I., Battisti, A., Dufour, B., Garcia-Ortiz, J.C., González-Muñoz, M., Moutou, F., Paolucci, P., Petrucco Toffolo, E., Rivière, J., Rodríguez-Mahillo, A.I. *et al.* (2015) Medical and veterinary impact of the urticating processionary larvae. In: Roques, A. (ed.) *Processionary Moths and Climate Change: An Update*. Springer, Dordrecht, the Netherlands and Éditions Quae, Versailles, France, pp. 359–410.

Müller, G., Landau Luescher, I. and Schmidt, M. (2011) New data on the incidence of household arthropod pests and new invasive pests in Zurich (Switzerland). In: Robinson, W.H. and de Carvalho Campos, A.E. (eds) *Proceedings of the 7th International Conference on Urban Pests*. Instituto Biológico, São Paulo, Brazil. pp. 99–104.

Müsken, H., Franz, J.T., Fernandez-Caldas, E. and Bergmann, K.C. (1998) Psocoptera (dust lice): new indoor allergens? *Allergologie* 21, 381–382.

Neghme, A., Schenone, H., Villarroel, F. and Rojas, A. (1991) Programa antitriatomico experimental de Santiago. *Boletin Chileno de parasitologia* 46, 47–57.

Nelder, M.P., Paysen, E.S., Zungoli, P.A. and Benson, E.P. (2006) Emergence of the introduced ant *Pachycondyla chinensis* (Formicidae: Ponerinae) as a public health threat in the southeastern United States. *Journal of Medical Entomology* 43, 1094–1098.

Patil, M.P., Niphadkar, P.V. and Bapat, M.M. (2001) *Psocoptera* spp. (book louse): a new major household allergen in Mumbai. *Annals of Allergy Asthma and Immunology* 87, 151–155.

Petrucco-Toffolo, E., Zovi, D., Perin, C., Paolucci, P., Roques, A., Battisti, A. and Horvath, H. (2014) Size and dispersion of urticating setae in three species of processionary moths. *Integrative Zoology* 9, 320–327.

Przybilla, B. and Ruëff, F. (2010) Hymenoptera venom allergy. *Journal der Deutschen Dermatologischen Gesellschaft* 8, 114–129.

Rabitsch, W. (2010) True bugs (Hemiptera, Heteroptera). *Biorisk* 4, 407–433.

Rabitsch, W. (2011) The hitchhiker's guide to alien ant invasions. *BioControl* 56, 551–572.

Ramsey, S. and Losey, J.E. (2012) Why is *Harmonia axyridis* the culprit in coccinellid biting incidents? An analysis of means, motive, and opportunity. *American Entomologist* 58, 166–170.

Rasplus, J.-Y. and Roques, A. (2010) Dictyoptera (Blattodea, Isoptera), Orthoptera, Phasmatodea and Dermaptera. *BioRisk* 4, 807–831.

Rasplus, J.Y., Villemant, C., Paiva, M.R., Delvare, G. and Roques, A. (2010) Hymenoptera. *BioRisk* 4, 669–776.

Reynaud, P. (2000) La punaise *Oxycarenus lavaterae*. Elle est responsable de pullulations spectaculaires à Paris. *Phytoma-La Défense des Végétaux* 528, 30–33

Rodriguez-Mahillo, A.I., Carballeda- Sangiao, N., Vega, J.M., Garcia-Ortiz, J.C., Roques, A., Moneo, I. and González-Muñoz, M. (2015) Diagnostic use of recombinant Tha p 2 in the allergy to *Thaumetopoea pityocampa*. *Allergy* 70, 1332–1335.

Rome, Q., Dambrine, L., Onate, C., Muller, F., Villemant, C., García-Pérez, A.L., Maia, M., Carvalho Esteves, P. and Bruneau, E. (2013) Spread of the invasive hornet *Vespa velutina* Lepeletier, 1836, in Europe in 2012 (Hym., Vespidae). *Bulletin de la Société Entomologique de France* 118, 21–22.

Romero, A., Potter, M.F., Potter, D.A. and Haynes, K.F. (2007) Insecticide resistance in the bed bug: a factor in the pest's sudden resurgence? *Journal of Medical Entomology* 44, 175–178.

Roques, A. (2010) Taxonomy, time and geographic patterns. *Biorisk* 4, 11–26.

Roques, A., Auger-Rozenberg, M.-A., Blackburn, T.M., Garnas, J.R., Pyšek, P., Rabitsch, W., Richardson, D.M., Wingfield, M.J., Liebhold, A.M. and Duncan, R.P. (2016) Temporal and interspecific variation in rates of spread for insect species invading Europe during the last 200 years. *Biological Invasions* 18, 907–920.

Roxo, E., Campos, A.E.C., Alves, M.P., Couceiro, A.P.M.R., Harakava, R., Ikuno, A.A., Ferreira, V.C.A., Baldassi, L., Almeida, E.A., Spada, D.T.A. *et al.* (2010) Ants' role (Hymenoptera: Formicidae) as potential vectors of mycobacteria dispersion. *Arquivos do Instituto Biológico (São Paulo)* 77, 359–362.

Roy, H.E., Brown, P.M.J, Adriaens, T., Berkvens, N., Borges, I., Clusella-Trullas, S., Comont, R.F., De Clercq, P., Eschen, R., Estoup, A. *et al.* (2016) The harlequin ladybird, *Harmonia axyridis*: global perspectives on invasion history and ecology. *Biological Invasions* 18, 997–1044.

Rust, M.K. and Su, N.-Y. (2012) Managing social insects of urban importance. *Annual Review of Entomology* 57, 355–375.

Rustin, M.H.A. and Munro, D.D. (1984) Papular urticaria caused by *Dermestes maculatus* DeGeer. *Clinical and Experimental Dermatology* 9, 317–321.

Schindler, S., Staska, B., Adam, M., Rabitsch, W. and Essl, F. (2015) Alien species and public health impacts in Europe: a literature review. *Neobiota* 27, 1–23.

Schneider, N. (2010) Psocids (Psocoptera). *BioRisk* 4, 793–805.

Seebens, H., Blackburn, T., Dyer, E., Genovesi, P., Hulme, P.E., Jeschke, J.M., Pagad, S., Pysek, P., Winter, M., Arianoutsou, M. *et al.* (2017) No saturation in the accumulation of alien species worldwide. *Nature Communications* 8:14435, 1–9.

Sekeyová, Z., Socolovschi, C., Špitalská, E., Kocianová, E., Boldiš, V., Diaz, M.Q., Berthová, L., Bohácsová, M., Valáriková, J., Fournier, P.E. *et al.* (2013) Update on Rickettsioses in Slovakia. *Acta Virologica* 57, 180–199.

Service, M. (2012) *Medical Entomology for Students*, 5th edn. Cambridge University Press, New York.

Siderhurst, M.S., Jang, E.B., Hara, A.H. and Conant, P. (2007) n-Butyl (E)-7,9-decadienoate: sex pheromone component of the nettle caterpillar, *Darna pallivitta*. *Entomologia Experimentalis et Applicata* 125, 63–69.

Srisong, H., Daduang, S. and Lopata, A.L. (2016) Current advances in ant venom proteins causing hypersensitivity reactions in the Asia-Pacific region. *Molecular Immunology* 69, 24–32.

Vega, J.M., Moneo, I., Garcia Ortiz, J.C., Sanchez Palla, P., Sanchis, M.E, Vega, J., Gonzalez-Muñoz, M., Battisti, A. and Roques, A. (2011) Prevalence of cutaneous reactions to pine processionary moth (*Thaumetopoea pityocampa*) in an adult population. *Contact Dermatitis* 64, 220–228.

Wheeler, A.G. Jr (1992) *Leptoglossus occidentalis*. A new conifer pest and house nuisance in Pennsylvania. *Regulatory Horticulture* 18, 29–30.

Yoder, J.A., Chambers, M.J., Condon, M.R., Benoit, J.B. and Zettler, L.W. (2008) The giant Madagascar hissing-cockroach (*Gromphadorhina portentosa*) as a source of antagonistic moulds: concerns arising from its use in a public setting. *Mycoses* 51, 95–98.

6 The Invasive Mosquitoes of Medical Importance

Roberto Romi[1]*, Daniela Boccolini[1], Marco Di Luca[1], Jolyon M. Medlock[2], Francis Schaffner[3,4], Francesco Severini[1] and Luciano Toma[1]

[1]*Istituto Superiore di Sanità, Rome, Italy;* [2]*Public Health England, Salisbury, United Kingdom;* [3]*University of Zurich, Switzerland; and* [4]*Francis Schaffner Consultancy, Riehen, Switzerland*

Abstract

Mosquitoes (Diptera: Culicidae) are the most important group of blood-sucking insects that are vectors of human diseases. This chapter focuses mainly on six species belonging to the *Aedes, Culex* and *Anopheles* genera, which, closely adapted to human habitats for thousands of years, have exploited human activities to spread and establish in areas far from their origin, becoming invasive. The mechanisms leading to the introduction and establishment of invasive mosquito species and the risk that they represent for human health in newly colonized areas are extensively described. In particular, this chapter focuses on the three powerful and widespread arbovirus disease vectors, *Ae. aegypti, Ae. albopictus* and *Ae. japonicus*, with shorter references to *Ae. koreicus* and other alien species recently recorded in Europe. The disease vectors belonging to the genus *Culex* are represented by the *Cx. pipiens* complex, whose history of travel between continents dates back centuries. The detailed story of the journey undertaken by species belonging to the *Anopheles gambiae* complex in the 19th century between Africa and Brazil represents the only significant example of invasion by Anophelinae mosquitoes. The last section of the chapter deals with the surveillance and control activities to be implemented in order to prevent the introduction and spread of invasive mosquito species.

6.1 Introduction

The Culicidae family, namely mosquitoes, belongs to the order of Diptera (Arthropoda: Insecta), which represents the main taxonomic group within the Arthropods of medical importance. Among these, mosquitoes are the most important family of blood-sucking insects that are vectors of human diseases. The major vector-borne diseases, such as malaria, yellow fever and dengue, that have an important role in the history of human epidemics, are transmitted by mosquitoes (Beaty and Marquardt, 1996).

This chapter focuses on some species of mosquitoes in the genera *Aedes, Anopheles* and *Culex*, which, closely adapted to human habitats for thousands of years, have exploited human activities to spread and establish in areas far from their origin, becoming invasive

---

* E-mail: roberto.romi@iss.it

(Lounibos, 2002). Other non-haematophagous Diptera, considered recently as invasive species but of little importance for human health, such as *Hermetia illucens* (Muscidae: Stratiomyidae), the Black soldier fly, and *Cochliomya hominivorax* (Muscidae: Calliforidae), the New World screwworm fly, are not considered in this context (Alexander, 2006; Vilà *et al.*, 2009; Concha *et al.*, 2016).

The processes and mechanisms leading to the introduction and establishment of invasive mosquito species, and the risk that they represent for human health in newly colonized areas, have been investigated extensively (Lounibos, 2011). The ability to adapt to different environments is the result of a complex strategy of survival, due to a set of genetic and behavioural characteristics, and is mainly shared by the invasive species that include most of the powerful arbovirus disease vectors (Juliano and Lounibos, 2005).

However, the first noticeable introduction of invasive mosquitoes dates back to the 16th century and was mainly related to the trade of slaves from West Africa to the Americas (Black, 2015). In the four centuries that followed, successive waves of invasion of the two most common species *Aedes aegypti* and *Culex pipiens* were sustained by global shipping transport (Powell and Tabachnick, 2013). In the 19th century, the *Anopheles gambiae* complex, which is native to Africa and includes the most important malaria vectors of the world, demonstrated its capacity to successfully invade new areas and cause significant epidemics of malaria (Lounibos, 2002, 2011).

After the Second World War and in particular over recent decades, other species have spread overseas, facilitated by the continuous increase in the worldwide trade of goods (Guerra, 1993), especially those suitable for the transportation of pre-imaginal stages of mosquitoes, such as scrap tyres (Reiter, 2010a). Several species were recorded in new geographic areas, for example *Aedes albopictus*, the Asian tiger mosquito, which in only a few years has successfully colonized the USA and Europe (Kraemer *et al.*, 2015). Recently, *Aedes japonicus* (Kampen and Werner, 2014) and currently *Aedes koreicus* (Medlock *et al.*, 2015) have shown a similar propensity to expand.

The spread of these species represents an important health risk in the newly invaded countries, where the local transmission of tropical diseases was thought unlikely until only a few years ago (Tatem *et al.*, 2006). Recently, *Ae. albopictus* has been proved to be the vector that caused large outbreaks of chikungunya virus. Repeated outbreaks have occurred in the Indian Ocean Islands (Weaver and Lecuit, 2015) and one notable outbreak in the north of Italy, the first outbreak in Europe (Rezza *et al.*, 2007). Furthermore, cases of autochthonous dengue have been reported recently in Europe and the USA (Añez and Rios, 2013; Schaffner and Mathis, 2014). The recent re-introduction into Europe and the continuous spreading northward in the USA of *Ae. aegypti* (Powell and Tabachnick, 2013), coupled with the establishment of new foci of *Ae. japonicus* (Kampen and Werner, 2014) on both continents have elevated the concern for public health posed by these invasive mosquitoes.

Furthermore, in the two last decades, the common and ubiquitous *Culex pipiens*, now considered as native all over the world, has been implicated in the transmission of arboviruses in temperate countries (Fonseca *et al.*, 2006), triggering outbreaks of human West Nile disease in the USA and Europe (Petersen *et al.*, 2013; Rizzoli *et al.*, 2015). Such incidences remind us of the global importance of mosquitoes as potential vectors of human diseases, rather than purely a source of annoyance.

With regard to other culicine species that have only occasionally been recorded in Europe, of which most have still to be confirmed as established (i.e. *Culex vishnui*, *Aedes triseriatus* and *Aedes atropalpus*), we refer to Medlock *et al.* (2015). With particular regard to two anopheline malaria vectors in the New World (i.e. *Anopheles darlingi* and *Anopheles quadrimaculatus*), these are not considered as true invaders in this context, and we refer to Juliano and Lounibos (2016).

6.2 *Aedes aegypti*

Aedes aegypti Linnaeus, 1762 is an extraordinary example of globalization by a vector mosquito. Able to exploit human habits by passive dispersion of its eggs, this species has expanded rapidly from its tropical zone to invade and establish in more temperate climes. Originally described from specimens collected in Cairo, Egypt, *Ae. aegypti* occurs worldwide with a cosmopolitan range extending throughout the tropical and subtropical world. *Ae. aegypti* is similar morphologically to *Aedes albopictus* (Skuse, 1894) and they share many biological traits. *Ae. aegypti* belongs to the *Stegomyia* subgenus, is renowned as a container-breeding mosquito, similar to *Ae. albopictus*, and is notorious for its ability to transmit arboviruses of human health concern. The main macroscopic difference distinguishing the adult form of these two vector species is the aspect of the thorax: *Ae. aegypti* has white scales on the dorsal surface shaped as a lyre, while *Ae. albopictus* only has a white stripe down the midline of the top of the thorax. *Ae. aegypti* eggs can survive desiccation for months and hatch once submerged by water, but despite its similarity with *Ae. albopictus*, its eggs are less adapted to overwintering in a temperate climate.

Genetically, *Ae. aegypti* is not homogeneous as a species or in its role as a disease vector. It exhibits genetic, morphological and ecological variation, with two morphological forms, or subspecies, having been described. These can be separated on the basis of the extent of white scaling on the first abdominal tergite (Brown *et al.*, 2011). The darker and presumably ancestral *Ae. aegypti formosus* was reportedly confined to East Africa, where it tended to breed in forested habitats and was predominantly zoophilic. On the other hand, the lighter *Ae. aegypti aegypti* was distributed throughout the tropics in West Africa and outside Africa, where it exploited larval habitats in artificial containers, with a greater tendency for anthropophily.

Aedes aegypti reached the New World by means of slave ships; presumably the casks used for onboard storage of water acted as a desirable breeding ground (Darsie and Ward, 2005). This mosquito was common in Florida until the invasion of *Ae. albopictus* in 1985, when the population of *Ae. aegypti* declined dramatically, but it remains in high densities in some urban areas. In the USA, the distribution of this species has changed in the last two centuries, but it has never disappeared. Today it occurs in 23 states, from south California to Connecticut, and is thought to be expanding north-eastward (Center for Disease Control, 2016).

Aedes aegypti was very abundant in the early 20th century in southern Europe and in the most important harbour cities of the Mediterranean Basin. In Italy, the species was very common until the 1940s; the last findings were in 1944 in Genoa (Toma *et al.*, 2011). The reasons for its disappearance from the Mediterranean region are still unclear (Schaffner and Mathis, 2014). The species seems to have disappeared from the Mediterranean Basin but sporadic European findings have been reported (see Table 6.1). Since the early 2000s, *Ae. aegypti* has started to again colonize the eastern Black Sea regions (parts of southern Russia, Georgia and Turkey) (Medlock *et al.*, 2015; Akiner *et al.*, 2016) and in 2004 the species was for the first time recorded in the Autonomous Region of Madeira, Portugal (Almeida *et al.*, 2007), where in 2012–2013 it was responsible for a significant outbreak of dengue (European Centre for Disease Prevention and Control (ECDC), 2014).

Also named the 'yellow fever mosquito', *Ae. aegypti* is known for its relevant worldwide role as the main vector of yellow fever virus (YFV), dengue virus (DENV), Zika virus (ZIKV), chikungunya virus (CHIKV) and others. YFV remains a significant public health problem in subtropical areas of South America and West Africa (World Health Organization (WHO), 2016a) but it does not occur in Asia, despite the presence of *Ae. aegypti*. Even if the topic is still debated and studied, several explanations for this issue have been proposed to explain the absence of yellow fever in Asia. The colonization out of Africa was unidirectional westward, as far as we know. In fact, YFV was unknown or very rare in East Africa and absent in

Asia, thus indicating that East African *Ae. aegypti* are not sufficiently anthropophilic and/or not competent to transmit yellow fever. Probably the limited adaptation to humans limited their ability to survive aboard ships for periods long enough to facilitate an Asian migration on ships. Another hypothesis is a cross-immunity between DENV and YFV, supporting DENV infections with cross protection against YFV in presence of both *Ae. albopictus* and *Ae. aegypti* in Asia (Saunders *et al.*, 1998; Mutebi and Barrett, 2002; Amaku *et al.*, 2011; Powell and Tabachnick, 2013).

However, human infections with DENV have become an increasingly grave concern for public health globally, with DENV currently considered the most important arbovirosis in terms of both morbidity and mortality (WHO, 2016a). *Aedes aegypti* is the main vector of ZIKV, but other *Aedes* species can also transmit the virus. Viral circulation and a few outbreaks have been documented in tropical Africa and in some areas of South-east Asia and South America, including Brazil. There is scientific consensus that ZIKV is a cause of microcephaly, congenital nervous system malformations and Guillain-Barré syndrome (WHO, 2016b).

6.3 *Aedes albopictus*

Aedes albopictus (Skuse, 1894), currently named Asian tiger mosquito because of its evident median white strip on the thorax and the white patches on its dark body and legs, was first described from specimens collected in Calcutta. In 2004, a new classification of Aedini tribe was proposed, elevating the *Stegomyia* subgenus to the genus level and subsequently changing *Aedes* to *Stegomyia albopicta*. This is a controversial matter, and the use of *St. albopicta* rather than *Ae. albopictus* (hereafter *Ae. albopictus*) is still debated (Reinert *et al.*, 2004; Schaffner and Aranda, 2005). However, a consensus might be arising since it was suggested to keep monophyletic groups in Aedini at subgenus level, and thus to maintain the use of *Aedes* for all *Stegomyia* species (Wilkerson *et al.*, 2015).

Aedes albopictus originated in the forests of tropical and subtropical areas of South-east Asia, where it was originally mainly zoophilic, and showed a rapid adaptation to human-caused environmental changes. Until the 20th century, the range of *Ae. albopictus* extended from southern China to Japan and most of the Indian and western Pacific Ocean (Huang, 1979). Since the early 1900s, the species has gradually expanded its range due to the passive transport of viable eggs and larvae through commercial trading routes and its establishment in new geographic localities has been permitted by its ability to adapt to different environmental and climatic conditions. However, over the last four decades *Ae. albopictus* has dramatically spread from its native range to now invading more than 50 other countries around the globe, largely through the international trade in used tyres, the primary means for its dissemination (Knudsen *et al.*, 1996).

In 1985, its introduction in the USA from Japan represented the first focus of *Ae. albopictus* established in the western hemisphere. The species has spread to other parts of the country and it is now reported in 40 states (Hahn *et al.*, 2016). As expected, it has since moved rapidly towards Central and South America (Rai, 1991). In 1979, *Ae. albopictus* was first recorded in Europe, in Albania, where it was evidently introduced through a shipment of goods from China (Adhami and Reiter, 1998). In 1990, a few adult *Ae. albopictus* were found in Italy (at the port of Genoa) (Sabatini *et al.*, 1990). In Italy, the species quickly spread throughout the country by passive transport and by successive waves of reintroduction (Romi *et al.*, 2006). Its establishment was further helped by its adaptation to the cold winters of temperate Europe (Romi *et al.*, 2008). To date, this species is largely distributed worldwide, and is firmly established in the Americas, Caribbean, Africa, in almost all countries bordering the Mediterranean sea, including parts of Turkey, Lebanon, Israel, Syria, and some countries of continental Europe (Table 6.1) from which it is gradually moving north-wards (Medlock *et al.*, 2012). A global geographic database of known occurrences of *Ae. albopictus* between 1960 and 2014 has recently been compiled by Kraemer *et al.* (2015).

Aedes albopictus is currently considered the most invasive mosquito in the world. The great ecological plasticity that allowed this species to adapt rapidly to exploit human habi-tats and to establish firmly in newly invaded countries is the result of a number of biological characteristics and environmental factors. These include the availability of a wide range of hosts for blood feeding; the ability to rapidly complete a full blood meal, often taking several blood meals within the same gonotrophic cycle; the egg-laying of the same batch over sev-eral days and in different containers; the particular structure of their eggs that allows them to hatch only in part at the first wetting stimulus and to overcome dryness or overwintering periods; the ability to colonize unusual breeding sites, such as the catch basins of street drains, rich in organic matter, in competition with *Culex pipiens* larvae; the ability of a mos-quito population to self-reduce their entity, according to the trophic resources available in the area. Finally, further evidence of its adaptability has highlighted its ability to change its behaviour; factors previously considered uncommon for this species, such as nocturnal tro-phic activity, indoor resting and the use of indoor breeding sites (Hawley, 1988; Juliano and Lounibos, 2005; Romi *et al.*, 2008). On its establishment in new areas, *Ae. albopictus* becomes a major pest, but its potential role as vector of human diseases is the cause of most concern.

According to the existing literature, the ability of *Ae. albopictus* to act as a vector has been experimentally demonstrated for more than 20 different arboviruses (Mitchell, 1995; Gratz, 2004). The tiger mosquito ranks second only to *Ae. aegypti* in importance as a vector of DENV, and in long-colonized regions is a proven vector of CHIKV (Paupy *et al.*, 2009) including in recently colonized regions, such as Italy (Rezza *et al.*, 2007). More recently, sev-eral studies on vector competence have demonstrated susceptibility of *Ae. albopictus* popula-tions to ZIKV (Chouin-Carneiro *et al.*, 2016; Di Luca *et al.*, 2016a). *Ae. albopictus* was also proven to be able to transmit heartworms, *Dirofilaria immitis* and *Dirofilaria repens* (Cancrini *et al.*, 2007).

6.4 *Aedes japonicus*

Aedes japonicus japonicus (Theobald, 1901) (hereafter *Ae. japonicus*), also known as the Asian rock pool or Asian bush mosquito, is part of a complex of sibling species. This mosquito originates from Korea, Japan, Taiwan, southern China and Russia (south-eastern Siberia) (Tanaka *et al.*, 1979). The spread beyond its native range started in the 1990s, once again following the movement of egg-infested used tyres, but only in recent times has it spread into large parts of North America (the USA and Canada) and Central Europe, where its dis-tribution is expanding (Kaufman and Fonseca, 2014; Kampen and Werner, 2014) (Table 6.1). The success of its invasion is due to various factors: its ability to withstand both long-distance dispersal and winter temperatures in temperate regions, as well as its tolerance of high organic matter content in water in natural and artificial containers (Versteirt *et al.*, 2009). It also emerges before other competing species and is active for longer. It has less specialized requirements for aquatic habitats, compared with *Ae. albopictus*, and this could facilitate its further spread (Schaffner *et al.*, 2003). In Asia, *Ae. japonicus* is not considered an important disease vector, but in the invaded countries concern is mounting over it becom-ing a pest problem or being involved in the transmission of arboviruses such as West Nile virus (WNV) and even CHIKV and DENV (Schaffner *et al.*, 2013b).

6.5 *Aedes koreicus*

Aedes koreicus (Edwards, 1917) was first reported in Europe within an industrial zone of Belgium (2008), with little evidence of spread (Versteirt *et al.*, 2012). Since 2010, it has been reported in northern Italy, southern Switzerland, southern Germany and southern Russia (Table 6.1). Nothing is known on the initial introductions into these countries, although international trade has been suggested as a possible route (Versteirt *et al.*, 2012). In parts of far-eastern Russia, *Ae. koreicus* has been suggested as a possible vector for Japanese encephalitis virus (JEV), but this was never confirmed from the field in Korea (Medlock *et al.*, 2012). There are indications that the species could play a role in the transmission of the nematode *Dirofilaria immitis* (Montarsi *et al.*, 2015).

6.6 Presence of Alien Invasive *Aedes* Mosquito in Europe and the Concern for Public Health

Since the first reports of *Ae. albopictus* from Albania (1979) and Italy (1990), concerns have been raised over the incursion and possible spread and establishment of invasive *Aedes* mosquitoes in Europe and their potential role in local transmission of arboviruses. Remarkably, it took 17 years before invasive *Aedes* became involved in local European transmission of pathogens to humans, when over 200 CHIKV infection cases were reported in Ravenna, Italy, in 2007 (Rezza *et al.*, 2007). Since then, CHIKV cases have occasionally been reported from France, and local DENV cases from Croatia and France, and a large DENV outbreak on Madeira, Portugal involving more than 2100 probable cases (Jenero-Margan *et al.*, 2011; Sousa *et al.*, 2012; Marchand *et al.*, 2013; Schaffner and Mathis, 2014).

Climate change predictions suggest that *Ae. albopictus* will continue spreading beyond its current range (ECDC, 2009, 2012), and indeed there are signs of adaptation to colder climates (Paupy *et al.*, 2009). It is a significant biting nuisance that has the potential to become a serious health threat as a known field vector of CHIKV, DENV, ZIKV and dirofilarial worms. Its recent involvement in the localized transmission of CHIKV and DENV in France, Croatia and Italy highlights the importance of mosquito surveillance.

All these recent occurrences of autochthonous CHIKV and DENV have occurred in areas where *Ae. aegypti* or *Ae. albopictus* have also established. In previous centuries, DENV was endemic on the European continent, with major DENV epidemics in Greece and Turkey in the 1920s (Schaffner and Mathis, 2014). Furthermore, other arboviruses such as ZIKV could also be transmitted by these species, with imported cases frequently introduced into localities where specific criteria for local mosquito-borne transmission of ZIKV are viable (Septfons *et al.*, 2016). There are no climatic reasons why *Ae. aegypti*, if introduced, could not survive across southern Europe (Reiter, 2010b). If this were to happen, it may increase the risk of disease transmission of DENV, CHIKV, YFV and ZIKV.

Historically, *Ae. aegypti* has moved between continents via ships, and this dispersal method presents the highest risk of its introduction from Madeira into continental Europe (Almeida *et al.*, 2007). In southern and eastern Europe, its introduction from the eastern Black Sea regions to the Balkans and central Europe is more likely via road or sea traffic (Akiner *et al.*, 2016). The list of current countries invaded by these four species of *Aedes* mosquito appears in Table 6.1.

The presence and continued spread of alien invasive *Aedes* species remains a concern for European public health. Active surveillance for invasive mosquitoes now occurs across much of Europe, and both the ECDC and the WHO Regional Office for Europe are promoting

Table 6.1. European countries and outermost regions (within the European continent) where invasive *Aedes* mosquito species have been reported, with year of first observation and present status of the colonization, after 1978 and up to July 2016. Light grey: localized or eliminated; medium grey: spreading; dark grey: widespread.

Country	Ae. aegypti	Ae. albopictus	Ae. japonicus	Ae. koreicus
Albania		1979		
Austria		2012	2011	
Belgium		2000	2002	2008
Bosnia & Herzegovina		2005		
Bulgaria		2011		
Croatia		2004	2013	
Czech Republic		2012		
France (mainland)		1999	2000	
Germany		2007	2008	2015
Georgia	2007	2015		
Greece		2003		
Hungary		2014	2012	
Italy	1972	1990	2015	2011
Malta		2009		
Montenegro		2001		
Netherlands (mainland)	2010	2005	2012	
Portugal (Madeira)	2004			
Romania		2012		
Russia	2001	2011		2013
Serbia		2009		
Slovakia		2012		
Slovenia		2002	2011	
Spain		2004		
Switzerland		2003	2009	2013
Turkey	2015	2011		

integrated surveillance of pathogens and vectors by providing guidelines and training (Schaffner *et al.*, 2013a; WHO, 2013). In addition, VectorNet gathers data on invasive *Aedes* and publishes regularly updated online European mosquito maps (ECDC, 2018).

6.7 *Culex pipiens* Complex

Members of the *Culex pipiens* complex are a group of species closely related both morphologically and evolutionarily and characterized by a long history of association with humans (Vinogradova, 2000). In spite of their morphological similarities, these species and their hybrids exhibit physiological and behavioural differences, affecting their feeding behaviour and vectorial capacity in the transmission of pathogens to animals and humans. In particular, *Cx. pipiens* complex is considered to be involved in the transmission of filarial worms, avian plasmodia and several arboviruses, including WNV, which recently has become endemic in North America, the Mediterranean Basin and southern central Europe and

Russia, causing thousands of human cases (Petersen *et al.*, 2013; Rizzoli *et al.*, 2015). Consequently, the discrimination of vector species and the evaluation of their involvement in pathogen circulation is becoming an important issue for risk assessment and for the adoption of correct public health strategies.

Although taxonomy and phylogeny remain controversial (Harbach *et al.*, 2012), four species are recognized as members of the complex. *Culex pipiens* Linnaeus, 1758 and *Culex quinquefasciatus* Say, 1823, named the northern and southern house mosquitoes, are the most invasive species within the group, becoming ubiquitous in temperate and tropical regions, respectively. Where their distributions overlap (North America, Argentina, Madagascar, Japan and Republic of South Korea), these species hybridize extensively. The other two members of the complex are *Culex australicus* Dobrotworsky & Drummond 1953 and *Culex globocoxitus* Dobrotworsky, 1953, with a distribution limited to Australia.

The nominal species, *Cx. pipiens*, includes two subspecies: *Cx. pipiens pipiens*, occurring throughout temperate regions of the world, and *Cx. pipiens pallens* Coquillett, 1898, limited to temperate East Asia, which partially hybridizes with *Cx. quinquefasciatus*. In addition, *Cx. p. pipiens* has two epidemiologically distinct forms or biotypes, *pipiens* and *molestus*, which differ dramatically in a number of behavioural and physiological characteristics. In North Europe and Russia, the two forms occur in distinct habitats, whereas in southern Europe and North Africa, they thrive in sympatry (Di Luca *et al.*, 2016b). In such circumstances, *molestus* and *pipiens* interbreed and their hybrids, exhibiting intermediate ecological features, are considered of great epidemiological importance (Ciota *et al.*, 2013). Fonseca and colleagues (2004) dated the origin of the *molestus* form at 10,000 years ago, after the advent of agriculture in North Africa or during the Pleistocene glaciations. From Africa, the *pipiens* form spread throughout the temperate areas of Europe and Asia, while the *molestus* form was introduced only to cities in Japan, the Republic of South Korea and Australia (Vinogradova, 2000). In the early 16th century, *Cx. pipiens* reached the New World through human trade of slaves and goods. Continuous transatlantic journeys between Europe and Africa have probably allowed repeated introductions to the Americas of separate *molestus* and *pipiens* populations and/or of hybrids from southern Europe, events confirmed by the current high abundance of hybrids in the USA aboveground *pipiens* populations (Huang *et al.*, 2008).

The domestication of *Cx. quinquefasciatus* dates back to the advent of agriculture and the flowering of different civilizations in South-east Asia. The species then spread throughout tropical and subtropical Asia, reaching eastern Africa. When *Cx. quinquefasciatus* colonized the New World remains unclear (Fonseca *et al.*, 2006). Certainly, its arrival from West Africa with the slave trade (Vinogradova, 2000) is ruled out because the species was absent before 1942 in this part of the world, which was reached only recently by sea from the New World (Fonseca *et al.*, 2006). *Cx. quinquefasciatus* invaded Australia with or immediately after the First Fleet colony in 1788 or after the opening of Australian ports to American whalers in 1831, while its arrival in New Zealand dates back prior to 1848 aboard sailing vessels or in the first half of the 20th century. Even more recent seems the expansion of the species to the smaller Pacific islands from Australia, linked to events during the Second World War. Of note, the introduction of *Cx. quinquefasciatus* in Hawaii (the first mosquito species to have reached the islands on a ship from Mexico as early as 1826 and the only vector of *Plasmodium relictum* occurring there) contributed to the decline and extinction of native species of Hawaiian birds (Lounibos, 2002).

The success of *Cx. pipiens* and *Cx. quinquefasciatus* in colonizing all temperate and tropical regions around the world during the whole Age of Discovery stems from their ability to exploit water sources with high organic content and in close association with humans. Inside the large sailing vessels, these mosquitoes were able to exploit the heavily polluted bilge waters as larval breeding sites and the presence of humans and animals as a constant source of blood. Furthermore, these extreme conditions may have contributed to select only those mosquito populations able to survive and to mate in confined spaces during these long jour-neys (Farajollahi *et al.*, 2011).

6.8 *Anopheles gambiae* Complex

The *Anopheles gambiae sensu lato* (*s.l.*), recognized in the 1960s, defines a complex of eight morphologically indistinguishable species, which includes the three major malaria vectors in sub-Saharan Africa, namely *Anopheles gambiae sensu stricto* (*s.s.*) Giles 1902, *Anopheles arabiensis* Patton 1905, *Anopheles coluzzii* Coetzee & Wilkerson 2013 (White, 1974; Coetzee *et al.*, 2013).

Accurate identification of the species, paramount to optimize any malaria-control pro-grammes, has been achieved over recent years using polytene chromosome patterns; gas chromatography of cuticular components; allozyme electrophoresis analysis and more recently using DNA-based methods that have greatly improved the efficiency of species dif-ferentiation among the complex (White *et al.*, 2011).

Anopheles gambiae s.l. shows a short developmental life cycle of about 6–10 days at tem-peratures of 25–28°C. It breeds in a broad range of habitats, in temporary ground-pools as well as in more extensive water surfaces (rice fields, river edges or swamp margins), usually in close association with anthropic environments. Adult females show nocturnal biting activity and mostly indoor-feeding behaviour (endophagy), human feeding preferences and indoor-resting habits during blood meal digestion (Gillies and Coetzee, 1987).

The geographical range of the members is mainly tropical Africa, from the coast to the humid rainforests and to the fringes of the desert. In particular, the more widespread spe-cies of the complex, *An. gambiae s.s.*, *An. arabiensis* and *An. coluzzii*, are distributed through-out sub-Saharan Africa including Madagascar, also occurring in sympatry in large areas. *An. arabiensis* extends its distribution range across the south-western Arabian peninsula (Coetzee *et al.*, 2000).

Regarding their medical importance, *An. gambiae s.s.*, *An. arabiensis* and *An. coluzzii* are highly susceptible to human *Plasmodium* parasites and, coupled with their ecological adapta-tion and their strong association with human populations, are considered the most efficient malaria vectors in the world (White *et al.*, 2011). Although historically the invasive potential of *An. gambiae s.l.*, due to its biological characteristics, may be considered lower than *Aedes* and *Culex* species (Lambrechts *et al.*, 2011), where it was accidentally introduced the impact on human health has always been dramatic, associated with large-scale malaria epidemics.

The first evidence of the invasive nature of *An. gambiae s.l.* was recorded in 1866 in Mauritius, probably introduced by ship from the African mainland or Madagascar. Large malaria outbreaks were recorded on the island in 1867 and, despite vector control efforts during the years that followed, malaria became endemic and caused recurrent epidemics with high mortality (Ross, 1908; WHO, 2012).

However, the most devastating invasion of a disease vector in recent times was the establishment of *An. gambiae s.l.* in Brazil in the 1930s. It was assumed that *An. gambiae s.l.* was introduced by shipping traffic, begun in the 1920s, between Brazil and Senegal, showing the ability of this mosquito to survive long international trips. Over the years, *An. gambiae s.l.* gradually spread further north-east, from the coast towards the mainland, following water courses (Lounibos, 2002, 2011). Extensive malaria epidemics occurred from 1930 to 1938, with a serious increase in human malaria cases accompanied by a 20–25% increase in the death rate. At that time, native *Anopheles* species had been responsible for transmitting malaria in north-eastern Brazil, but the arrival of a new, more efficient vector, such as *An. gambiae s.l.*, changed malaria transmission from endemic to epidemic. A control cam-paign launched by the Brazilian government led to the eradication of *An. gambiae s.l.* in the

1940s; after its complete elimination, malaria incidence dropped immediately (Soper, 1966; Lambrechts *et al.*, 2011; Griffing *et al.*, 2015). In total, the costs were estimated as 16,000 lives and the equivalent of about US$3 billion in healthcare, drugs and the vector eradication programme (Killeen *et al.*, 2002). At the time of its invasion *An. gambiae s.l.* was not recognized as a species complex, and both *An. gambiae s.s.* and *An. arabiensis* were speculated to be the invader based on eco-epidemiological data (White, 1974; Levine *et al.*, 2004). The results of a recent molecular investigation by Parmakelis *et al.* (2008), applied to museum specimens, has identified *An. arabiensis* as the true invader species of north-eastern Brazil.

Another documented invasion by *An. gambiae s.l.* occurred in 1941 in Egypt associated with the increasing traffic from Sudan, due to wartime difficulties of shipping in the Mediterranean Sea. In 1942–43 in Upper Egypt, malaria epidemics were disastrous. Successful *An. gambiae s.l.* eradication came in 1945, about three years after the invasion occurred (Soper, 1966).

6.9 The Control of Invasive Mosquitoes

Active surveillance of invasive mosquitoes is necessary to ensure that their arrival and spread can be prevented. In some cases, their early detection and prompt control has facilitated their eradication. However, if an invasive species becomes well established in a region, then the implementation of a control plan, which integrates all the available methods, is paramount. Plans should be developed centred on an entomological surveillance programme, organized around four major components: (i) an in-depth knowledge of the bionomics of the target species (biology, ecology and behaviour); (ii) a monitoring system for collecting field data; (iii) a dedicated software program for the management and storage of all collected data; and (iv) an operational team that can rapidly respond to deliver control interventions. This brief overview of mosquito control methods will only consider the role that invasive species can potentially have on public health as disease vectors in newly invaded areas, rather than their role as nuisance species. The purpose of controlling mosquitoes is to drastically reduce the abundance of a vector population, selecting the most suitable integrated control strategies based on mosquito density, in order to minimize their vector potential. Until new experimental techniques (such as the use of genetically modified females or of sterile males) become available and are field-tested, control efforts should focus on those methods that are currently being implemented. Integrated vector management programmes should adapt the current control methods to the target species, ensuring maximum impact while limiting the negative impact on the environment. The main components of a vector management programme aimed at controlling invasive mosquitoes, as potential arbovirus vectors, include:

- Source reduction. This involves the elimination of all potential removable outdoor water containers, which if left can be rapidly exploited by invasive mosquitoes for egg laying and larval development. This can be very difficult to achieve because it is based on public education, community participation and local compliance. A range of strategies can be employed to inform the local population using a variety of media. Messages should be targeted on the risks associated with these mosquitoes as potential vectors of arboviruses and their impact of public health, and should provide clear instruction on good practices to prevent the production of larval breeding sites.
- Interruption of human–vector contact. Suggested mainly for endemic areas, this activity is also based on public education and aims to protect people from mosquito bites indoors by promoting personal and/or community protection using bed nets and screening doors and windows with nets or curtains (pyrethroid-impregnated netting

should be promoted to enhance their efficiency). Repellents may also be used to provide some protection outdoors.

- Larval control. In locations where aquatic habitats are not removable or are natural larval breeding sites, a range of biological products are available. These products can be selective and very effective, without environmental hazard, and include entomo-pathogen bacteria (such as *Bacillus thuringiensis israelensis* and/or *Bacillus sphaericus*). Additionally, certain chemicals (semi-selective molecules) are available for larval control (insect growth regulators and similar molecules). Although very effective in control, they represent a moderate hazard for the environment. Alternative products (effective on street drains only) include different kinds of oils, which form a film on the water surface preventing larvae from breathing.
- Adult control. The use of adulticidal control should be added in emergency situations only. Available products are mainly pyrethrum-derived products (pyrethrins and pye-throids). Treatments with these synthetic non-selective molecules may not be very effective and represent a severe environmental hazard. Most commonly, it includes space spraying and treatment of low vegetation (outdoor), fogging and residual wall treatments (indoor).

6.10 Conclusions

This chapter highlights the importance of mosquitoes to global public health. It also illustrates the role that humans have played, and continue to play, in the worldwide dissemination of mosquito vectors to new localities. Outbreaks of mosquito-borne diseases have shaped human history, and our desire to travel the world, explore new territories, trade and travel between distant lands has facilitated the greatest global migration of any species other than humans. Mosquitoes are not only one of the greatest enemies of humans, but they have also exploited our behaviour and our habitat to further enhance their talent to exploit humans for their survival, and in doing so they have enabled a range of parasites and pathogens to take further advantage of their haematophagic tendencies. Despite the wealth of evidence for the global dissemination of mosquitoes and the consequent disease outbreaks of the last few centuries, humans are still afflicted by mosquito-borne diseases, newly emerging in certain regions. The recent continental-wide outbreaks of Chikungunya and Zika viruses in the Americas are testament to the success of the mosquito and the limitations of healthcare in arresting their spread and in controlling the pathogens they vector.

References

Adhami, J. and Reiter, P. (1998) Introduction and establishment of *Aedes* (*Stegomyia*) *albopictus* Skuse (Diptera: Culicidae) in Albania. *Journal of the American Mosquito Control Association* 14, 340–343.

Akiner, M.M., Demirci, B., Babuadze, G., Robert, V. and Schaffner, F. (2016) Spread of the invasive mosquitoes *Aedes aegypti* and *Aedes albopictus* in the Black Sea region increases risk of chikungunya, dengue, and Zika outbreaks in Europe. *PLoS Neglected Tropical Diseases* 10(4), e0004664.

Alexander, J.L. (2006) Screwworms. *American Veterinary Medical Association* 228(3), 357–367.

Almeida, A.P., Gonçalves, Y.M., Novo, M.T., Sousa, C.A., Melim, M. and Gracio, A.J. (2007) Vector monitoring of *Aedes aegypti* in the Autonomous Region of Madeira, Portugal. *Eurosurveillance* 12(46), pii=3311.

Amaku, M., Coutinho, F.A.B. and Massad, E. (2011) Why dengue and yellow fever coexist in some areas of the world and not in others? *Biosystems* 106, 111–120.

Añez, G. and Rios, M. (2013) Dengue in the United States of America: a worsening scenario? *Bio Medical Research International* doi: 10.1155/2013/678645.

Beaty, B.J. and Marquardt, W.C. (1996) *The Biology of Disease Vectors*. University Press of Colorado, Denver, Colorado.

Black, J. (2015) *The Atlantic Slave Trade in World History*. Utledge Publishing, New York.

Brown, J.E., McBride, C.S., Johnson, P., Ritchie, S., Paupy, C., Bossin, H., Luthomiah, J., Fernandez-Salas, I., Ponlawat, A., Cornel, A.J. *et al.* (2011) Worldwide patterns of genetic differentiation imply multiple 'domestications' of *Aedes aegypti*, a major vector of human diseases. *Proceedings of the Royal Society B* 278, 2446–2454.

Cancrini, G., Scaramozzino, P., Gabrielli, S., Di Paolo, M., Toma, L. and Romi, R. (2007) *Aedes albopictus* and *Culex pipiens* implicated as natural vectors of *Dirofilaria repens* in Central Italy. *Journal of Medical Entomology* 44(6), 1064–1066.

Center for Disease Control (2016) Estimated range of *Aedes aegypti* and *Aedes albopictus* in the United States, 2017. Available at: https://www.cdc.gov/zika/pdfs/zika-mosquito-maps.pdf (accessed 19 December 2017).

Chouin-Carneiro, T., Vega Rua, A., Vazeille, M., Yebakima, A., Girod, R., Goindin, D., Dupont Rouzeyrol, M., Lourenço de Oliveira, R. and Failloux, A.B. (2016) Differential susceptibilities of *Aedes aegypti* and *Aedes albopictus* from the Americas to Zika virus. *PLoS Neglected Tropical Diseases* 10(3), e0004543.

Ciota, A.T., Chin, P.A. and Kramer, L.D. (2013) The effect of hybridization of *Culex pipiens* complex mosquitoes on transmission of West Nile virus. *Parasites & Vectors* 6, 305.

Coetzee, M., Craig, M. and le Sueur, D. (2000) Distribution of African malaria mosquitoes belonging to the *Anopheles gambiae* complex. *Parasitology Today* 16, 74–77.

Coetzee, M., Hunt, R.H., Wilkerson, R., Della Torre, A., Coulibaly, M.B. and Besansky, N.J. (2013) *Anopheles coluzzii* and *Anopheles amharicus*, new members of the *Anopheles gambiae* complex. *Zootaxa* 3619, 246–274.

Concha, C., Palavesam, A., Guerrero, F.D., Sagel, A., Li, F., Osborne, J.A., Hernandez, Y., Pardo, T., Quintero, G., Vasquez, M. *et al.* (2016) A transgenic male-only strain of the New World screwworm for an improved control program using the sterile insect technique. *Bio Medical Central Biology* 2016 14, 72.

Darsie, R.F. and Ward, R.A. (2005) *Identification and Geographical Distribution of the Mosquitoes of North America, North of Mexico*. University of Florida Press, Gainesville, Florida, USA.

Di Luca, M., Severini, F., Toma, L., Boccolini, D., Romi, R., Remoli, M.E., Sabbatucci, M., Rizzo, C., Venturi, G., Rezza, G. *et al.* (2016a) Experimental studies of susceptibility of Italian *Aedes albopictus* to Zika virus. *Eurosurveillance* 21(18), pii=30223.

Di Luca, M., Toma, L., Boccolini, D., Severini, F., La Rosa, G., Minelli, G., Bongiorno, G., Montarsi, F., Arnoldi, D., Capelli, G. *et al.* (2016b) Ecological distribution and CQ11 genetic structure of *Culex pipiens* complex (Diptera: Culicidae) in Italy. *PLoS ONE* 11(1), e0146476.

European Centre for Disease Prevention and Control (ECDC) (2009) *Development of Aedes albopictus Risk Maps*. ECDC Technical Report. ECDC, Stockholm.

European Centre for Disease Prevention and Control (ECDC) (2012) *The Climatic Suitability for Dengue Transmission in Continental Europe*. ECDC Technical Report. ECDC, Stockholm.

European Centre for Disease Prevention and Control (ECDC) (2014) *Dengue Outbreak in Madeira, Portugal, March 2013*. ECDC, Stockholm. Available at: https://ecdc.europa.eu/en/publications-data/dengue-outbreak-madeira-portugal-march-2013 (accessed 19 December 2017).

European Centre for Disease Prevention and Control (ECDC) (2018) *Mosquito Maps*. Available at: https://ecdc.europa.eu/en/disease-vectors/surveillance-and-disease-data/mosquito-maps (accessed February 2018).

Farajollahi, A., Fonseca, D.M., Kramer, L.D. and Kilpatrick, A.M. (2011) 'Bird biting' mosquitoes and human disease: a review of the role of *Culex pipiens* complex mosquitoes in epidemiology. *Infection, Genetics and Evolution* 11(7), 1577–1585.

Fonseca, D.M., Keyghobadi, N., Malcolm, C.A., Mehmet, C., Schaffner, F., Mogi, M., Fleischer, R.C. and Wilkerson, R.C. (2004) Emerging vectors in the *Culex pipiens* complex. *Science* 303, 1535–1538.

Fonseca, D.M., Smith, J.L., Wilkerson, R.C. and Fleischer, R.C. (2006) Pathways of expansion and multiple introductions illustrated by large genetic differentiation among worldwide populations of the southern house mosquito. *The American Journal of Tropical Medicine and Hygiene* 74(2), 284–289.

Gillies, M.T. and Coetzee, M. (1987) *A Supplement to the Anophelinae of Africa South of the Sahara*. Publication no. 55. South African Institute of Medical Research, Johannesburg, South Africa.

Gratz, N.G. (2004) Critical review of the vector status of *Aedes albopictus*. *Medical and Veterinary Entomology* 18, 215–227.

Griffing, S.M., Tauil, P.L., Udhayakumar, V. and Silva-Flannery, L. (2015) A historical perspective on malaria control in Brazil. *Memórias do Instituto Oswaldo Cruz* 110, 701–718.

Guerra, F. (1993) The European–American exchange. *History, Philosophy, Life Sciences Journal* 15, 313–327.

Hahn, M.B., Eisen, R.J., Eisen, L., Boegler, K.A., Moore, C.G., McAllister, J., Savage, H.M. and Mutebi, J.P. (2016) Reported Distribution of *Aedes* (*Stegomyia*) *aegypti* and *Aedes* (*Stegomyia*) *albopictus* in the United States, 1995–2016 (Diptera: Culicidae). *Journal of Medical Entomology* 53(5), 1169–1175.

Harbach, R.E., Kitching, I.J., Lorna Culverwell, C., Dubois, J. and Linton Y.M. (2012) Phylogeny of mosquitoes of tribe Culicini (Diptera: Culicidae) based on morphological diversity. *Zoologica Scripta* 41(5), 499–514.

Hawley, A.H. (1988) The biology of *Aedes albopictus*. *Journal of the American Mosquito Control Association* 4, 2–39.

Huang, Y.M. (1979) Medical entomology studies – XI. The subgenus *Stegomyia* of *Aedes* in the Oriental region with keys to the species (Diptera: Culicidae). *Contributions of the American Entomological Institute* 15, 1–76.

Huang, S., Molaei, G. and Andreadis, T.G. (2008) Genetic insights into the population structure of *Culex pipiens* (Diptera: Culicidae) in the Northeastern United States by using microsatellite analysis. *The American Journal of Tropical Medicine and Hygiene* 79(4), 518–527.

Jenero-Margan, I., Aleraj, B., Krajcar, D., Lesnikar, V., Klobucar, A., Pem-Novosel, I., Kurecic-Filipovic, S., Komparak, S., Martic, R., Duricic, S. *et al.* (2011) Autochthonous dengue fever in Croatia, August–September 2010. *Eurosurveillance* 16(9), pii=19805.

Juliano, S.A. and Lounibos, L.P. (2005) Ecology of invasive mosquitoes: effects on resident species and on human health. *Ecology Letters* 8, 558–574.

Juliano, S.A. and Lounibos, L.P. (2016) Invasions by mosquitoes: the roles of behaviour across the life cycle. In: Weis, J.S. and Sol, D. (eds) *Biological Invasions and Animal Behaviour*. Cambridge University Press, Cambridge, pp. 245–265.

Kampen, H. and Werner, D. (2014) Out of the bush: the Asian bush mosquito *Aedes japonicus japonicus* (Theobld1901) (Diptera, Culicidae) becomes invasive. *Parasites & Vectors* 7, 59–68.

Kaufman, M.G. and Fonseca, D.M. (2014) Invasion biology of *Aedes japonicus japonicus* (Diptera: Culicidae). *Annual Review of Entomology* 59, 31–49.

Killeen, G.F., Fillinger, U., Kiche, I., Gouagna, L.C. and Knols, B.G. (2002) Eradication of *Anopheles gambiae* from Brazil: lessons for malaria control in Africa? *The Lancet Infectious Diseases* 2, 618–627.

Knudsen, A.B., Romi, R. and Majori, G. (1996) Occurrence and spread in Italy of *Aedes albopictus*, with implications for its introduction into other parts of Europe. *Journal of the American Mosquito Control Association* 12, 177–183.

Kraemer, M.U.G., Sinka, M.E., Duda, K.A., Mylne, A.Q.N., Shearer, F.M., Barker, C.M., Moore, C.G., Carvalho, R.G., Coelho, G.E., Van Bortel, W. *et al.* (2015) The global distribution of the arbovirus vectors *Aedes aegypti* and *Ae. albopictus*. *Elife* 4, e08347.

Lambrechts, L., Cohuet, A. and Robert, V. (2011) Malaria vectors. In: Simberloff, D. and Reimánek, M. (eds) *Encyclopedia of Biological Invasions*. University of California Press, Berkeley, California, pp. 442–445.

Levine, R.S., Peterson, A.T. and Benedict, M.Q. (2004) Geographic and ecologic distributions of the *Anopheles gambiae* complex predicted using a genetic algorithm. *American Journal of Tropical Medicine and Hygiene* 70, 105–109.

Lounibos, L.P. (2002) Invasions by insect vectors of human disease. *Annual Review of Entomology* 47, 233–266.

Lounibos, L.P. (2011) Mosquitoes. In: Simberloff, D. and Reimánek, M. (eds) *Encyclopedia of Biological Invasions*. University of California Press, Berkeley, California, USA, pp. 462–466.

Marchand, E., Prat, C., Jeannin, C., Lafont, E., Bergmann, T., Flusin, O., Rizzi, J., Roux, N., Busso, V., Deniau, J. *et al.* (2013) Autochthonous case of dengue in France, October 2013. *Eurosurveillance* 18(50), 20661.

Medlock, J.M., Hansford, K.M., Schaffner, F., Versteirt, V., Hendrickx, G., Zeller, H. and Van Bortel, W. (2012) A review of the invasive mosquitoes in Europe: ecology, public health risks, and control options. *Vector-Borne and Zoonotic Diseases* 12(6), 435–447.

Medlock, J.M., Hansford, K.M., Versteirt, V., Cull, B., Kampen, H., Fontenille, D., Hendrickx, G., Zeller, H., Van Bortel, W. and Schaffner, F. (2015) An entomological review of invasive mosquitoes in Europe. *Bulletin of Entomological Research* 105(6), 637–663.

Mitchell, C.J. (1995) Geographic spread of *Aedes albopictus* and potential for involvement in arbovirus cycles in the Mediterranean Basin. *Journal of Vector Ecology* 20, 44–58.

Montarsi, F., Ciocchetta, S., Devine, G., Ravagnan, S., Mutinelli, F., Frangipane Di Regalbono, A., Otranto, D. and Capelli, G. (2015) Development of *Dirofilaria immitis* within the mosquito *Aedes (Finlaya) koreicus*, a new invasive species for Europe. *Parasites & Vectors* 8, 177.

Mutebi, J.P. and Barrett, A.D.T. (2002) The epidemiology of yellow fever in Africa. *Microbes Infections* 4, 1459–1468.

Parmakelis, A., Russello, M.A., Caccone, A., Marcondes, C.B., Costa, J., Forattini, O.P., Sallum, M.A., Wilkerson, R.C. and Powell, J.R. (2008) Historical analysis of a near disaster: *Anopheles gambiae* in Brazil. *American Journal of Tropical Medicine and Hygiene* 78, 176–178.

Paupy, C., Delatte, H., Bagny, L., Corbel, V. and Fontenille, D. (2009) *Aedes albopictus*, an arbovirus vector: from the darkness to the light. *Microbes & Infection* 11(14–15), 1177–1185.

Petersen, L.R., Brault, A.C. and Nasci, R.S. (2013) West Nile virus: review of the literature. *Journal of the American Medical Association* 310, 308–315.

Powell, J.R. and Tabachnick, W.J. (2013) History of domestication and spread of *Aedes aegypti* – a review. *Memorias Do Instituto Oswaldo Cruz* 108 (Suppl. I), 11–17.

Rai, K.S. (1991) *Aedes albopictus* in the Americas. *Annual Review of Entomology* 36, 459–484.

Reinert, J.F., Harbach, R.E. and Kitching, I.J. (2004) Phylogeny and classification of *Aedini* (Diptera: Culicidae), based on morphological characters of all life stages. *Zoological Journal of the Linnean Society* 142, 289–368.

Reiter, P. (2010a) The standardized freight container: vector of vectors and vector-borne diseases. *Revue Scientifique et Technique* 29, 57–64.

Reiter, P. (2010b) Yellow fever and dengue: a threat to Europe? *Eurosurveillance* 15(10), 19509.

Rezza, G., Nicoletti, L., Angelini, R., Romi, R., Finarelli, A.C., Panning, M., Cordioli, P., Fortuna, C., Boros, S., Magurano, F. *et al.* (2007) Infection with chikungunya virus in Italy: an outbreak in a temperate region. *The Lancet* 370, 1840–1846.

Rizzoli, A., Jiménez-Clavero, M.A., Barzon, L., Cordioli, P., Figuerola, J., Koraka, P., Martina, B., Moreno, A., Nowotny, N., Pardigon, N. *et al.* (2015) The challenge of West Nile virus in Europe: knowledge gaps and research priorities *Eurosurveillance* 20(20), pii=21135.

Romi, R., Severini, F. and Toma, L. (2006) Cold acclimation and overwintering of female *Aedes albopictus* in Roma. *Journal of the American Mosquito Control Association* 22(1), 149–151.

Romi, R., Toma, L., Severini, F. and Di Luca, M. (2008) Twenty years of the presence of *Aedes albopictus* in Italy – from the annoying pest mosquito to the real disease vector. *European Infectious Disease* 2(2), 98–101.

Ross, D. (1908) *Report on the Prevention of Malaria in Mauritius*. Waterlow and Sons, London.

Sabatini, A., Raineri, V., Trovato, G. and Coluzzi, M. (1990) *Aedes albopictus* in Italia e possibile diffusione della specie nell'area mediterranea. *Parassitologia* 32, 301–304.

Saunders, E.J., Marfin, A.A., Tukei, P.M., Kuria, G., Ademba, G., Agata, N.N., Ouma, J.O., Cropp, C.B., Karabatsos, N., Reiter, P., Moore, P.S. and Guber, D.J. (1998) First recorded outbreak of yellow fever in Kenya, 1992–1993. Epidemiologic investigations. *American Journal of Tropical Medicine and Hygiene* 59, 644–649.

Schaffner, F. and Aranda, C. (2005) European SOVE – MOTAX group: Technical Note.

Schaffner, F. and Mathis, A. (2014) Dengue and dengue vectors in the WHO European region: past, present, and scenarios for the future. *The Lancet Infectious Diseases* 14(12), 1271–1280.

Schaffner, F., Chouin, S. and Guilloteau, J. (2003) First record of *Ochlerotatus (Finlaya) japonicus japonicus* (Theobald, 1901) in metropolitan France. *Journal of the American Mosquito Control Association* 19(1), 1–5.

Schaffner, F., Bellini, R., Petri , D., Scholte, E.-J., Zeller, H. and Marrama Rakotoarivony, L. (2013a) Development of guidelines for the surveillance of invasive mosquitoes in Europe. *Parasites and Vectors* 6, 209.

Schaffner, F., Medlock, J.M. and Van Bortel, W. (2013b) Public health significance of invasive mosquitoes in Europe. *Clinical Microbiology and Infection* 19, 685–692.

Septfons, A., Leparc-Goffart, I., Couturier, E., Franke, F., Deniau, J., Balestier, A., Guinard, A., Heuzé, G., Liebert, A., Mailles, A. *et al.* (2016) Travel-associated and autochthonous Zika virus infection in mainland France, 1 January to 15 July 2016. *Eurosurveillance* 21(32), pii=30315.

Soper, F.L. (1966) Paris Green in the eradication of *Anopheles gambiae*: Brazil, 1940; Egypt, 1945. *Mosquito News* 26, 470–476.

Sousa, C.A., Clairouin, M., Seixas, G., Viveiros, B., Novo, M.T., Silva, A.C., Escoval, M.T. and Economo-poulou, A. (2012) Ongoing outbreak of dengue type 1 in the Autonomous Region of Madeira, Portugal: preliminary report. *Eurosurveillance* 17(49), 15–18.

Tanaka, K., Mizusawa, K. and Saugstad, E. (1979) A revision of the adult and larval mosquitoes of Japan (including the Ryukyu Archipelago and the Ogasawara islands) and Korea (Diptera: Culicidae). *Contributions of the American Entomological Institute* 16, 1–987.

Tatem, A.J., Rogers, D.J. and Hay, S.I. (2006) Global transport networks and infectious disease spread. *Advances in Parasitology* 62, 293–343.

Toma, L., Di Luca, M., Severini, F., Boccolini, D. and Romi, R. (2011) *Aedes aegypti*: risk of introduction in Italy and strategy to detect the possible re-introduction. *Veterinaria Italiana. Collana di monografie* 23, 18–26.

Versteirt, V., De Clercq, E.M., Fonseca, D.M., Pecor, J., Schaffner, F., Coosemans, M. and Van Bortel, W. (2009) Introduction and establishment of the exotic mosquito species *Aedes japonicus japonicus* (Diptera: Culicidae) in Belgium. *Journal of Medical Entomology* 46(6), 1464–1467.

Versteirt, V., Schaffner, F., Garros, C., Dekoninck, W., Coosemans, M. and Van Bortel, W. (2012) Bionomics of the established exotic mosquito species *Aedes koreicus* in Belgium, Europe. *Journal of Medical Entomology* 49(6), 1226–1232.

Vilà, M., Basnou, C., Pyøek, P., Josefsson, M., Genovesi, P., Gollasch, S., Nentwig, W., Olenin, S., Roques, A., Roy, D., Hulme, P.E. and Daisie partners (2009) How well do we understand the impacts of alien species on ecosystem services? A pan-European, cross-taxa assessment. *Frontiers in Ecology and the Environment* 8, 135–144.

Vinogradova, E.B. (2000) *Culex pipiens pipiens Mosquitoes: Taxonomy, Distribution, Ecology, Physiology, Genetics, Applied Importance and Control*. Pensoft, Sofia-Moscow, Russia.

Weaver, S.C. and Lecuit, M. (2015) Chikungunya virus and the global spread of a mosquito-borne disease. *New England Journal of Medicine* 372, 1231–1239.

White, G.B. (1974) *Anopheles gambiae* complex and disease transmission in Africa. *Transactions of the Royal Society of Tropical Medicine and Hygiene* 68, 278–301.

White, B.J., Collins, F.H. and Besansky, N.J. (2011) Evolution of *Anopheles gambiae* in relation to humans and malaria. *Annual Review in Ecology, Evolution, and Systematics* 42, 111–132.

Wilkerson, R.C., Linton, Y.-M., Fonseca, D.M., Schultz, T.R., Price, D.C. and Strickman, D.A. (2015) Making mosquito taxonomy useful: a stable classification of tribe Aedini that balances utility with current knowledge of evolutionary relationships. *PLoS ONE* 10(7), e0133602.

World Health Organization (WHO) (2012) *Eliminating Malaria. Case-Study 4. Preventing Reintroduction in Mauritius*. Available at: http://apps.who.int (accessed 19 December 2017).

World Health Organization (WHO) (2013) *Regional Framework for Surveillance and Control of Invasive Mosquito Vectors and Re-emerging Vector-borne Diseases 2014–2020*. WHO Regional Office for Europe, Copenhagen.

World Health Organization (WHO) (2016a) Media Center, Fact Sheets, Yellow Fever. Available at: http://www.who.int/mediacentre/factsheets/fs100/en/ (accessed 19 December 2017).

World Health Organization (WHO) (2016b) Statement on the third meeting of the International Health Regulations (2005) (IHR 2005). Emergency Committee on Zika virus and observed increase in neurological disorders and neonatal malformations. Available at: http://www.who.int/mediacentre/news/statements/2016/zika-third-ec/en/ (accessed 16 September 2016).

7

Invasive Freshwater Invertebrates and Fishes: Impacts on Human Health

Catherine Souty-Grosset[1]*, Pedro Anastácio[2], Julian Reynolds[3] and Elena Tricarico[4]

[1]Université de Poitiers, France; [2]Universidade de Évora, Portugal; [3]Trinity College, University of Dublin, Ireland; and [4]Department of Biology, University of Florence, Italy

Abstract

Inland waters are subject to more widespread biotic invasions than terrestrial ecosystems. During the last century, 756 aquatic species were introduced in Europe, frequently carrying new parasites for native fauna and humans. The consequences of such invasions are the loss of the invaders' original parasites, the introduction of new parasites, or new intermediate hosts or vectors for existing parasites. Many parasites are water-borne and need aquatic species to complete their transmission cycles. The list of *100 of the World's Worst Invasive Alien Species* (Lowe *et al.*, 2000) does not take into account human health problems, so a risk assessment of the consequences of invasive freshwater alien species requires more attention. Here we review the direct and indirect impacts of invasive freshwater alien species on human health. Direct impacts include the injuries or allergies and new contaminants (bacteria, toxins), and their role as intermediate hosts to human parasites. Indirect impacts include the effects of the chemicals needed to control these aliens, changes to ecosystem services making the invaded area less suitable for recreational human use and damage to cultivation/aquaculture affecting human well-being in developing countries. A clear management response is urgently needed to halt their spread and reduce or minimize the risk of human and wildlife disease.

7.1 Introduction

Inland waters are subject to more widespread biotic invasions than terrestrial ecosystems (Strayer, 2010). These invasions may be unintended, for example with ships ballast or migrating via canals, or intentional, for aquaculture or other human use (Savini *et al.*, 2010), freshwater fishes being the most frequently moved aquatic group at a global level (García-Berthou *et al.*, 2007). During the last century, 756 aquatic species were introduced in Europe (Nunes *et al.*, 2015), frequently carrying new parasites for native fauna and humans. Parasitized alien species can cause the loss of the invaders' original parasites or the introduction of new parasite species, or can act as new intermediate hosts or vectors for existing parasites.

* E-mail: catherine.souty@univ-poitiers.fr

Many parasites are water-borne and need aquatic species to complete their transmission cycles. Thus, risk assessment of the consequences of invasive species in freshwater ecosystems requires more attention, because they may have direct impacts on humans through causing injuries or allergies, carrying contaminants (bacteria, toxins), and by acting as intermediate hosts to human parasites (Mazza *et al.*, 2014). Invasive freshwater molluscs have been studied extensively, especially for their impacts on human health (Pointier, 1999). There is also an increasing awareness about the spread of new diseases or parasites with fish movement; for example, in Europe, over 100 new fish parasite species have been recorded as contaminants of introduced fish (Gherardi, 2010). Fish are frequently used for the evaluation of trace metal pollution because they are a potential source of environmental contaminants with adverse effects on human health (Barak and Mason, 1990). Despite the health benefits of regular consumption of fish and seafood, when fish are exposed to elevated metal levels, they can absorb the available metals from the environment via their gills and skin or through ingesting contaminated water and food. Therefore, metals can accumulate in their tissues and enter the food chain, causing health problems to humans (Ahmad and Othman, 2010). However, although the list of *100 of the World's Worst Invasive Alien Species* (Lowe *et al.*, 2000) selected species with heavy ecological and economic impacts, it did not fully take into account species causing human health problems, mentioning this for only seven species.

Many organisms introduced for ornamental purposes may be released into the wild. A third of the world's worst aquatic invaders are ornamental species (Padilla and Williams, 2004) with impacts on human health (i.e. venomous species or vectors for parasites or infections). Invasive ampullarid molluscs, such as the Apple snails *Pomacea* spp., are commonly used as aquarium pets (Duggan, 2010). They have strong agricultural and environmental impacts, but are also important vectors of the rat lungworm (*Angiostrongylus cantonensis*), a parasitic nematode that can cause eosinophilic meningoencephalitis in humans (Yang *et al.*, 2013) if ingested through undercooked meat. Other health issues related to apple snails are the possible injuries empty shells may cause to the feet of farmers and the poorly regulated application of heavy pesticides to control the species (Cowie, 2002). Even planorbid snails, common in the aquarium trade, such as *Biomphalaria* sp., are the intermediate hosts of *Schistosoma* blood flukes that can cause intestinal schistosomiasis, a disease that infects over 83 million people (Crompton, 1999).

We will illustrate some well-known examples of direct and indirect contamination from invasive freshwater invertebrates and fishes (see Table 7.1 for a summary).

7.2 Direct Effects on Human Health

7.2.1 Allergies and injuries

Crustacean shellfish is included in allergen labelling recommendations for pre-packaged foods. Crustacean and mollusc shellfish allergies are relatively common, associated with Tropomyosin allergen (anaphylactic, not neutralized by heating). In molluscs, the sharp shells of zebra mussel *Dreissena polymorpha* may cause injuries to bathers and paddlers (DAISIE, 2009). Among crustaceans, the waterflea *Cercopagis pengoi*, native to the Ponto-Caspian area and invasive in waterways of Eastern Europe and the Great Lakes of North America, can cause allergic reactions in fishermen when cleaning nets (Leppäkoski and Olenin, 2000). Generally, alien crustaceans increase allergies already present. Some amphipods may bite humans when fishing, diving and swimming, causing skin reactions (Dick *et al.*, 2002). The Asian silver carp *Hypophthalmichthys molitrix* threatens the safety of anglers and recreational boaters because it can jump at least 3 m out of the water, causing injuries or even death to boaters (Kolar *et al.*, 2007). Many invasive catfish sold in markets have

Table 7.1. Examples of introduced freshwater molluscs, crustaceans and fish potentially causing health problems to humans.

Species	Common name	Impact on human health	References
Molluscs			
Pomacea spp.	Apple snails	Vectors of the rat lungworm (*Angiostrongylus cantonensis*) – eosinophilic meningoencephalitis. Empty shells can injure the feet of farmers. Problems with poorly regulated application of dangerous pesticides	Cowie, 2002
Biomphalaria spp.	Planorbid snails	Intermediate hosts of *Schistosoma* digenetic trematodes	DAISIE, 2009
Lymnaea spp. (*natalensis, columella, truncatula*)	Pond snails	Intermediate hosts for liver flukes that cause fascioliasis – *Fasciola gigantica*	Mas-Coma *et al.*, 2009
Corbicula fluminea	Asian clam	Human waterborne pathogens *Cryptosporidium parvum, C. hominis, Giardia intestinalis*	Graczyk *et al.*, 2003
		Concentrate heavy metals and are consumed by fish, which may result in bioaccumulation	García and Protogino, 2005 (consumption by fish)
		The trematode *Echinostoma* sp. is the most referenced parasite	Huffman and Fried, 1990
Dreissena polymorpha	Zebra mussel	Human waterborne pathogens *Cryptosporidium parvum, C. hominis, Giardia intestinalis*	Graczyk *et al.*, 2003
		Accumulate toxins like polychlorinated biphenyls and polycyclic aromatic hydrocarbons and may increase human exposure to organic pollutants as these toxins are passed up the food chain	Bruner *et al.*, 1994
		Injuries to bathers have been documented	DAISIE, 2009
Limnoperna fortunei	Golden mussel	Concentrate heavy metals and are consumed by fish, which may result in bioaccumulation	García and Protogino, 2005 (consumption by fish)
Ferrissia fragilis	Fragile Ancylid	Human schistosomiasis	Sankarappan *et al.*, 2015
Crustaceans			
Cercopagis pengoi	Fishhook waterflea	Allergic reactions in fishermen when cleaning nets	Leppäkoski and Olenin, 2000
Dikerogammarus villosus	Killer shrimp	May bite humans, when fishing, diving and swimming, causing skin reaction	Platvoet, 2007
Eriocheir sinensis	Chinese mitten crab	Possible second intermediate host for human lung fluke parasite *Paragonimus westermani*	Gollasch *et al.*, 2008
		Potential to bioaccumulate contaminants that then may be passed up the food chain	Rudnick *et al.*, 2000
		Arsenic, selenium and DDT derivatives were detected in crabs from California	Veilleux and de Lafontaine, 2007
		Contaminated with *Vibrio parahaemolyticus*	Wagley *et al.*, 2009

continued

Table 7.1. *continued.*

Species	Common name	Impact on human health	References
Pacifascus leniusculus	Signal crayfish	Allergies towards shellfish may occur	Johnsen and Taugbøl, 2010
		Accumulate cyanotoxins in their tissues	Vasconcelos *et al.*, 2001
Cherax quadricarinatus		Potential disease vector for lethal microbes including *Vibrio cholerae*, enterococcids, and *Escherischia coli*	Edgerton *et al.*, 2002
		Hepatopancreas and intestines may accumulate high levels of cyanotoxins	Saker and Eaglesham, 1999
Procambarus clarkii	Red swamp crayfish	Several impacts on human health	Gherardi, 2010
		Tularemia, caused by the Gram-negative bacteria *Francisella tularensis*	Anda *et al.*, 2001
		Accumulate cyanotoxins in their tissues and may consume cyanobacteria	Vasconcelos *et al.*, 2001; Gherardi and Lazzara, 2006
		May function as a contamination vector for pesticides to humans	Lages *et al.*, 2009
		Able to bioaccumulate heavy metals, and often live in areas contaminated by sewage and toxic industrial residues	Geiger *et al.*, 2005
		Cancer risks from lifetime consumption of crayfish exposed to a herbicide mixture	Green and Abdelghani, 2004
Crayfish	Crayfish	Humans acquire *Paragonimus kellicotti* when they consume infected raw crayfish	Lane *et al.*, 2009
Several aquatic crustaceans		Close relationship between microsporidian taxa infecting aquatic crustaceans and the human pathogen *Enterocytozoon bieneusi*	Conn, 2014
Fish			
Ameirus melas	Black bullhead (catfish)	Injures humans – venomous [the same applies to many other catfish]	Elia *et al.*, 2007
		Accumulates organochlorine pesticides, PCBs and heavy metals	
Ictalurus punctatus	Channel catfish	Microcystins have been found	Zimba *et al.*, 2001
Micropterus salmoides	Largemouth bass	Levels of mercury especially harmful to developing children	Karouna-Renier *et al.*, 2011
Hypophthalmichthys molitrix	Silver carp	Collisions between boaters and jumping carp have caused human fatalities	A. Smith, Mississippi State University, 2015

poisonous spines (Rixon *et al.*, 2005). These spines frequently cause morbidity among anglers (Mann and Werntz, 1991), retained foreign bodies, respiratory compromise, arterial hypotension and cardiac dysrhythmias (Dorooshi, 2012). Other serious complications can occur, mostly resulting from infections, as in *Ictalurus punctatus* (Blomkalns and Otten, 1999).

7.2.2 Bacteria and their toxins

In crustaceans, a waterborne outbreak of tularemia (caused by the Gram-negative bacteria *Francisella tularensis*) was associated with fishing the invasive red swamp crayfish *Procambarus clarkii*. Crayfish stomach and hepatopancreas harboured *F. tularensis*: in 1997, an outbreak affected 585 patients in central Spain (Anda *et al.*, 2001). Vibrios are also Gram-negative bacteria that form part of the autochthonous microbial flora of estuarine environments. Those of primary human health significance are *Vibrio cholerae*, *V. parahaemolyticus* and *V. vulnificus*. *Vibrio cholerae* may be picked up from fresh waters, directly into cuts or through consumption. *Vibrio parahaemolyticus* food-poisoning has been identified in many countries (Blake *et al.*, 1980), most commonly from those that have both high temperature and traditions of raw food consumption. Like other crayfish, the redclaw *Cherax quadricarinatus* is a potential disease vector for lethal microbes including *V. cholerae*, enterococcids and *Escherischia coli* (Edgerton *et al.*, 2002). In the Thames estuary, Wagley *et al.* (2009) investigated the presence of likely human pathogens in the invasive catadromous Chinese mitten crab *Eriocheir sinensis*, for the first time detecting *V. parahaemolyticus* from *E. sinensis* in Europe. Bacterial recovery was concomitant with an increasing water temperature during the summer months.

The occurrence of harmful cyanobacterial blooms in surface waters is often accompanied by the production of a variety of cyanotoxins that are generally classified according to the human organs on which they act: hepatotoxins (liver), neurotoxins (nervous system) and dermatotoxins (skin). Microcystins are stable and resistant to degradation even during cooking (Drobac *et al.*, 2013). Their presence has been reported worldwide and liver-toxic microcystins are commonly found in 50–75% of cyanobacterial blooms. Contamination by cyanotoxins from accidental ingestion by humans (from drinking water, recreational contact and consumption of aquatic organisms) causes abdominal pain, vomiting, diarrhoea, skin irritation and muscle tremors. Aquatic organisms could contribute to food chain transfer of cyanotoxins and constitute a health risk (Ettoumi *et al.*, 2011).

Cyanobacteria may be associated with the invasive South American colonial bryozoan *Pectinatella magnifica* in reservoirs, harbouring and releasing cyanotoxins into the water (Joo *et al.*, 1992). Invasive crayfish such as *C. quadricarinatus* (Saker and Eaglesham, 1999), *Pacifastacus leniusculus* and *P. clarkii* (Vasconcelos *et al.*, 2001) may accumulate high levels of cyanotoxins in their hepatopancreas and intestines and transfer them to more sensitive organisms, man included. *P. clarkii* hatchlings were found to be resistant to cyanobacteria and their toxins, surviving cyanobacteria densities during acute exposures, while juveniles tolerated toxic cyanobacteria better than non-toxic ones. *Procambarus clarkii* accumulated up to 2.9 µg microcystin/dry crayfish weight with a depuration pattern similar to that observed for mussels. The consumption of Cyanobacteria by *P. clarkii* is an increasing concern for human health (Gherardi and Lazzara, 2006), even if depuration is possible before human consumption (Tricarico *et al.*, 2008).

Finally, being often at the top of the food chain, fish are the most exposed to cyanotoxins, which may be accumulated in liver, muscles, gills, guts and kidneys. These contaminants can be magnified along the food chain: fish consuming zebra mussels will accumulate these organic pollutants (Vanderploeg *et al.*, 2001). In China, Zhao *et al.* (2006) suggested that Nile tilapia fed on toxic cyanobacteria are not suitable for human food. In the laboratory, fish fed with toxic cyanobacteria accumulated more microcystins than in natural waters, mainly due to a low food choice and restricted living area. Microcystins have also been found in edible catfish *Ictalurus punctatus* (Zimba *et al.*, 2001). Microcystin concentration in muscle is highest in omnivorous fish (such as the invasive goldfish *Carassius auratus*), followed by phytoplanktivorous and carnivorous fish (Zhang *et al.*, 2009).

7.2.3 Fungi and oomycetes

Parasitic or saprophytic fungi may be carried to new areas by invasive organisms. Oomycetes (water moulds) are known to cause major mortalities in plants (e.g. potato blight) and animals, potentially affecting human health. Microsporidia may be hosted by all major animal groups, their spores being released through faeces or urine. The spores may be consumed by humans and may cause serious disease, especially in immuno-compromised individuals. The close relationship between microsporidian taxa infecting aquatic crustaceans and the human pathogen *Enterocytozoon bieneusi* raises interesting questions regarding the ecological and evolutionary linkage between these parasites and the potential role of invertebrates (and water) as a source of zoonotic infections in humans (Conn, 2014).

7.2.4 Contaminants: pesticides and herbicides, heavy metals

Aquatic ecosystems are constantly exposed to chemical contaminants from industrial, agricultural and domestic sources. The deleterious effects of pesticides on human health have increased due to their toxicity and persistence in the environment and ability to enter the food chain. Many aquatic invaders are able to concentrate these compounds, with consequences for human health if consumed by people.

Zebra mussels accumulate organic toxins such as polychlorinated biphenyls and polycyclic aromatic hydrocarbons in their tissues, and may increase human exposure as these toxins are accumulated and transferred along the food chain (Bruner *et al.*, 1994). They can concentrate organic pollutants within their tissues to levels more than 300,000 times greater than concentrations in the environment, also depositing these pollutants in their pseudofaeces. García and Protogino (2005) demonstrated that in South America the invasive molluscs *Corbicula fluminea* and *Limnoperna fortunei*, which accumulate heavy metals, are consumed by native fishes, often caught for human consumption. Even the Chinese mitten crab *E. sinensis* has the potential to bioaccumulate inorganic and organic contaminants such as arsenic, selenium and DDT derivatives that then may be passed up the food chain (Rudnick *et al.*, 2000; Veilleux and de Lafontaine, 2007). The red swamp crayfish *P. clarkii* may function as a contamination vector for pesticides to humans using it as a food source (Lages *et al.*, 2009), potentially causing health problems (e.g. thyroid modification by the contact fungicide propineb, along with reproductive impairment and carcinogenesis; neurological effects by cymoxanil), especially if ingestion occurs over long time periods. Bioconcentration studies have analysed the uptake of arsenic by crayfish after long-term exposure to herbicides such as 4-dichlorophenoxyaceticacid and monosodium methanearsonate. The hepatopancreas bioconcentrated the highest amount of arsenic. Assessment of the human health risk associated with consuming crayfish showed an exposure dose approximately twice the reference dose for arsenic. Cancer risks have been detected, resulting from a lifetime consumption of crayfish exposed to the herbicide mixture (Green and Abdelghani, 2004).

Heavy metals accumulation has also been studied in crayfish (Sneddon and Richert, 2011). *Procambarus clarkii* bioaccumulates heavy metals nickel, lead and zinc in tissues and organs at a significantly higher rate than the indigenous European crayfish *Austropotamobius pallipes* (Gherardi *et al.*, 2002), and often lives in areas contaminated by sewage and toxic industrial residues (Geiger *et al.*, 2005), increasing the probability of transferring contaminants to their consumers, including man. In Spain, several heavy metals released into the River Guadiamar by a mine spill (Toja *et al.*, 2003) are accumulated by *P. clarkii* and the potential to seriously affect human health is widely documented (Oyarzun and Higueras, 2005). Indeed, Cd, Pb and As have been registered by the CIS (2006) as carcinogenic to humans. In China, the world's largest producer and consumer of crayfish, *P. clarkii* has been reported to accumulate high levels of metals. Peng *et al.* (2016a) estimated that the bioaccessibility of metals in crayfish was variable (68–95%) and metal-specific. Their results suggest a need to consider crayfish when assessing human dietary exposure to metals and associated health risks, especially for high crayfish-consuming populations, such as in China, the USA and Sweden. Peng *et al.* (2016b) showed methylmercury (MeHg) accumulation in crayfish, raising global concern about human exposure. The potential health risk of MeHg exposure through eating crayfish depends largely on consumption rates.

Finally, bioaccumulation of Cu, Pb, Ni and Cr was observed in two widespread invasive fish species, *Oreochromis niloticus* and *Cyprinus carpio*, in the Indian Yamuna River (Singh *et al.*, 2014). The calculated hazard index for Pb was more than one, which indicated human health concern. The carcinogenic risk value for Ni was higher than for all other metals studied. Results indicated heavy metals bioaccumulate in fish muscle. Thus, regular environmental monitoring of heavy metal contamination in fish is advocated for assessing food safety, since eating fish contaminated because of a degraded environment may be associated with health risks.

7.2.5 Intermediate hosts to human parasites

One of the most significant ecological and economic impacts of aquatic invaders is associated with their role as vectors for the introduction of parasites into invaded areas. It is estimated that molluscs are among the most notorious invaders in terms of transmitting parasitic diseases to humans (Prenter *et al.*, 2004). The stage in the intermediate host is often a multiplicative form. Fish-borne trematodiasis infecting humans is well known (Hung *et al.*, 2013) and fishes are mainly the second intermediate host, while freshwater snails are very often the first intermediate ones. Freshwater crustaceans are the second intermediate hosts for human lung flukes, whereas fish may transmit liver or intestinal flukes. Freshwater-food-associated infections by another group of helminths, the acanthocephalans, are also known but rarely reported in humans. Both trematode and acanthocephalan infections are associated with social-cultural and behavioural factors, in particular the consumption of raw or undercooked seafood.

Freshwater snails: schistosomiasis, fasciolasis, angiostrongyliasis and echinostomiasis

Major human parasites such as *Schistosoma* blood flukes and food-borne trematodes affecting the liver, lungs and intestines of humans are spread through the introduction of aquatic host snails from long-established foci in other areas. Freshwater snails are intermediate hosts of these infections. People acquire schistosomiasis through repeated contact with fresh water during fishing, farming, swimming, washing, bathing and recreational activities. Humans harbour the sexual stages of the parasites and snails host the asexual stages; people serve as vectors by contaminating the environment. Transfer of the infection requires no direct contact between snails and people.

Lymnaea sp. serve as intermediate hosts for the liver flukes that cause fascioliasis; *L. natalensis* for *Fasciola gigantica*, the most common fasciolid in sub-Saharan Africa, and *L. truncatula* for *F. hepatica* (Mas-Coma *et al.*, 2009). The North American *L. columella*, invasive in Africa, is susceptible to both *F. hepatica* and *F. gigantica*, so its role in the transmission of fascioliasis in Africa needs investigation. Human *Fasciola* infections are uncommon in Africa, except in Egypt, where fascioliasis is regarded as an emerging disease: 830,000 people are thought to be infected, most in the Nile Delta, with prevalence up to 17% (Soliman, 2008). The severe pathology caused by the immature flukes as they migrate through the liver has been recognized as a public health problem in Egypt.

Eosinophilic meningitis (angiostrongyliasis) caused by *Angiostrongylus cantonensis* has been emerging as an infectious disease in mainland China since the late 1970s (Lv *et al.*, 2009). Two snail species, *Achatina fulica* and the South American golden apple snail *Pomacea canaliculata*, are closely associated with angiostrongyliasis. These snails were imported to China in 1931 and 1981 respectively, and have rapidly extended their geographic ranges; they are well established in 6 and 11 provinces, respectively, and are now listed as invasive species by the Chinese government. The parasite can cause meningoencephalitis, and thus paralysis and death, in humans (Mochida, 1988). Golden apple snails can also play host to various trematodes that cause skin irritations (Keawjam *et al.*, 1993). Through a national survey, Lv *et al.* (2009) confirmed that *P. canaliculata* and *A. fulica* were the predominant intermediate hosts of *A. cantonensis* in China. Infected snails were also found in markets and restaurants. This survey indicated that health education, rigorous food inspection and surveillance are needed to prevent recurrent angiostrongyliasis outbreaks. The biggest outbreak in China thus far attributed to *P. canaliculata* took place in the capital Beijing in 2006. Humans become infected by ingesting third-stage larvae in raw or undercooked intermediate host molluscs or in paratenic hosts (e.g. freshwater prawns, crabs, fish).

The Asian freshwater clam *Corbicula fluminea* is one of the hosts for some parasite forms that cause severe diseases in man and is still a public health problem even in endemic areas. Pathway transmission is by eating clams raw or barely cooked. The trematode *Echinostoma* sp. is the most referenced parasite within *Corbicula* sp. Echinostomiasis is spread over Southeast Asia and the Far East, the prevalence of infection ranging from 44% in the Philippines to 5% in mainland China, and from 50% in northern Thailand to 9% in Korea (Huffman and Fried, 1990). Echinostomiasis occurs in numerous other countries (Chai, 2009) and is also associated with other non-native molluscs.

Crustaceans: paragonimiasis and onchocersiasis

Some crustaceans are potential vectors for trematode, cestode and acanthocephalan parasites (spiny headed worms) from parts of the globe where most human infections occur by consumption of undercooked crab meat (East Asia), such as the freshwater crabs *Eriocheir*, *Potamon* and *Potamiscus*, second intermediate hosts for the trematode *Paragonimus* spp. in North America and East Asia (Lane *et al.*, 2009). Among the species reported to infect humans, the most common is *Paragonimus westermani*, the oriental lung fluke, in the Far East. Other species of *Paragonimus* are encountered in Asia, America and Africa. There have been rare reports of paragonimiasis from North America as a zoonosis caused by *Paragonimus kellicotti*, acquired when humans consume infected raw crayfish (Lane *et al.*, 2009).

Human infection with *P. westermani* occurs by eating inadequately cooked or pickled crab or crayfish harbouring encysted metacercariae of the parasite. The metacercariae penetrate the intestinal wall into the peritoneal cavity and then into the lungs, where they become encapsulated and develop into adults. The eggs are excreted or are swallowed and passed with faeces. In the external environment, the eggs hatch as ciliated miracidia that enter the first intermediate host, a snail, where they multiply. Cercariae emerge from the snail and invade the second intermediate host, a crustacean, crab or crayfish, where they encyst as metacercariae. Ingestion of these late larval forms closes the cycle (Fig. 7.1). Symptoms such as diarrhoea, fever and eosinophilia may mark the acute phase of invasion and migration. Extrapulmonary locations of the adult worms result in more severe manifestations, especially involving the brain.

Fish as vectors of diphyllobothriasis and clonorchiasis

Diphyllobothriasis, caused by the cestode *Diphyllobothrium latum*, occurs where lakes and rivers coexist with human consumption of raw or undercooked freshwater fish, in Europe,

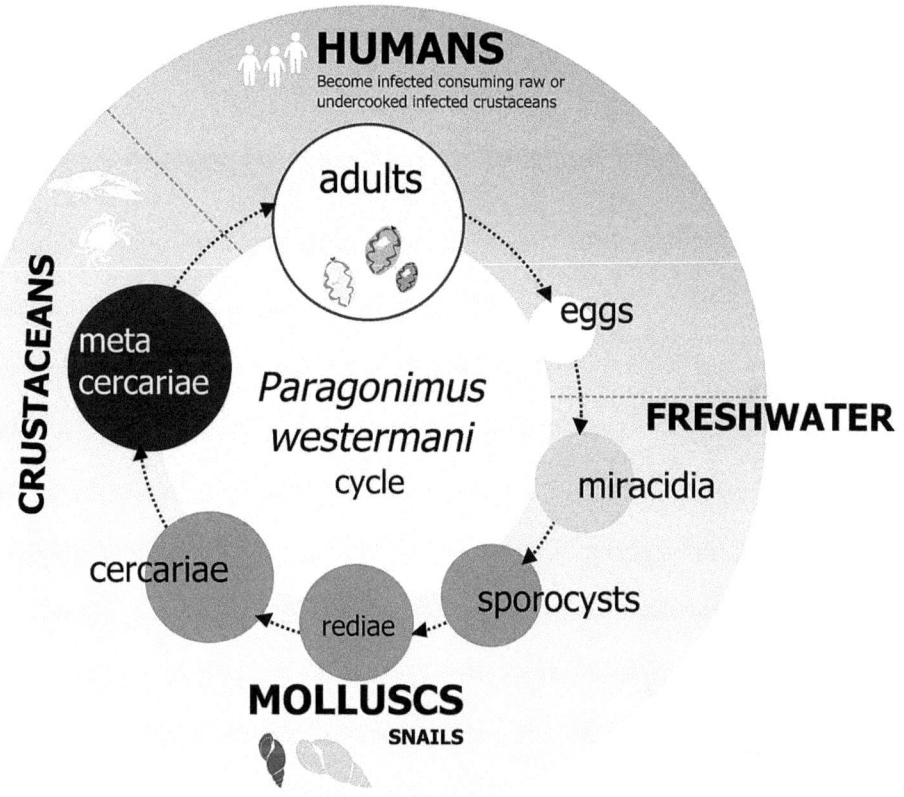

Fig. 7.1. Adult lung flukes *Paragonimus westermani* live encysted in mammalian (man included) lungs and release their eggs through the host's faeces or mucus. See text for life-cycle details. (Souty-Grosset.)

North America, South-east Asia, Uganda and Chile. The life cycle of this tapeworm is human faeces–copepod–fish–man (Fig. 7.2). After ingestion by the first intermediate host, a copepod crustacean, the ciliated coracidium develops into a procercoid larva. Following ingestion of the copepod by a suitable freshwater fish, the second intermediate host, the larva develops into a plerocercoid larva or sparganum. If a larger fish eats the smaller infected fish, the sparganum may migrate into the flesh of the larger fish, eaten finally by humans.

In European fresh waters, over 100 new fish parasite species have been recorded (Gollasch *et al.*, 2008), but only a few species are capable of infecting humans. The invasive Asian fish tapeworm *Bothriocephalus acheilognathi*, spreading actively with its host the grass carp, recently infected humans in Europe (Yera *et al.*, 2013). Among important helminths acquired by humans from fish are anisakid nematodes, particularly *Anisakis simplex* and *Pseudoterranova decipiens*. *Anisakis simplex* is most common in freshwater and anadromous fish such as wild salmon, while *P. decipiens* is also encountered in eels (Möller *et al.*, 1991). Nico *et al.* (2011) explored the potential for imported Asian swamp eels to become pathways for human helminths into North American food markets. Sokolov *et al.* (2014) reviewed parasites of the invasive Chinese sleeper fish, *Percottus glenii*, from its non-native range, noting that the trematode parasite *Isthmiophora melis* may be regarded as potentially dangerous to human health. However, Reshetnikov and Chibilev (2009) also hypothesized that this fish can depress natural centres of human opisthorchiidosis caused by the trematodes *Opisthorchis felineus*, *Metorchis bilis* and *Pseudamphistomum truncatum*, through active predation on

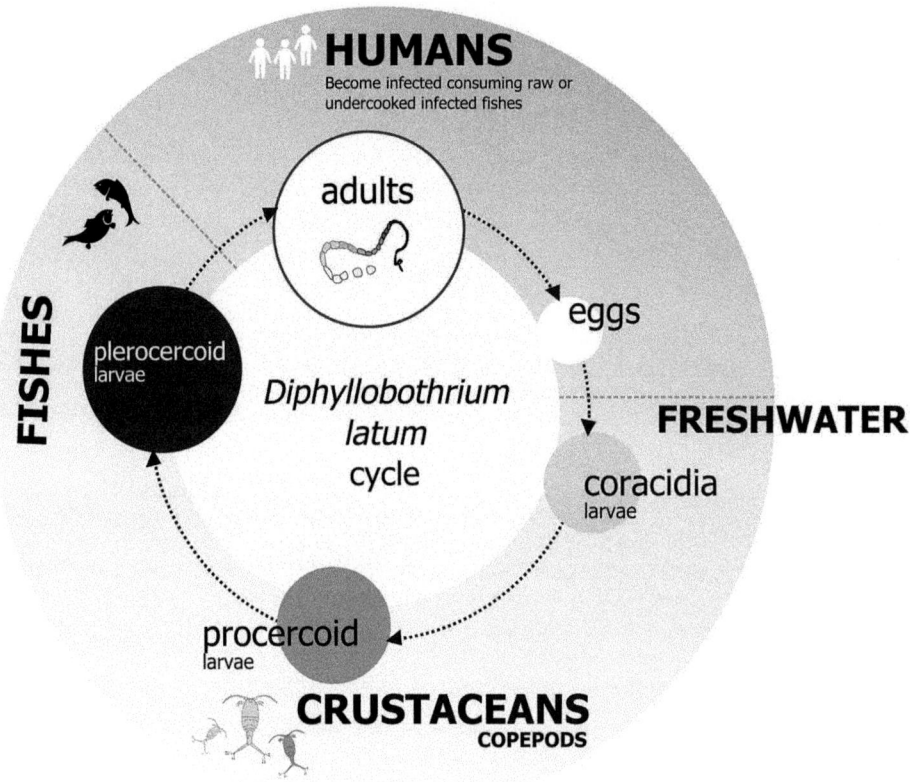

Fig. 7.2. Adult tapeworms *Diphyllobothrium latum* live in intestines of humans and other mammals and release their eggs in the host's faeces. See text for life-cycle details. (Souty-Grosset.)

intermediate hosts (bithyniid molluscs and young cyprinid fish). The oriental liverfluke, *Clonorchis sinensis*, is of major socio-economic importance in parts of Asia, including China, Japan, Korea, Taiwan and Vietnam (Lun *et al.*, 2005). About 35 million people are infected globally, of whom approximately 15 million are in China, and there is evidence for an aetiological connection between clonorchiasis and bile duct cancer in humans. The parasite is transmitted via snails to native or introduced freshwater cyprinid fish, and then to humans.

7.3 Indirect Effects on Human Health

This category of effects includes chemicals needed to control invasive species, where, due to their low selectivity and/or residues, they may have side effects on humans, but also some potential effects of invaders on the incidence of disease. Cascading effects on food webs are also analysed. Invasive species can change ecosystem services, making affected areas less suitable for recreational human use (e.g. fishing, boating). Some invasive species such as common carp and crayfish, due to their bioturbatory feeding or burrowing activity, can alter water quality as well as water flow regulation (Gherardi *et al.*, 2011). These changes in waters can affect its potability for human consumption, alter flooding regimes, damage canal banks and obstruct intakes of pumps used for irrigation and potable water. Zebra mussels can also alter water potability through favouring blooms of Cyanobacteria and macrophytes that can produce microcystins (Vanderploeg et al., 2001). The water is thus not drinkable and it is necessary to ban its use.

Aquatic invaders can affect human well-being in developing countries (reviewed in Mazza et al., 2014). *Pomacea canaliculata*, introduced into the Philippines in 1980 to increase the income of local people, escaped from rearing facilities, became established and now has damaging effects on rice seedlings over 426,000 ha, with costs of US$27.8–45.3 million in 1990 alone. Control efforts use a combination of molluscides and hand-picking snails. Chemical control is frequently applied to invasive fish species, chiefly using rotenone, antymicin, saponins and 3-trifluoromethyl-4-nitrophenol (Ling, 2009). Rotenone is a natural toxin of plant origin, long used as a fish poison (Ling, 2003). Controversy has been raised regarding possible involvement of rotenone in Parkinson's disease. However, this link was established using intravenous injection of rotenone; ingesting rotenone effectively detoxifies it (Ott, 2006). Nevertheless, prolonged occupational exposure to rotenone dust may cause non-lethal symptoms such as headaches, sore throats and other cold symptoms, also sores on mucous membranes, skin rashes and eye irritation (Ling, 2003).

In chemical control of crustacean invaders, organophosphates, organochlorines, pyrethroids, rotenone and surfactants are often used, posing problems of bioaccumulation and biomagnification in the food chain (Hänfling et al., 2011), with therefore a possibility of ultimately affecting human health. Attempts to control crustacean invaders may sometimes lead to dangerous decisions. One example is the attempt to control *P. clarkii* invasion in Portuguese rice fields during the early 1990s using the toxic chemicals parathion and dimethoate (Jørgensen et al., 1997). The leading route of human exposure to methyl parathion is by inhalation, but there is also significant risk from dermal contact and by ingestion (Edwards and Tchounwou, 2005), so individuals living near application or disposal sites are vulnerable to methyl parathion exposure. Known methyl parathion health effects include headaches, nausea, diarrhoea, difficulty breathing, lack of coordination and mental confusion (Edwards and Tchounwou, 2005). Mollusc invasions have also been widely controlled using chemical methods. One of the major methods used for *Dreissena polymorpha* is chlorination, which is considered a source of relevant carcinogens (Burger et al., 2013).

Apart from molluscs, crustaceans and fish, other aquatic invaders have indirect impacts on human health and well-being, chiefly through food-web modifications. For example, the freshwater medusa *Craspedacusta sowerbii* Lancester, originally from the Yangtze Basin in China, is now widespread. *Craspedacusta* may compete with planktivore fish for zooplankters, leading to increased phytoplankton density and associated cascading effects in water quality (Gasith et al., 2011). Myxozoa, highly modified cnidarians, parasitize freshwater fish, particularly in aquaculture (Anderson et al., 1999). While no direct harmful effects on humans have been noted, a skin test for allergens using *Kudoa*, a myxosporean parasite of marine fish, showed skin reactions in 4 out of 15 patients complaining of gastrointestinal and/or allergic symptoms related to ingesting fish (Martinez de Velasco et al., 2008).

7.4 Management and Conclusions

Peeler et al. (2011) argued, using examples of disease emergence in aquatic animals in Europe, that the introduction of aquatic aliens drives disease emergence by both extending the geographic range of parasites and pathogens, and by increasing threats to both biodiversity and human health (Daszak et al., 2000). Alien species have impacts on human health through transmission of diseases and parasites to humans, bioaccumulation of noxious substances or becoming a health hazard due to contamination with pathogens or parasites (Mazza et al., 2014; Van der Veer and Nentwig, 2015). In addition to transmitting novel pathogens to native hosts, invasive species can alter disease transmission through dilution or amplification of native pathogens, or via indirect effects on native host immunity (Poulin *et al.*, 2010). Paradoxically, some invaders may control the transmission of diseases: the mosquitofish *Gambusia holbrookii* has been introduced in many areas to control malaria through consumption of the larvae of the transmitting mosquitoes (Krumholz, 1948), while *P. clarkii* has been proposed as a means of controlling the abundance of freshwater snail species responsible for human parasites in Kenya (Lodge *et al.*, 2005).

It is important to monitor the spread and impacts of invasive species and pathogens in developing national data and research networks. 'Citizen science' can also play a role; individuals can report new invasions and record phenological changes associated with invasions or disease outbreaks. Management also involves hygiene and correct food preparation. For example, fish, crabs, crayfish and other crustaceans must be well cooked; human faeces should be kept out of watercourses and buried; raw watercress or other wetland plants should not be eaten from areas where snails, crabs, crayfish or other crustaceans and crustacean-eating animals are found. However, the human health significance of potentially pathogenic bacterial strains from Chinese mitten crabs is currently uncertain and requires clarification.

In many cases, invasive species can themselves be used as monitors for pathogen invasion. For example, assessment of human waterborne parasites was performed in Irish river basin districts using zebra mussels as bioindicators (Graczyk *et al.*, 2003); they are recognized biomonitors for the human waterborne pathogens *Cryptosporidium parvum*, *C. hominis*, *Giardia intestinalis* and microsporidia in surface waters (Lucy *et al.*, 2008). Bioinvasion scientists should search for other models that foster integration of invasive species studies with biomonitoring and surveillance for human and animal disease agents, thus helping to ensure good health for all (Conn, 2014).

A clear management response is to control the international pet trade, responsible for many invasions. Aquatic fish and invertebrates are traded worldwide through the internet (e.g. Mazza *et al.*, 2015), especially in the USA, Western Europe and the Far East. Aquatic pets are often released into the wild, resulting in alterations of food webs and transmission of diseases. These environmental and human health implications must be considered and acted on by international bodies, including the World Trade Organization. Understanding the interactions of invasive species, disease vectors and pathogens with other drivers of ecosystem change is critical to human health and economic well-being. Such situations require further monitoring by bioinvasion scientists, as international trade and thus the international movement of infective agents and their hosts increases (Conn, 2009). According to Johnson and Paull (2011), managing freshwater ecosystems to reduce or minimize human and wildlife disease risk – arguably one of the most significant ecosystem services – will require enhanced incorporation of ecological approaches alongside medical and veterinary tools. As Conn states, we need 'One Health strategies for integrating human, animal, and environmental monitoring and surveillance to better prepare for or prevent geographic spread of major human health threats associated with aquatic systems' (2014, p. 383).

References

Ahmad, A.K. and Othman, S.M. (2010) Heavy metal concentrations in sediments and fishes from Lake Chini, Pahang, Malaysia. *Journal of Biological Sciences* 10, 93–100.

Anda, P., Segura del Pozo, J., Díaz García, J.M., Escudero, R., García Peña, F.J., López Velasco, M.C., Sellek, R.E., Jiménez Chillarón, M.R., Sánchez Serrano, L.P. and Martínez Navarro, J.F. (2001) Waterborne outbreak of tularemia associated with crayfish fishing. *Emerging Infectious Diseases* 7, 575–582.

Anderson, C., Canning, E.U. and Okamura, B. (1999) Molecular data implicate bryozoans as hosts for PKX (Phylum Myxozoa) and identify a clade of bryozoan parasites within the Myxozoa. *Parasitology* 119(6), 555–561.

Barak, N.A.E. and Mason, C.F. (1990) Mercury, cadmium and lead concentrations in five species of freshwater fish from Eastern England. *Science of the Total Environment* 92, 257–263.

Blake, P.A., Weaver, R.E. and Hollis, D.G. (1980) Diseases of humans (other than cholera) caused by vibrios. *Annual Review of Microbiology* 34, 341–367.

Blomkalns, A.L. and Otten, E.J. (1999) Catfish spine envenomation: a case report and literature review. *Wilderness & Environmental Medicine* 10, 242–246.

Bruner, K.A., Fisher, S.W. and Landrum, P.F. (1994) The role of the zebra mussel, *Dreissena polymorpha*, in contaminant cycling: I. The effect of body size and lipid content on the bioconcentration of PCBs and PAHs. *Journal of Great Lakes Research* 20(4), 725–734.

Burger, M., Catto, J.W., Dalbagni, G., Grossman, H.B., Herr, H., Karakiewicz, P., Kassouf, W., Kiemeney, L.A., La Vecchia, C. and Shariat, S. (2013) Epidemiology and risk factors of urothelial bladder cancer. *European Urology* 63, 234–241.

Chai, J.-Y. (2009) Echinostomes in humans. In: Toledo. R. and Fried. B. (eds) *The Biology of Echinostomes*. Springer-Verlag, New York, pp. 147–183.

CIS (2006) *Occupational Safety and Health*. ILO/BIT/OIT–CIS, Geneva, Switzerland. Available at: http://www.ilo.org/public/english/protection/safework/cis/products/icsc/ (accessed 19 December 2017)

Conn, D.B. (2009) Presidential address: parasites on a shrinking planet. *Journal of Parasitology* 95, 1253–1263.

Conn, D.B. (2014) Aquatic invasive species and emerging infectious disease threats: a one health perspective. *Aquatic Invasions* 9(3), 383–390.

Cowie, R.H. (2002) Apple snails (Ampullariidae) as agricultural pests: their biology, impacts and management. In: Barker, G.M. (ed.) *Molluscs as Crop Pests*. CAB International, Wallingford, UK, p. 145.

Crompton, D.W.T. (1999) How much human helminthiasis is there in the world? *The Journal of Parasitology* 85, 397–403.

DAISIE (2009) *Handbook of Alien Species in Europe*. Springer, Dordrecht, the Netherlands.

Daszak, P., Cunningham, A.A. and Hyatt, A.D. (2000) Emerging infectious diseases of wildlife – threats to biodiversity and human health. *Science* 287, 443–449.

Dick, J.T.A., Platvoet, D. and Kelly, D.W. (2002) Predatory impact of the freshwater invader *Dikerogammarus villosus* (Crustacea: Amphipoda). *Canadian Journal of Fisheries and Aquatic Sciences* 59, 1078–1084.

Doroshi, G. (2012) Catfish stings: a report of two cases. *Journal of Research in Medical Sciences: The Official Journal of Isfahan University of Medical Sciences* 17, 578.

Drobac, D., Tokodi, N., Simeunovic, J., Baltic, V., Stanic, D. and Svircev, Z. (2013) Effects of cyanotoxins in humans. *Archives of Industrial Hygiene and Toxicology* 64, 305–316.

Duggan, I. (2010) The freshwater aquarium trade as a vector for incidental invertebrate fauna. *Biological Invasions* 12, 3757–3770.

Edgerton, B.F., Evans, L.H., Stephens, F.J. and Overstreet, R.M. (2002) Synopsis of freshwater crayfish diseases and commensal organisms. *Aquaculture* 206(1–2), 57–135.

Edwards, F.L. and Tchounwou, P. (2005) Environmental toxicology and health effects associated with methyl parathion exposure – a scientific review. *International Journal of Environmental Research and Public Health* 2(3), 430–441.

Elia, A.C., Dorr, A.J. and Galarini, R. (2007) Comparison of organochlorine pesticides, PCBs, and heavy metal contamination and of detoxifying response in tissues of *Ameiurus melas* from Corbara, Alviano, and Trasimeno Lakes, Italy. *Bulletin of Environmental Contamination and Toxicology* 78(6), 463–468.

Ettoumi, A., Khalloufi, F.E., Ghazali, I.E., Oudra, B., Amrani, A., Nasri, H. and Bouaicha, N. (2011) Bioaccumulation of cyanobacterial toxins in aquatic organisms and its consequences for public health. In: Kattel, G. (ed.) *Zooplankton and Phytoplankton: Types, Characteristics and Ecology*. Nova Science, New York, pp. 1–34.

García, M.L. and Protogino, L.C. (2005) Invasive freshwater molluscs are consumed by native fishes in South America. *Journal of Applied Ichthyology* 21, 34–38.

García-Berthou, E., Boix, D. and Clavero, M. (2007) Non-indigenous animal species naturalized in Iberian inland waters. In: Gherardi, F. (ed.) *Biological Invaders in Inland Waters: Profiles, Distribution, and Threats*. Springer, Dordrecht, The Netherlands, pp. 123–140.

Gasith, A., Gafny, S., Hershkovitz, Y., Goldstein, H. and Galil, B.S. (2011) The invasive freshwater medusa *Craspedacusta sowerbii* Lankester, 1880 (Hydrozoa: Olindiidae) in Israel. *Aquatic Invasions* 6(1), S147–S152.

Geiger, W., Alcorlo, P., Baltanas, A. and Montes, C. (2005) Impact of an introduced Crustacean on the trophic webs of Mediterranean wetlands. *Biological Invasions* 7, 49–73.

Gherardi, F. (2010) The invasive freshwater crayfish and fishes of the world. Invited review paper. *OIE Scientific and Technical Review* 29(2), 241–254.

Gherardi, F. and Lazzara, L. (2006) Effects of the density of an invasive crayfish (*Procambarus clarkii*) on pelagic and surface microalgae in a Mediterranean wetland. *Archiv Fur Hydrobiologie* 165(3), 401–414.

Gherardi, F., Barbaresi, S., Vaselli, O. and Bencini, A. (2002) A comparison of trace metal accumulation in indigenous and alien freshwater macro-decapods. *Marine and Freshwater Behaviour and Physiology* 35, 179–188.

Gherardi, F., Britton, J.R., Mavuti, K.M., Pacini, N., Grey, J., Tricarico, E. and Harper, D.M. (2011) A review of allodiversity in Lake Naivasha, Kenya: developing conservation actions to protect Africa's freshwater lakes from the impacts of alien species. *Biological Conservation* 144, 2585–2596.

Gollasch, S., Miossec, L., Peeler, E. and Cowx, I.G. (2008) Spread of novel pathogens and diseases caused by the introduction of alien species. Report to EC of Environmental Impacts of Alien Species in Aquaculture, FP6 2005-SSP-5A.

Graczyk, T.K., Conn, D.B., Marcogliese, D.J., Graczyk, H. and De Lafontaine, Y. (2003) Accumulation of human waterborne parasites by zebra mussels (*Dreissena polymorpha*) and Asian freshwater clams (*Corbicula fluminea*). *Parasitology Research* 89(2), 107–112.

Green, R.M. and Abdelghani, A.A. (2004) Toxicity of a mixture of 2,4-dichlorophenoxyaceticacid and mono-sodium methanearsonate to the red swamp crawfish, *Procambarus clarkii*. *International Journal of Environmental Research and Public Health* 1, 35–38.

Hänfling, B., Edwards, F. and Gherardi, F. (2011) Invasive alien Crustacea: dispersal, establishment, impact and control. *BioControl* 56, 573–595.

Huffman, J.E. and Fried, B. (1990) Echinostoma and echinostomiasis. In: Baker, J.R. and Muller, R. (eds) *Advances in Parasitology*. Academic Press, New York, pp. 215–269

Hung, N., Madsen, H. and Fried, B. (2013) Global status of fish-borne zoonotic trematodiasis in humans. *Acta Parasitologica* 58(3), 231–258.

Johnsen, S.I. and Taugbøl, T. (2010) NOBANIS – Invasive Alien Species Fact Sheet *Pacifastacus leniusculus*. Online Database of the European Network on Invasive Alien Species – NOBANIS. Available at: www.nobanis.org (accessed December 2017).

Johnson, P.T.J. and Paull, S. (2011) The ecology and emergence of disease in fresh waters. *Freshwater Biology* 56, 638–657.

Jørgensen, S.E., Marques, J.C. and Anastácio, P.M. (1997) Modelling the fate of detergents and pesticides in a rice field. *Ecological Modelling* 104, 205–213.

Joo, G.J., Ward, A.K. and Ward, G.M. (1992) Ecology of *Pectinatella magnifica* (Bryozoa) in an Alabama oxbow lake: colony growth and association with algae. *Journal of the North American Benthological Society* 11(3), 324–333.

Karouna-Renier, N.K., Snyder, R.A., Lange, T., Gibson, S., Allison, J.G., Wagner, M.E. and Rao, K.R. (2011) Largemouth bass (*Micropterus salmoides*) and striped mullet (*Mugil cephalus*) as vectors of contaminants to human consumers in northwest Florida. *Marine Environmental Research* 72(3), 96–104.

Keawjam, R.S., Poonswad, P., Upatham, E.S. and Banpavichit, S. (1993) Natural parasitic infection of the golden apple snail, *Pomacea canaliculata*. *Southeast Asian Journal of Tropical Medicine Public Health* 24, 170–177.

Kolar, C.S., Chapman, D.C., Courtenay, W.R. Jr, Housel, C.M., Williams, J.D. and Jennings, D.P. (2007) *Bigheaded Carps: A Biological Synopsis and Environmental Risk Assessment*. American Fisheries Society Special Publication 33, Bethesda, Maryland, USA.

Krumholz, L.A. (1948) Reproduction in the western mosquitofish, *Gambusia affinis affinis* (Baird and Girard), and its use in mosquito control. *Ecological Monograph* 18, 1–4.

Lages, N., Balcao, V.M. and Nunes, B.A. (2009) Risk assessment of human consumption of potentially contaminated red swamp crayfish (*Procambarus clarkii*): a conceptual approach. *Revista da Faculdade de Ciências da Saúde* 6, 332–342.

Lane, M.A., Barsanti, M.C., Santos, V., Yeung, M., Lubner, S.J. and Weil, G.J. (2009) Human paragonimiasis in North America following ingestion of raw crayfish. *Clinical Infectious Diseases* 49(6), 55–61.

Leppäkoski, E. and Olenin, S. (2000) Non-native species and rates of spread: lessons from the brackish Baltic Sea. *Biological Invasions* 2, 151–163.

Ling, N. (2003) Rotenone: a review of its toxicity and use for fisheries management. *Science For Conservation* 211, 40.

Ling, N. (2009) Management of invasive fish. In: Clout, M.N. and Williams, P.A. (eds) *Invasive Species Management: A Handbook of Principles and Techniques.* Oxford University Press, Oxford, UK, pp. 185–204.

Lodge, D.M., Rosenthal, S.K., Mavuti, K.M., Muohi, W., Ochieng, P., Stevens, S.S. and Mkoji, G.M. (2005) Louisiana crayfish (*Procambarus clarkii*) (Crustacea: Cambaridae) in Kenyan ponds: non-target effects of a potential biological control agent for schistosomiasis. *African Journal of Aquatic Science* 30(2), 119–124.

Lowe, S., Browne, M., Boudjelas, S. and De Poorter, M. (2000) *100 of the World's Worst Invasive Alien Species. A Selection from the Global Invasive Species Database.* The Invasive Species Specialist Group (ISSG) a specialist group of the Species Survival Commission (SSC) of the World Conservation Union (IUCN). First published as special lift-out in *Aliens* 12, December 2000.

Lucy, F.E., Graczyk, T.K., Minchin, D., Tamang, L. and Miraflor, A. (2008) Biomonitoring of surface and coastal water for *Cryptosporidium, Giardia* and human virulent microsporidia using molluscan shellfish. *Parasitology Research* 103, 1369–1375.

Lun, Z.R., Gasser, R.B., Lai, D.H., Li, A.X., Zhu, X.Q., Yu, X.B. and Fang, Y.Y. (2005) Clonorchiasis: a key foodborne zoonosis in China. *The Lancet Infectious Diseases* 5(1), 31–41.

Lv, S., Zhang, Y., Liu, H.X., Hu, L., Yang, K., Steinmann, P., Chen, Z., Wang, L.Y., Utzinger, J. and Zhou, X.N. (2009) Invasive snails and an emerging infectious disease: results from the first national survey on *Angiostrongylus cantonensis* in China. *PLoS Neglected Tropical Diseases* 3(2), e368.

Mann, J.W. and Werntz, J.R. (1991) Catfish stings to the hand. *The Journal of Hand Surgery* 16, 318–321.

Martinez de Velasco, G., Rodero, M., Cuellar, C., Chivatom, T., Mateos, J.M. and Laguna, R. (2008) Skin prick test of *Kudoa anytigens* on patients with gastrointestinal and/or allergic symptoms relative to fish ingestion. *Parasitology Research* 103(3), 713–715.

Mas-Coma, S., Valero, M.A. and Bargues, M.D. (2009) Fasciola, lymnaeids and human fascioliasis, with a global overview on disease transmission, epidemiology, evolutionary genetics, molecular epidemiology and control. *Advances in Parasitology* 69, 41–146.

Mazza, G., Tricarico, E., Genovesi, P. and Gherardi, F. (2014) Biological invaders are threats to human health: an overview. *Ethology Ecology and Evolution* 26(2–3), 112–129.

Mazza, G., Aquiloni, L., Inghilesi, A.F., Giuliani, C., Lazzaro, L., Ferretti, G., Lastrucci, L., Foggi, B. and Tricarico, E. (2015) Aliens just a click away: the online aquarium trade in Italy. *Management of Biological Invasions* 6(3), 253–261.

Mochida, O. (1988) Spread of freshwater *Pomacea* snails (Pilidae, Mollusca) from Argentina to Asia. *Micronesica* 3, 51–62.

Möller, H., Holst, S., Lüchtenberg, H. and Petersen, F. (1991) Infection of eel *Anguilla anguilla* from the River Elbe estuary with two nematodes, *Anguillicola crassus* and *Pseudoterranova decipiens*. *Diseases of Aquatic Organisms* 11, 193–199.

Nico, L.G., Sharp, P. and Collins, T.M. (2011) Imported Asian swamp eels (Synbranchidae: Monopterus) in North American live food markets: potential vectors of non-native parasites. *Aquatic Invasions* 6, 69–76.

Nunes, A.L., Tricarico, E., Panov, V., Katsanevakis, S. and Cardoso, A.C. (2015) Pathways and gateways of freshwater invasions in Europe. *Aquatic Invasions* 10, 359–370.

Ott, K.C. (2006) Rotenone. A brief review of its chemistry, environmental fate, and the toxicity of rotenone formulations. New Mexico Council of Trout Unlimited. Available at: http://www.newmexicotu.org/Rotenone%20summary.pdf (accessed December 2017).

Oyarzun, R. and Higueras, P. (2005) Minerales, metales, compuestos químicos, y seres vivos: una difícil pero inevitable convivencia. Available at: http://med.se-todo.com/biolog/1995/index.html (accessed February 2018).

Padilla, D.K. and Williams, S.L. (2004) Beyond ballast water: aquarium and ornamental trades as sources of invasive species in aquatic ecosystems. *Frontiers in Ecology and the Environment* 2, 131–138.

Peeler, E.J., Oidtmann, B.C., Midtlyng, P.J., Miossec, L. and Gozlan, R.E. (2011) Non-native aquatic animals introductions have driven disease emergence in Europe. *Biological Invasions* 13, 1291–1303.

Peng, Q., Ben, K., Greenfield, B.K., Dang, F. and Zhong, H. (2016a) Human exposure to methylmercury from crayfish (*Procambarus clarkii*) in China. *Environmental Geochemistry and Health* 38, 169–181.

Peng, Q., Nunes, L.M., Ben, K., Greenfield, B.K., Dang, F. and Zhong, H. (2016b) Are Chinese consumers at risk due to exposure to metals in crayfish? A bioaccessibility-adjusted probabilistic risk assessment. *Environment International* 88, 261–268.

Platvoet, D. (2007) *Dikerogammarus villosus* (Sowinsky, 1894), an amphipod with a bite. PhD thesis. University of Amsterdam, The Netherlands.

Pointier, J.P. (1999) Invading freshwater gastropods: some conflicting aspects for public health. *Malacologia* 41, 403–411.

Poulin, R., Paterson, R.A., Townsend, C.R., Tompkins, D.M. and Kelly, D.W. (2010) Biological invasions and the dynamics of endemic diseases in freshwater ecosystems. *Freshwater Biology* 56, 676–688.

Prenter, J., MacNeil, C., Dick, J.T.A. and Dunn, A.M. (2004) Roles of parasites in animal invasions. *Trends in Ecology and Evolution* 19, 385–390.

Reshetnikov, A.N. and Chibilev, E.A. (2009) Distribution of the fish rotan (*Perccottus glenii* Dybowski, 1877) in the Irtysh River basin and analysis of possible consequences for environment and people. *Contemporary Problems of Ecology* 2(3), 224–228.

Rixon, C., Duggan, I.C., Bergeron, N.M.N., Ricciardi, A. and Macisaac, H.J. (2005) Invasion risks posed by the aquarium trade and live fish markets on the Laurentian Great Lakes. *Biodiversity Conservation* 14, 1365–1381.

Rudnick, D., Halat, K. and Resh, V. (2000) Distribution, ecology and potential impacts of the Chinese mitten crab (*Eriocheir sinensis*) in San Francisco Bay. University of California Water Resources Center contribution no. 206, Riverside, California, USA.

Saker, M. and Eaglesham, G.K. (1999) The accumulation of cylindrospermopsin from the cyanobacterium *Cylindrospermopsis raciborskii* in tissues of the redclaw crayfish *Cherax quadricarinatus*. *Toxicon* 37(7), 1065–1077.

Sankarappan, A., Chellapandian, B., Vimalanathan, A.P., Mani, K., Sundaram, D. *et al.* (2015) Vector ecology of human schistosomiasis in south India and description of a new species of the genus *Ferrissia* (Mollusca: Gastropoda: Planorbidae). *Journal of Vector Borne Diseases* 52(3), 201–207.

Savini, D., Occhipinti-Ambrogi, A., Marchini, A., Tricarico, E., Gherardi, F., Olenin, S. and Gollasch, S. (2010) The top 27 alien animal species intentionally introduced by European aquaculture and related activities: stocking, sport fishery and ornamental purposes. *Journal of Applied Ichthyology* 26, 1–7.

Singh, A.K., Srivastava, S.C., Verma, P., Ansari, A. and Verma, A. (2014) Hazard assessment of metals in invasive fish species of the Yamuna River, India in relation to bioaccumulation factor and exposure concentration for human health implications. *Environmental Monitoring and Assessment* 186(6), 3823–3836.

Sneddon, J. and Richert, J.C. (2011) *Metals in Crawfish, Aquaculture and the Environment.* InTech. Available at: http://www.intechopen.com/books/aquaculture-and-the-environment-a-shared-destiny/metals-in-crawfish (accessed 19 December 2017).

Sokolov, S.G., Reshetnikov, A.N. and Protasova, E.N. (2014) A checklist of parasites in non-native populations of rotan *Perccottus glenii* Dybowski, 1877 (Odontobutidae). *Journal of Applied Ichthyology* 30, 574–596.

Soliman, M.F.M. (2008) Epidemiological review of human and animal fascioliasis in Egypt. *Journal of Infection in Developing Countries* 2(3), 182–189.

Strayer, D.L. (2010) Alien species in fresh waters: ecological effects, interactions with other stressors, and prospects for the future. *Freshwater Biology* 55, 152–174.

Toja, J., Alcalá, E., Martín, J., Solà, C., Plans, M., Burgos, A., Plazuelo, A. and Prat, N. (2003). *Ciencia y Restauración del río Guadiamar*. Consejería de Medio Ambiente, Junta de Andalucía, Sevilla, 78–92.

Tricarico, E., Bertocchi, S., Brusconi, S., Casalone, E., Gherardi, F., Giorgi, G., Mastromei, G. and Parisi, G. (2008) Depuration of a cyanobacterial toxin from the red swamp crayfish *Procambarus clarkii* and assessment of its food quality. *Aquaculture* 285, 90–95.

Vanderploeg, H.A., Liebig, J.R., Carmichael, W.W., Agy, M.A., Johengen, T.H., Fahnenstiel, G.L. and Nalepa, T.F. (2001) Zebra mussel (*Dreissena polymorpha*) selective filtration promoted toxic *Microcystis* blooms in Saginaw Bay (Lake Huron) and Lake Erie. *Canadian Journal of Fisheries and Aquatic Sciences* 58(6), 1208–1221.

Van der Veer, G. and Nentwig, W. (2015) Environmental and economic impact assessment of alien and invasive fish species in Europe using the generic impact scoring system. *Ecology of Freshwater Fish* 24, 646–656.

Vasconcelos, V.M., Oliveira, S. and Teles, F.O. (2001) Impact of a toxic and non-toxic strain of *Microcystis aeruginosa* on the crayfish *Procambarus clarkii*. *Toxicon* 39, 1461–1470.

Veilleux, E. and de Lafontaine, Y. (2007) *Biological Synopsis of the Chinese Mitten Crab (Eriocheir sinensis)*. Canadian Manuscript Report of Fisheries and Aquatic Sciences 2812, Nanaimo, Canada.

Wagley, S., Koofhethile, K. and Rangdale, R. (2009) Prevalence and potential pathogenicity of *Vibrio parahaemolyticus* in Chinese mitten crabs (*Eriocheir sinensis*) harvested from the River Thames estuary, England. *Journal of Food Protection* 72, 60–66.

Yang, T.-B., Wu, Z.-D. and Lun, Z.-R. (2013) The apple snail *Pomacea canaliculata*, a novel vector of the rat lungworm, *Angiostrongylus cantonensis*: its introduction, spread, and control in China. *Hawai'i Journal of Medicine & Public Health* 72, 23.

Yera, H., Kuchta, R., Brabec, J., Peyron, F. and Dupouy-Camet, J. (2013) First identification of eggs of the Asian fish tapeworm *Bothriocephalus acheilognathi* (Cestoda: Bothriocephalidea) in human stool. *Parasitology International* 62, 268–271.

Zhang, D.W., Xie, P., Liu, Y.Q. and Qiu, T. (2009) Transfer, distribution and bioaccumulation of microcystins in the aquatic food web in Lake Taihu, China, with potential risks to human health. *Science of the Total Environment* 407, 2191–2199.

Zhao, M., Xie, S.Q., Zhu, X.M., Yang, Y.X., Gan, N.Q. and Song, L.R. (2006) Effect of dietary cyanobacteria on growth and accumulation of microcystins in Nile tilapia (*Oreochromis niloticus*). *Aquaculture* 261, 960–966.

Zimba, P.V., Khoo, L., Gaunt, P.S., Brittain, S. and Carmichael, W.W. (2001) Confirmation of catfish, *Ictalurus punctatus* (Rafinesque) mortality from *Microcystis* toxins. *Journal of Fish Diseases* 24, 41–47.

8

Risks for Human Health Related to Invasive Alien Reptiles and Amphibians

Olivier S.G. Pauwels[1]* and Nikola Pantchev[2]

[1]*Institut Royal des Sciences Naturelles de Belgique, Brussels, Belgium and* [2]*IDEXX Laboratories, Ludwigsburg, Germany*

Abstract

More than 100 amphibian and reptile species have established populations outside their natural geographical range, mostly as a consequence of the international pet trade. About 40 zoonoses are associated with reptiles and amphibians. The main zoonotic risks from alien invasive reptiles and amphibians are salmonellosis and probably also vibriosis from a bacteriological point of view, pentastomids, sparganosis and potentially trichinellosis from a parasitical point of view, and West Nile virus. There are also new and emerging pathogens, e.g. atypical *Brucella* spp., with zoonotic potential. Transmission of pathogens from introduced reptile and amphibian species to humans is limited by the important physiological differences between them and humans, the secretive or shy habits of most introduced species and the rarity of direct contact (with the notable exception of a few exotic species eaten by humans). Locally, alien reptiles include venomous species and large species able to inflict bites of medical concern. In certain areas some species (mainly anuran amphibians) are generating noise pollution affecting human well-being. Given the continued increase of invasive alien population establishments with time, the spread of alien arthropod vectors and aggravating factors such as climate change, it is expected that alien reptiles and amphibians and their associated pathogens will generate more public health concern in the future.

8.1 Introduction

The establishment of viable populations of exotic reptiles and amphibians (R&A) is closely linked to the development of the international pet trade. An extreme example is found in Florida, the region housing the highest density of exotic R&A, where the accidental introduction or voluntary release of 180 taxa, most of them imported for the pet trade, has caused the establishment of 63 taxa (Krysko *et al.*, 2016). The popularity of R&A as pets keeps growing (Pietzsch *et al.*, 2006; Nardoni *et al.*, 2008; Martinho and Heatley, 2012; Wang *et al.*, 2013). For instance, people in the USA, some of the biggest participants of the international wildlife trade, import nearly five million live amphibians every year (Kolby and Daszak, 2016). Nearly 9 million R&A were kept as pets in the USA in 2000 (Pantchev and Tappe, 2011).

---

* E-mail: osgpauwels@yahoo.fr

To a lesser extent, accidental shipping in cargo has generated a number of alien R&A establishments, especially among anthropophilic species such as various house geckos. A striking example of accidental transport is the harmless minute flowerpot blind snake *Indotyphlops braminus*, probably originating from India or Sri Lanka, carried inside ornamental plant pots and other exported goods, and which established itself in more than 80 countries (Wallach, 2009), a colonization success explained by its parthenogenetic reproduction. Several R&A species were imported to act as a biological control of agricultural pests, such as the cane toad *Rhinella marina* (Fig. 8.1), native to Central and South America, which has been introduced to more than 40 countries worldwide (Selechnik *et al.*, 2017). Several other species colonizations have resulted from escapes from farms breeding them for human consumption, as in the case of the bullfrog *Lithobates catesbeianus* (Giovanelli *et al.*, 2007).

The Global Register of Introduced and Invasive Species (http://www.griis.org/) lists more than 110 R&A species. Among all exotic R&A which established populations, the Global Invasive Species Database of the International Union for Conservation of Nature (IUCN, see http://www.iucngisd.org/gisd/) has classified 11 amphibians and 32 reptiles as invasive. Two reptile and three amphibian species were listed by the IUCN in the *100 World's Worst Invasive Alien Species* (the brown tree snake *Boiga irregularis*, the red-eared slider *Trachemys scripta* (Fig. 8.2), the bullfrog, the cane toad and the Caribbean tree frog *Eleutherodactylus coqui*) (Lowe *et al.*, 2000).

The vast majority of scientific literature dedicated to invasive R&A focuses on the cane toad and on the mildly venomous brown tree snake. The latter, originating from Australia, New Guinea and the Solomon Islands, was accidentally introduced on Guam where it exterminated more than half of the island's native birds and lizards, and generated numerous

Fig. 8.1. Cane toads (*Rhinella marina*) hiding behind a flower pot at a motel in Normanton in northern Queensland, Australia. This invasive species, originating from Central and South America, can reach high densities in human settlements. (G. Shea.)

Fig. 8.2. An exotic invasive red-eared slider *Trachemys scripta elegans* (right) and an indigenous Western Caspian turtle *Mauremys rivulata* basking at Kournas Lake, Crete. (F. Lavail and J. Maran.)

ecological, economic and human health impacts (Smith *et al.*, 2016). The R&A that have successfully established outside their respective natural geographical ranges are phylogenetically very diverse, as are the pathogens (bacteria, fungi, parasites and viruses) they carry, and thus they pose various risks to human health, from zoonoses to bites and envenoming.

8.2 Zoonoses and Other Health Issues

8.2.1 Pathogens

Paleoparasitological studies on human coprolites suggest that transmissions of zoonotic agents from R&A to humans have occurred since ancient times (Sianto *et al.*, 2012). About 40 zoonoses are associated with R&A (Pantchev and Tappe, 2011; Warwick *et al.*, 2012). The zoonotic role of R&A in the transmission of bacteria, parasites and other pathogens is however still poorly known globally (Pantchev and Tappe, 2011; Zając *et al.*, 2016). Handling of R&A for food consumption and consumption when the meat is insufficiently cooked greatly increase the risks for pathogen transmissions (e.g. Magnino *et al.*, 2009; Gilbert *et al.*, 2014; Mühldorfer *et al.*, 2016). Several R&A species from alien populations have become regular or occasional food items, for example the cane toad in Australia, the bullfrog and freshwater turtles (e.g. Liner, 2005), and thus privileged pathways for zoonotic disease transmission.

Several R&A were shown as reservoirs for viruses able to cause severe diseases in humans, including the West Nile Fever arbovirus in crocodilians (Ariel, 2011; Warwick *et al.*, 2012). Reptiles could represent potential amplifying hosts of the West Nile Fever virus because they are known to develop a viraemia of long duration and that can overwinter

(Dauphin *et al.*, 2004). It has been shown that some mosquitoes can be vectors for equine encephalitis virus between humans and snakes (Ariel, 2011). However, 'the risk of transfer of viruses between reptiles and humans is negligible due to the thermoregulatory differences between reptilian and mammalian hosts which would limit the suite of pathogens able to grow in both temperature regimes' (Ariel, 2011, p. 8).

Reptiles, especially turtles, have long been known as sources of bacteria capable of causing arthritis, gastroenteritis, meningitis, wound infections and septicaemia in humans (Chiodini and Sundberg, 1981). Numerous R&A species asymptotically excrete salmonellae, and salmonellosis is by far the most important zoonosis transmitted by these two animal groups to humans, as is abundantly illustrated in the literature (e.g. Nardoni *et al.*, 2008; Van Meervenne *et al.*, 2009; Pantchev and Tappe, 2011; Drake *et al.*, 2012; Pees *et al.*, 2013; Zając *et al.*, 2016). The risk of *Salmonella* transmission to humans is directly linked to the level of contact and exposure (Van Meervenne *et al.*, 2009; Hydeskov *et al.*, 2012); for example, transmission from captive pet bearded dragons to children has been demonstrated (Pees *et al.*, 2013). Cane toads, often living in direct proximity to humans, have been shown to be important reservoirs for *Salmonella* that are potentially implicated in human salmonellosis (Thomas *et al.*, 2001; Drake *et al.*, 2012). African anthropophilic *Agama* lizards have established populations in several regions outside their natural range, including in Florida, and often live in high densities in direct proximity to humans (Fig. 8.3). They are known as vectors of salmonellosis and transmission to humans from an exotic *Agama* population has been documented (see references in Sancho and Pauwels, 2015). The anthropophilic and

Fig. 8.3. An *Agama picticauda* on fish for sale in Mayonami market, south-western Gabon. Transmission of salmonellosis to humans by alien populations of this invasive species has been documented. (O.S.G. Pauwels.)

now near-cosmopolitan Asian house gecko *Hemidactylus frenatus* and other introduced house geckos potentially play an important role in the epidemiology of human salmonellosis in several regions (Callaway *et al.*, 2011; Jiménez *et al.*, 2015). In Costa Rica, for example, all *Salmonella* isolates obtained by Jiménez *et al.* (2015) from the lower gut of Asian house geckos are associated with human salmonellosis. Introduced terrapins, including *Trachemys scripta*, represent a risk of transmission of *Salmonella* to humans (Hidalgo-Vila *et al.*, 2008; Martínez-Silvestre, 2013). However, this does not necessarily represent an additional risk compared with the local herpetofauna. For example, all *Salmonella* serovars identified in imported exotic reptiles during a recent study in New Zealand (Kikillus, 2010) had already been previously reported both in humans and reptiles within the country.

Among the most anthropophilic exotic reptiles are the house geckos of the genus *Hemidactylus*, frequently found inside houses, including in kitchens, where they hunt insects attracted by light, food, etc. Their oral cavity can contain Gram-positive *Staphylococcus* bacteria strains, which could cause serious infections if they reached the bloodstream (Das *et al.*, 2011). However, geckos of this genus generally have teeth too small and jaws too weak to cause severe injuries; their bites leave at most some puncture marks without any other symptoms (Das *et al.*, 2011). *Hemidactylus* droppings regularly come into contact with food and drinking water, and have been shown to contain various bacteria of public health importance, including *Enterobacter*, *Salmonella* and *Staphylococcus* (Nwachukwu *et al.*, 2014).

Chlamydiosis, caused by Chlamydiaceae bacteria, is a zoonotic disease that can be found in diverse animals, including R&A; possible transmission from R&A to humans should be evaluated, as identical *Chlamydia* genotypes were found in snakes and their owners (Sariya *et al.*, 2015; Taylor-Brown *et al.*, 2015). Apart from *Salmonella* and *Chlamydia* spp., other bacterial infections involving *Borrelia*, *Campylobacter*, *Clostridium*, *Enterobacter*, *Enterococcus*, *Escherichia*, *Klebsiella*, *Peptococcus*, *Proteus*, *Pseudomonas*, *Serratia*, *Shigella*, *Streptococcus* (streptococcosis), *Yersinia* and others can be linked to R&A (Rana *et al.*, 2010; Warwick *et al.*, 2012; Gilbert *et al.*, 2014), but their zoonotic significance is poorly understood. The genus *Borrelia* includes three main distinct phylogenetic lineages: Lyme borreliosis borreliae, relapsing fever borreliae and the recently described reptile-associated borreliae, the latter not known to cause human infections (Kalmár *et al.*, 2015). The zoonotic and tick-transmitted rickettsial agent *Anaplasma phagocytophilum* can be occasionally found in reptiles but they do not seem to be primary reservoir hosts (Nieto *et al.*, 2009). *Campylobacter fetus* was found in a variety of reptiles, including in *Trachemys scripta* (Wang *et al.*, 2013). A distinct genetic variant of *C. fetus* causing infection in humans with underlying disease has been found in reptiles and in humans having direct or indirect contact with reptiles (Gilbert *et al.*, 2014). Rare but relevant amphibian and/or reptile-related zoonotic bacterial diseases involve the genera *Burkholderia*, *Mycobacterium* (e.g. *M. marinum*; Bouricha *et al.*, 2014) and *Vibrio* (Magnino *et al.*, 2009; Martinho and Heatley, 2012; Warwick *et al.*, 2012). A case of importation of *Burkholderia pseudomallei* from Central America to Europe through a green iguana imported for the pet trade has been documented by Elschner *et al.* (2014). Amphibians are potential reservoirs for *Brucella* bacteria, and their zoonotic potential for brucellosis in humans should be carefully evaluated (Mühldorfer *et al.*, 2016; Soler-Lloréns *et al.*, 2016). There are known cases of *Mycobacterium marinum* infection of humans following handling of captive reptiles and amphibians or their contaminated tank water (Warwick *et al.*, 2012; Bouricha *et al.*, 2014), and reptiles can also be infected by the zoonotic *M. haemophilum* (Hernandez-Divers and Shearer, 2002). Other *Mycobacterium* found in reptiles might present a zoonotic risk, especially concerning immunocompromised persons, but the role of reptiles as vectors is not yet fully understood (Ullmann *et al.*, 2016).

Major R&A-borne mycotic zoonotic infections involve *Candida*, *Coccidioides* (coccidiomycosis), *Cryptococcus* (cryptococcosis), *Fusarium*, *Rhodotorula*, etc. (Warwick *et al.*, 2012). Amphibian-borne adiaspiromycosis (*Chrysosporium*) is recorded as a minor zoonotic mycotic infection (Warwick *et al.*, 2012). The zoonotic capabilities of the *Chrysosporium* spp. found in reptiles for human fungal infections are not yet properly evaluated (Mitchell and Walden, 2013; Kahraman *et al.*, 2015). Reptiles could play a possible role as carriers of opportunistic fungi affecting humans, including *Acremonium, Alternaria, Aspergillus, Cladosporium, Exophiala, Mucor, Paecylomyces, Penicillium, Rhizopus, Scopulariopsis* (Kostka *et al.*, 1997; Nardoni *et al.*, 2008).

Consumption of R&A as food, if they are not properly cooked, represents a risk of parasitosis of humans by various worms (Magnino *et al.*, 2009). Sparganosis, caused by cestodes (tapeworms) of the genus *Spirometra*, is well known in regions where R&A are commonly eaten, especially in China (Dorny *et al.*, 2009; Pantchev and Tappe, 2011). Other ways of contamination include consumption of unfiltered water with procercoid-containing copepods and direct contact between plerocercoid-containing raw meat and an open wound. This zoonosis is serious and occasionally deadly. Other parasitoses resulting from the consumption of improperly cooked R&A meat involve the cestodes of the genus *Diphyllobothrium* (diphyllobothriasis), nematodes of the genus *Gnathostoma* and trematodes of the genera *Alaria* and *Echinostoma* (echinostomiasis) (Graczyk and Fried, 1998; Dorny *et al.*, 2009; Warwick *et al.*, 2012). Trichinellosis, a common zoonosis originating from raw meat consumption, has been discussed, especially with regard to turtles and crocodiles, but further evidence-based studies are required to prove reptile-to-human transmissions (Magnino *et al.*, 2009; Pantchev and Tappe, 2011).

The zoonotic potential of reptile pentastomid parasites of the genera *Armillifer* and *Porocephalus* (tongue worms) has been documented (Dorny *et al.*, 2009; Wolf *et al.*, 2014). Most human infections in Africa are caused by *Armillifer armillatus*, and in Asia by *A. agkistrodontis* (Chen *et al.*, 2010), but severe cases are not common. Most exposed people are those with close and regular contact with snakes (e.g. transmission has been proved within a snake farm in Gambia; Tappe *et al.*, 2011), including people consuming reptile meat that has not been cooked properly, or close contact with snake excretions, such as in python tribal totemism in Africa (Pantchev and Tappe, 2011; Tappe *et al.*, 2011; Vanhecke *et al.*, 2016). These vermiform organisms form a unique group comprising approximately 120 species which are obligate endoparasites; 91% of the pentastome hosts are reptiles and 5% of the pentastome species are capable of causing zoonotic infections. In geographical areas with particular ethnic traditions, such as in West and Central Africa, and East Asia, 8–45% of the human population can be affected.

Reptiles and amphibians imported for the pet trade are regularly parasitized by ticks, which themselves could affect humans or transmit human pathogens, among them several species of *Amblyomma* and *Aponomma*, known reservoirs and vectors of several *Rickettsia* species affecting humans (Pietzsch *et al.*, 2006). Corn *et al.* (2011) collected seven tick species from 51 wild-caught exotic reptiles in Florida belonging to eight species; several of these tick species represent a potential zoonotic risk and three of them had never been found in the USA. Corn *et al.* (2011) suggested that the number of tick species associated with exotic free-living reptile species must be much higher. Generally, little is known about the ecology of these ticks, even in their respective areas of origin, and their potential as vectors of human diseases is often poorly understood. In Japan, for example, there is no quarantine imposed for imported pet R&A and no laws regarding their ectoparasitic ticks, simply because the risks posed to humans and domestic animals by these ticks are unknown (Goka *et al.*, 2013). Additional pathogens with a zoonotic potential have been detected in R&A, including free-living amoebae, *Cryptosporidium* protozoans (cryptosporidiosis) and *Hymenolepis* tapeworms, but the associated transmission and health risks are currently often poorly understood or hypothetical (Pantchev and Tappe, 2011; Díaz *et al.*, 2013; Wolf *et al.*, 2014).

Apart from their role as zoonotic agents, parasites can also help in differentiating wild caught from legally bred imported R&A, based on the life cycle of the detected stages (Moré *et al.*, 2013).

8.2.2 Envenoming

On Guam, where it has been introduced, the brown tree snake can locally reach very high densities (up to 100 individuals per ha in some forested areas) and bite incidents are not rare; large individuals could in theory cause serious human envenomations (Mackessy *et al.*, 2006). Surprisingly, a high proportion of victims on Guam are bitten while they are sleeping at home and some bites even seem to result from attempts to feed on small children (Fritts *et al.*, 1994). Toxins released through monitor lizard bites are still poorly known, but can sometimes have serious effects, including death (Vikrant and Singh Verma, 2013).

8.2.3 Bites and physical injuries

In Florida, established alien species include the spectacled caiman *Caiman crocodilus*, the Nile monitor *Varanus niloticus*, the Burmese python *Python bivittatus* and the Northern African rock python *P. sebae* (Krysko *et al.*, 2016). All of them can reach large sizes and are capable of inflicting serious bites if cornered or handled; the largest pythons can even kill humans (Campbell, 2003; Rodda *et al.*, 2009). It is to be noted that reptiles do not carry the rabies virus (Das *et al.*, 2011).

8.2.4 Noise pollution

The Tokay gecko *Gekko gecko* has been introduced in several areas outside its natural range, like in Florida (Krysko *et al.*, 2016), and can generate noise pollution and locally affect human well-being because its loud territorial vocalizations are often heard at night in urban areas. In Hawaii in the early 1980s, when the exotic Tokay gecko population was becoming established, residents called Honolulu Zoo to ask it to identify these nocturnal vocalizations (McKeown, 1996). Similarly, the first record of a population of the invasive frog *Eleuthero-dactylus johnstonei* in São Paulo, Brazil was made because residents called the environmental authorities to complain about its noisy calls. In French Guiana, this introduced frog is known to disturb the sleep of local residents, and in Brazil there is even a documented case of hospitalization because of chronic stress due to this exotic frog's noise (Melo *et al.*, 2014). On Hawaii, the other invasive exotic frog *Eleutherodactylus coqui*, originating from Puerto Rico, has reached some of the highest densities ever attained by terrestrial amphibians (up to 133,000 per ha); the noise pollution they generate is so important that it has a significant negative impact on property values in some areas (Kaiser and Burnett, 2006). Other anuran amphibians can be a source of noise pollution when they form choruses in the breeding season, especially in areas where no amphibians existed before the arrival of the exotic species, like in Hawaii (Kaiser and Burnett, 2006).

8.3 Conclusions

Zoonotic risks strongly depend on the frequency and types of contacts between humans and alien invasive R&A. Most of the knowledge gathered on pathogen transmission is based on the contamination of humans through captive exotic pet R&A or R&A used as food in their countries of origin. Close interactions with free-living alien R&A are much rarer. Only a few of them are anthropophilic and spontaneously enter into regular contact with humans, their households or their food or water. Avoiding handling exotic R&A with bare hands, and hand washing afterwards, effectively decreases the risk of *Salmonella* and other pathogen transmission (Hydeskov *et al.*, 2012). If wild-caught alien invasive R&A are used as food, their meat should be cooked properly. Alien R&A (as well as indigenous species) should be kept away from food and food utensils, drinking water, and young children, pregnant women and immunodeficient persons (Pantchev and Tappe, 2011). Avoiding bites from snakes and other reptiles, indigenous or exotic, is common sense. The identity, diversity and potential role of ticks introduced with invasive R&A as reservoirs and vectors of human pathogens is still poorly known. Transmission of pathogens to humans in geographical areas where alien R&A are newly established will depend on which alien species and associated pathogens were introduced, on local human habits towards R&A, and also on ecological conditions and potential arthropod vectors (indigenous or introduced) such as ticks and mosquitoes. Sometimes pathogens found in the original geographical range of alien invasive R&A species are not found in the populations inhabiting newly conquered areas, due to the absence of these pathogens in the founder population or the lack of vectors in the introduced area (Selechnik *et al.*, 2017). The human behaviour factor can be addressed by awareness campaigns about the health risks. However, from an ecological point of view, consumption of alien R&A as food or other products should not be discouraged if it does not represent a specific public health concern.

The number of established alien R&A populations is increasing, and will keep increasing with the wildlife traffic and possibly in the longer term with climate change (Li *et al.*, 2013), increasing in parallel the health hazards they represent. Release of unwanted exotic R&A pets should be prohibited in all cases. Quarantine for imported R&A for the pet trade is not observed in numerous countries (Goka *et al.* 2013) and should be encouraged (Pasmans *et al.*, 2008). Risk assessments for tick-borne pathogens (*Anaplasma*, *Rickettsia*, etc.) carried by ticks found on imported exotic R&A should be undertaken. The problem of exotic invasive R&A-related health risks should be tackled through early detection and rapid response as soon as potentially invasive exotic R&A species are detected (Campbell, 2007). Stopping the invasion of exotic R&A has to be adapted to the species targeted, their ecology and the specificities of the impacted zone (Florance *et al.*, 2011; Tingley *et al.*, 2012; Smith *et al.*, 2016; Southwell *et al.*, 2016).

Acknowledgements

We are grateful to Tobias Eisenberg (Hessian State Laboratory), Philippe Le Gall (Institute of Research for Development), Patrocinio Morrondo (University of Santiago de Compostela) and Volker Schmidt (University of Leipzig) for providing useful literature, and to Fred Lavail, Jérôme Maran and Glenn Shea for their photographs.

References

Ariel, E. (2011) Viruses in reptiles. *Veterinary Research* 42, 100.

Bouricha, M., Castan, B., Duchene-Parisi, E. and Drancourt, M. (2014) *Mycobacterium marinum* infection following contact with reptiles: *Vivarium granuloma. International Journal of Infectious Diseases* 21, 17–18.

Callaway, Z., Buttner, P. and Speare, R. (2011) *Salmonella* Virchow and *Salmonella* Weltevreden in a random survey of the Asian House Gecko, *Hemidactylus frenatus*, in houses in Northern Australia. *Vector-Borne Zoonotic Diseases* 11, 621–625.

Campbell, T.S. (2003) Species profile: Nile monitors (*Varanus niloticus*) in Florida. *Iguana* 10, 119–120.

Campbell, T.S. (2007) The role of early detection and rapid response in thwarting amphibian and reptile introductions in Florida. *Managing Vertebrate Invasive Species*. Paper 6. Available at: http://digitalcommons. unl.edu/nwrcinvasive/6 (accessed December 2017).

Chen, S.-H., Liu, Q., Zhang, Y.-N., Chen, J.-X., Li, H., Chen, Y., Steinmann, P. and Zhou, X.-N. (2010) Multi-host model-based identification of *Armillifer agkistrodontis* (Pentastomida), a new zoonotic parasite from China. *PLoS Neglected Tropical Diseases* 4(4), e647.

Chiodini, R.J. and Sundberg, J.P. (1981) Salmonellosis in reptiles: a review. *American Journal of Epidemiology* 113(5), 494–499.

Corn, J.L., Mertins, J.W., Hanson, B. and Snow, S. (2011) First reports of ectoparasites collected from wild-caught exotic reptiles in Florida. *Journal of Medical Entomology* 48(1), 94–100.

Das, M., Brahma, R.K. and Purkayastha, J. (2011) More in our mind than in their mouth? A preliminary inspection inside the oral cavity of two house geckos: *Hemidactylus frenatus* Schlegel, 1836 and *Hemidactylus aquilonius* McMahan & Zug, 2007. *Herpetology Notes* 4, 303–306.

Dauphin, G., Zientara, S., Zeller, H. and Murgue, B. (2004) West Nile: worldwide current situation in animals and humans. *Comparative Immunology, Microbiology & Infectious Diseases* 27, 343–355.

Díaz, P., Rota, S., Marchesi, B., López, C., Panadero, R., Fernández, G., Díez-Baños, P., Morrondo, P. and Poglayen, G. (2013) *Cryptosporidium* in pet snakes from Italy: molecular characterization and zoonotic implications. *Veterinary Parasitology* 197, 68–73.

Dorny, P., Praet, N., Deckers, N. and Gabriel, S. (2009) Emerging food-borne parasites. *Veterinary Parasitology* 163, 196–206.

Drake, M., Amadi, V., Zieger, U., Johnson, R. and Hariharan, H. (2012) Prevalence of *Salmonella* spp. in Cane Toads (*Bufo marinus*) from Grenada, West Indies, and their antimicrobial susceptibility. *Zoonoses and Public Health* 60(6), 437–441.

Elschner, M.C., Hnizdo, J., Stamm, I., El-Adawy, H., Mertens, K. and Melzer, F. (2014) Isolation of the highly pathogenic and zoonotic agent *Burkholderia pseudomallei* from a pet Green Iguana in Prague, Czech Republic. *BMC Veterinary Research* 10, 283.

Florance, D., Webb, J.K., Dempster, T., Kearney, M.R., Worthing, A. and Letnic, M. (2011) Excluding access to invasion hubs can contain the spread of an invasive vertebrate. *Proceedings of the Royal Society B* 278, 2900–2908.

Fritts, T.H., McCoid, M.J. and Haddock, R.L. (1994) Symptoms and circumstances associated with bites by the Brown Tree Snake (Colubridae: *Boiga irregularis*) on Guam. *Journal of Herpetology* 28(1), 27–33.

Gilbert, M.J., Kik, M., Timmerman, A.J., Severs, T.T., Kusters, J.G., Duim, B. and Wagenaar, J.A. (2014) Occurrence, diversity, and host association of intestinal *Campylobacter, Arcobacter*, and *Helicobacter* in reptiles. *Plos ONE* 9(7), e101599.

Giovanelli, J.G.R., Haddad, C.F.B. and Alexandrino, J. (2007) Predicting the potential distribution of the alien invasive American bullfrog (*Lithobates catesbeianus*) in Brazil. *Biological Invasions* 10, 585.

Goka, K., Okabe, K. and Tanako, A. (2013) Recent cases of invasive alien mites and ticks in Japan: why is a regulatory framework needed? *Experimental and Applied Acarology* 59, 245–261.

Graczyk, T.K. and Fried, B. (1998) Echinostomiasis: a common but forgotten food-borne disease. *American Journal of Tropical Medicine and Hygiene* 58(4), 501–504.

Hernandez-Divers, S.J. and Shearer, D. (2002) Pulmonary mycobacteriosis caused by *Mycobacterium haemophilum* and *M. marinum* in a royal python. *Journal of the American Veterinary Medical Association* 220(11), 1661–1663.

Hidalgo-Vila, J., Diaz-Panaguia, C., Perez-Santigosa, N., de Frutos-Escobar, C. and Herrero-Herrero, A. (2008) *Salmonella* in free-living exotic and native turtles and in pet exotic turtles from SW Spain. *Research in Veterinary Science* 85, 449–452.

Hydeskov, H.B., Guardabassi, L., Aalbæk, B., Olsen, K.E.P., Nielsen, S.S. and Bertelsen, M.F. (2012) *Salmonella* prevalence among reptiles in a zoo education setting. *Zoonoses and Public Health* 2012, 1–5.

Jiménez, R.R., Barquero-Calvo, E., Abarca, J.G. and Porras, L.P. (2015) Salmonella isolates in the introduced Asian House Gecko (*Hemidactylus frenatus*) with emphasis on *Salmonella* Weltevreden, in two regions in Costa Rica. *Vector-borne and Zoonotic Diseases* 15(9), 550–555.

Kahraman, B.B., Siğirci, B.D., Metiner, K., Ak, S., Koenhemsi, L., Or, M.E., Castellá, G. and Abarca, M.L. (2015) Isolation of *Chrysosporium guarroi* in a Green Iguana (*Iguana iguana*) in Turkey. *Journal of Exotic Pet Medicine* 24, 427–429.

Kaiser, B.A. and Burnett, K. (2006) Economic impacts of *E. coqui* frogs in Hawaii. *Interdisciplinary Environmental Review* 8(2), 1–11.

Kalmár, Z., Cozma, V., Sprong, H., Jahfari, S., D'Amico, G., Mărcutan, D.I., Ionică, A.M., Magdaş, C., Modrý, D. and Mihalca, A.D. (2015) Transstadial transmission of *Borrelia turcica* in *Hyalomma aegyptium* ticks. *PLoS ONE* 10(2), e0115520.

Kikillus, K.H. (2010) Exotic reptiles in the pet trade: are they a threat to New Zealand? PhD thesis. Victoria University of Wellington, New Zealand.

Kolby, J.E. and Daszak, P. (2016) The emerging amphibian fungal disease, chytridiomycosis: a key example of the global phenomenon of wildlife emerging infectious diseases. *Microbiology Spectrum* 4(3), EI10-0004-2015.

Kostka, V.M., Hoffmann, L., Balks, E., Eskens, U. and Wimmershof, N. (1997) Review of the literature and investigations on the prevalence and consequences of yeasts in reptiles. *Veterinary Record* 140, 282–287.

Krysko, K.L., Somma, L.A., Smith, D.C., Gillette, C.R., Cueva, D., Wasilewski, J.A., Enge, K.M., Johnson, S.A., Campbell, T.S., Edwards, J.R. *et al.* (2016) New verified nonindigenous amphibians and reptiles in Florida through 2015, with a summary of over 152 years of introductions. *Reptiles & Amphibians* 23(2), 110–143.

Li, Y., Cohen, J.M. and Rohr, J.R. (2013) Review and synthesis of the effects of climate change on amphibians. *Integrative Zoology* 8, 145–161.

Liner, E.A. (2005) *The Culinary Herpetologist.* Bibliomania!, Salt Lake City, USA.

Lowe, S., Browne, M., Boudjelas, S. and De Poorter, M. (2000) *100 of the World's Worst Invasive Alien Species. A Selection from the Global Invasive Species Database.* World Conservation Union (IUCN), Gland, Switzerland.

Mackessy, S.P., Sixberry, N.M., Heyborne, W.H. and Fritts, T. (2006) Venom of the Brown Treesnake, *Boiga irregularis*: ontogenetic shifts and taxa-specific toxicity. *Toxicon* 47, 537–548.

Magnino, S., Colin, P., Dei-Cas, E., Madsen, M., McLauchlin, J., Nöckler, K., Prieto Maradona, M., Tsigarida, E., Vanopdenbosch, E. and Van Peteghem, C. (2009) Biological risks associated with consumption of reptile products. *International Journal of Food Microbiology* 134, 163–175.

Martínez-Silvestre, A. (2013) Reptiles as invasive species: effects on environment and public health. In: *Abstracts. Symposium on Freshwater Turtles Conservation*, Parque Biológico, Gaia, p. 15.

Martinho, F. and Heatley, J.J. (2012) Amphibian mycobacteriosis. *Veterinary Clinics: Exotic Animal Practice* 15, 113–119.

McKeown, S. (1996) *Reptiles and Amphibians in the Hawaiian Islands.* Diamond Head Publishing, Los Osos, California.

Melo, M.A., Lyra, M.L., Brischi, A.M., Geraldi, V.C. and Haddad, C.F.B. (2014) First record of the invasive frog *Eleutherodactylus johnstonei* (Anura: Eleutherodactylidae) in São Paulo, Brazil. *Salamandra* 50(3), 177–180.

Mitchell, M.A. and Walden, M.R. (2013) *Chrysosporium* anamorph *Nannizziopsis vriesii*. An emerging fungal pathogen of captive and wild reptiles. *Veterinary Clinics of North America Exotic Animal Practice* 16, 659–668.

Moré, G., Pantchev, N., Herrmann, D.C., Globokar Vrhovec, M., Öfner, S., Conraths, F.J. and Schares, G. (2013) Molecular identification of *Sarcocystis* spp. helped to define the origin of Green pythons (*Morelia viridis*) confiscated in Germany. *Parasitology* 141, 646–651.

Mühldorfer, K., Wibbelt, G., Szentiks, C.A., Fischer, D., Scholz, H.C., Zschöck, M. and Eisenberg, T. (2016) The role of 'atypical' *Brucella* in amphibians: are we facing novel emerging pathogens? *Journal of Applied Microbiology* 122(1), 40–53.

Nardoni, S., Papini, R., Marcucci, G. and Mancianti, F. (2008) Survey on the fungal flora of the cloaca of healthy pet reptiles. *Revue de Médecine Vétérinaire* 159(3), 159–165.

Nieto, N.C., Foley, J.E., Bettaso, J. and Lane, R.S. (2009) Reptile infection with *Anaplasma phagocytophilum*, the causative agent of granulocytic anaplasmosis. *Journal of Parasitology* 95(5), 1165–1170.

Nwachukwu, M.I., Duru, M.K.C., Nwachukwu, I.O. and Anomodu, C.K. (2014) Incidence of pathogenic bacteria in Wall Gecko dropping. *Intraspecific Journal of Microbiology and Life Science* 1(1), 1–6.

Pantchev, N. and Tappe, D. (2011) Pentastomiasis and other parasitic zoonoses from reptiles and amphibians. *Berliner und Münchener Tierärztliche Wochenschrift* 124(11–12), 528–535.

Pasmans, F., Blahak, S., Martel, A. and Pantchev, N. (2008) Introducing reptiles into a captive collection: the role of the veterinarian. *The Veterinary Journal* 175, 53–68.

Pees, M., Rabsch, W., Plenz, B., Fruth, A., Prager, R., Simon, S., Schmidt, V., Münch, S. and Braun, P.G. (2013) Evidence for the transmission of *Salmonella* from reptiles to children in Germany, July 2010 to October 2011. *Euro Surveillance* 18(46), pii=20634.

Pietzsch, M., Quest, R., Hillyard, P.D., Medlock, J.M. and Leach, S. (2006) Importation of exotic ticks into the United Kingdom via the international trade in reptiles. *Experimental and Applied Acarology* 38, 59–65.

Rana, S.W., Kumar, A., Walia, S.K., Berven, K., Cumper, K. and Walia, S.K. (2010) Isolation of Tn1546-like elements in vancomycin-resistant *Enterococcus faecium* isolated from wood frogs: an emerging risk for zoonotic bacterial infections to humans. *Journal of Applied Microbiology* 110, 35–43.

Rodda, G.H., Jarnevich, C.S. and Reed, R.N. (2009) What parts of the US mainland are climatically suitable for invasive alien pythons spreading from Everglades National Park? *Biological Invasions* 11, 241–252.

Sancho, V. and Pauwels, O.S.G. (2015) An accidental importation of an anthropophilic lizard (Squamata: Agamidae: *Agama lebretoni*) into Spain. *Bulletin of the Chicago Herpetological Society* 50(12), 218–219.

Sariya, L., Kladmanee, K., Bhusri, B., Thaijongrak, P., Tonchiangsai, K., Chaichoun, K. and Ratanakorn, P. (2015) Molecular evidence for genetic distinctions between Chlamydiaceae detected in Siamese crocodiles (*Crocodylus siamensis*) and known Chlamydiaceae species. *Japanese Journal of Veterinary Research* 63(1), 5–14.

Selechnik, D., Rollins, L.A., Brown, G.P., Kelehear, C. and Shine, R. (2017) The things they carried: the pathogenic effects of old and new parasites following the intercontinental invasion of the Australian cane toad (*Rhinella marina*). *International Journal for Parasitology: Parasites and Wildlife* 6(3), 375–385.

Sianto, L., Teixeira-Santos, I., Chame, M., Chaves, S.M., Souza, S.M., Ferreira, L.F., Reinhard, K. and Araujo, A. (2012) Eating lizards: a millenary habit evidenced by paleoparasitology. *BMC Research Notes* 5, 586.

Smith, J.B., Turner, K.L., Beasley, J.C., DeVault, T.L., Pitt, W.C. and Rhodes, O.E. (2016) Brown Tree Snake (*Boiga irregularis*) population density and carcass locations following exposure to acetaminophen. *Ecotoxicology* 25, 1556–1562.

Soler-Lloréns, P.F., Quance, C.R., Lawhon, S.D., Stuber, T.P., Edwards, J.F., Ficht, T.A., Robbe-Austerman, S., O'Callaghan, D. and Keriel, A. (2016) A *Brucella* spp. isolate from a Pac-Man Frog (*Ceratophrys ornata*) reveals characteristics departing from classical Brucellae. *Frontiers in Cellular and Infection Microbiology* 6, 116.

Southwell, D., Tingley, R., Bode, M., Nicholson, E. and Phillips, B.L. (2016) Cost and feasibility of a barrier to halt the spread of invasive cane toads in arid Australia: incorporating expert knowledge into model-based decision-making. *Journal of Applied Ecology* 54(1), 216–224.

Tappe, D., Meyer, M., Oesterlein, A., Jaye, A., Frosch, M., Schoen, C. and Pantchev, N. (2011) Transmission of *Armillifer armillatus* ova at snake farm, The Gambia, West Africa. *Emerging Infectious Diseases* 17(2), 251.

Taylor-Brown, A., Rüegg, S., Polkinghorne, A. and Borel, N. (2015) Characterisation of *Chlamydia pneumoniae* and other novel chlamydial infections in captive snakes. *Veterinary Microbiology* 178(1–2), 88–93.

Thomas, A.D., Forbes-Faulkner, J.C., Speare, R. and Murray, C. (2001) Salmonelliasis in wildlife from Queensland. *Journal of Wildlife Diseases* 37(2), 229–238.

Tingley, R., Phillips, B.L., Letnic, M., Brown, G.P., Shine, R. and Baird, S.J.E. (2012) Identifying optimal barriers to halt the invasion of cane toads *Rhinella marina* in arid Australia. *Journal of Applied Ecology* 50(1), 129–137.

Ullmann, L.S., Dias-Neto, R.d.N., Quevedo Cagnini, D., Seiti Yamatogi, R., Oliveira-Filho, J.-P., Nemer, V., Teixeira, R.H.F., Welker Biondo, A. and Araújo, J.P. (2016) *Mycobacterium genavense* infection in two species of captive snakes. *Journal of Venomous Animals and Toxins including Tropical Diseases* 22, 27.

Van Meervenne, E., Botteldoorn, N., Lokietek, S., Vatlet, M., Cupa, A., Naranjo, M., Dierick, K. and Bertrand, S. (2009) Turtle-associated *Salmonella septicaemia* and meningitis in a 2-month-old baby. *Journal of Medical Microbiology* 58, 1379–1381.

Vanhecke, C., Le-Gall, P., Le Breton, M. and Malvy, D. (2016) Human pentastomiasis in Sub-Saharan Africa. *Médecine et Maladies infectieuses* 46(6), 269–275.

Vikrant, S. and Singh Verma, B. (2013) Monitor lizard bite-induced acute kidney injury – a case report. *Renal Failure* 36, 444–446.

Wallach, V. (2009) *Ramphotyphlops braminus* (Daudin): a synopsis of morphology, taxonomy, nomenclature and distribution (Serpentes: Typhlopidae). *Hamadryad* 34(1), 34–61.

Wang, C.-M., Shia, W.-Y., Jhou, Y.-J. and Shyu, C.-L. (2013) Occurrence and molecular characterization of reptilian *Campylobacter fetus* strains in Taiwan. *Veterinary Microbiology* 164, 67–76.

Warwick, C., Arena, P.C. and Steedman, C. (2012) Visitor behaviour and public health implications associated with exotic pet markets: an observational study. *Journal of the Royal Society of Medicine Short Reports* 3, 63.

Wolf, D., Globokar Vrhovec, M., Failing, K., Rossier, C., Hermosilla, C. and Pantchev, N. (2014) Diagnosis of gastrointestinal parasites in reptiles: comparison of two coprological methods. *Acta Veterinaria Scandinavica* 56, 44.

Zając, M., Wasyl, D., Różycki, M., Bilska-Zając, E., Fafiński, Z., Iwaniak, W., Krajewska, M., Hoszowski, A., Konieczna, O., Fafińska, P. and Szulowski, K. (2016) Free-living snakes as a source and possible vector of *Salmonella* spp. and parasites. *European Journal of Wildlife Research* 62, 161.

9

Do Alien Free-ranging Birds Affect Human Health? A Global Summary of Known Zoonoses

Emiliano Mori[1]*, Saverio Meini[2], Diederik Strubbe[3], Leonardo Ancillotto[4], Paolo Sposimo[5] and Mattia Menchetti[6]

[1]Università di Siena, Italy; [2]Centro Veterinario Cimarosa, Livorno, Italy; [3]Ghent University, Belgium; [4]Università degli Studi di Napoli 'Federico II', Italy; [5]NEMO s.r.l., Florence, Italy; and [6]University of Florence, Italy

Abstract

Non-native birds are prominent among alien taxa, with at least 415 species established outside their natural distribution ranges. Impacts of introduced birds on human health have received little attention up to now, despite previous works suggesting that disease transmission is a major impact exerted by introduced bird species. Our synthesis reveals that at least 42 species of introduced birds may represent a hazard to human well-being. Among those, most are Psittaciformes, Columbiformes and Anseriformes, species that frequently occur in urban areas, partly because of their popularity as pets and ornamental species. The main zoonoses potentially brought by these birds include psittacosis, cryptococcosis, listeriosis and salmonellosis, transmitted by direct contact or via arthropod vectors (fleas, lice, ticks and mites). Many Galliformes introduced for hunting purposes can lead to salmonellosis and other gastroenteric diseases in humans. Non-native birds can threaten human health through bird-strikes around airports and through noise pollution by species sharing colonial roosts. While we found that alien birds can theoretically transmit several diseases to humans, empirical case studies of disease outbreaks linked to alien birds are rare or non-existent. The synergistic impacts of ongoing species introductions and global climate change may increase the risk of health hazards in the future. Therefore, sanitary monitoring of traded birds, mainly of the most synanthropic and game species released for human consumption would be prudent. Strict attention should be paid to alien bird populations already established within urban areas, to verify their role in affecting human health and well-being.

9.1 Introduction

Among introduced species, birds are prominent mainly because of their strong association with humans in a range of different contexts, i.e. many species of birds are traded and transported for a wide range of reasons, from the pet trade to hunting, meat consumption or

* E-mail: moriemiliano@tiscali.it

ornamental purposes in public areas (Duncan *et al.*, 2003; Cassey *et al.*, 2004; Mori *et al.*, 2014), with at least 415 species currently naturalized throughout the world (Evans *et al.*, 2016). While an increasing number of studies have assessed alien bird impacts on biodiversity, economy and human well-being (Kumschick and Nentwig, 2010; Kumschick *et al.*, 2013; Martin-Albarracin *et al.*, 2015; Evans *et al.*, 2016), a mechanistic understanding of the factors underlying bird impacts is still incomplete (Shirley and Kark, 2009), hindering the formulation of effective invasive species management strategies (Kumschick *et al.*, 2016). In their assessments on impact of non-native birds, Martin-Albarracin *et al.* (2015) found that, although only a few studies have shown disease transmission from alien birds to native biota or humans, those impacts may be (potentially) severe. When focusing on human health, according to Mazza *et al.* (2014), less than 5 alien bird genera/species are reported to be a hazard to human health. Nonetheless, as many alien birds have colonized anthropogenic landscapes and often reach their highest population densities in urban areas (Lever, 2005; Menchetti and Mori, 2014), contacts with humans may be frequent and intense (Fig. 9.1).

Consequently, disease transmission by alien birds is a topic worthy of further study. The likelihood of an alien species acting as a (potential) vector for human diseases will be an important factor in prioritizing control and eradication programmes (Genovesi and Shine, 2004). Below, we report a synthesis of currently available literature on disease transmission by alien birds. We first discuss the risks of direct contacts with pathogens, followed by the risk of vector-borne transmission and also discuss other mechanisms through which alien birds may threaten human health.

9.2 Potential Zoonoses Transmitted by Alien Free-ranging Birds

A number of potential zoonoses can be transmitted by alien introduced birds, through (i) direct contacts with pathogens or (ii) through vectors. A number of aetiological agents of human pathologies have been detected on introduced birds, despite only few clinical cases being attributed definitively to contacts with introduced birds (i.e. 6 out of 57 analysed cases, see following paragraphs; cf. Mazza *et al.*, 2014). Even if actual evidence on zoonoses transmitted by alien birds is limited, the following summary on the current knowledge on pathogens and vectors of human diseases detected on introduced birds will be essential for planning effective early warning programmes (Genovesi and Shine, 2004).

Fig. 9.1. Alien birds in urban areas: left, Canada geese, mallards and pigeons in Regent's Park, London, UK (Leonardo Ancillotto); right, a ring-necked parakeet in Reggio Emilia, Italy. (Stefano Manfredini.)

9.2.1 Direct contacts with pathogens

The literature on possible direct transmission of pathogens is dominated by studies on Psittaciformes, Columbiformes, Anseriformes and Galliformes. Among birds introduced through the pet trade, Psittaciformes (i.e. parrots and parakeets) include the highest number of naturalized species (Menchetti and Mori, 2014). The ring-necked parakeet *Psittacula krameri* and the monk parakeet *Myiopsitta monachus* show the highest numbers established outside their native ranges (Edelaar *et al.*, 2015; Menchetti *et al.*, 2016). Main populations of these species occur within or in the immediate surroundings of cities, urban parks and human settlements, thus increasing the possibility of encounters with humans (Clergeau and Vergnes, 2011; Menchetti and Mori, 2014). Parakeets are recorded as reservoirs of several bacterial and viral diseases, such as psittacosis (also known as ornithosis or avian chlamydiosis), avian influenza, salmonellosis, erysipelas, pseudotuberculosis, pasteurellosis and tuberculosis (Gismondi, 1991; Runde *et al.*, 2007; Menchetti and Mori, 2014). Therefore, alien parrot populations could pose a threat to human health (Fletcher and Askew, 2007; Runde *et al.*, 2007). Direct contact with infected birds (e.g. handling of plumage and tissues), contaminated equipment and droppings, as well as aerosols of the secretions and bites from infected birds, are the main pathways of pathology transmissions (through nostrils, mouth and eyes: Gismondi, 1991). The spread of psittacosis and avian influenza through non-native parakeets has been named as a potential risk for human health (Runde *et al.*, 2007), although no actual cases of such disease transmission are known. Psittacosis is a lung infection caused by *Chlamydophila psittaci* and can affect both birds and their owners (Gismondi, 1991; Gregory and Schaffner, 1997). Bacteria avoid the defence mechanisms of the lungs, triggering an infection process, variable in severity from mild flu-like illness to severe fever (40–41°C) and pneumonia, and abortion in pregnant women (Gregory and Schaffner, 1997). Psittacosis is not transmitted among humans and contact with infected birds is always required (Gregory and Schaffner, 1997). *Chlamydophila*-positive individuals have been observed in over 20 commonly traded parrot species, although we cannot rule out that other species could be affected (Menchetti and Mori, 2014). In Belgium, 114 ring-necked parakeets were trapped in 2014 at a roost site and one positive case of *Chlamydia psittaci* was detected, corresponding to a prevalence of 0.87% (Vangeluwe, 2014). Psittacosis may also be transmitted by introduced game species (e.g. the common pheasant *Phasianus colchicus*, the bobwhite *Colinus virginianus* and the chukar partridge *Alectoris chukar*), as well as by pigeons (e.g. *Columba livia*), waterfowls, geese and peacocks released for ornamental purposes in public or private gardens (Keymer, 1974; Erbeck and Nunn, 1999; Gonzalez, 2008; Šatrović *et al.*, 2010; Krawiec *et al.*, 2015). Alien common mynas *Acridotheres tristis* are reported to be potential reservoirs of psittacosis, listeriosis, salmonellosis and erysipelas (Circella *et al.*, 2011; Boseret *et al.*, 2013). Acariasis by introduced pigeons and mynas may cause severe dermatoses (Weber, 1979; Boseret *et al.*, 2013).

Many Psittaciformes and all the Anseriformes, including those released within urban park lakes (e.g. the Canada goose *Branta canadensis*, the Aegyptian goose *Alopochen aegyptiacus* and the swans *Cygnus* spp.) are natural asymptomatic hosts of avian influenza, mostly dangerous for poultry (Andreotti *et al.*, 2001). The highly pathogenic avian influenza (subtype H5N1) shows episodic global spread and may also infect humans (Kilpatrick *et al.*, 2006), as typical for infectious emerging diseases with wild reservoirs. The H5N1 virus subtype has its natural reservoir in wild Anseriformes and shorebirds (the former including species often traded as ornamental species or in game hunting, e.g. mallards *Anas platyrhynchos*), which are considered the main long-distance vectors (Keawcharoen *et al.*, 2008). Also, Newcastle disease, caused by a *Paramyxovirus* potentially transmitted by introduced parrots (mainly monk parakeets: Nelson *et al.*, 1952; Fitzwater, 1988), may affect human health (Mustaffa-Babjee *et al.*, 1976), leading to acute attacks and respiratory failures. Introduced parrots were also identified as being responsible for the introduction of the aetiological agent of nocardiosis (the Gram-positive bacterium *Neocardia asteroids*), detected in introduced free-ranging rainbow lorikeets *Trichoglossus haematodus* in Western Australia (Raidal, 1997). Avian influenza, harmful to birds and humans, has been detected in an unnamed parrot species introduced in the UK from Suriname (Fletcher and Askew, 2007; http://news. bbc.co.uk/2/hi/uk_news/4365956.stm), although no transmission of this disease to humans by parrots has been detected yet. The influenza virus H9N2 has been isolated from the respiratory organs of two ring-necked parakeets imported from Pakistan to Japan (Mase *et al.*, 2001), where naturalized populations are known to occur (Menchetti *et al.*, 2016).

Galliformes often harbour a number of *Salmonella* pathogens. Outbreaks of bacterial salmonellosis due to *Salmonella pullorum* and *S. typhimurium* frequently occur in introduced common pheasants, red-legged partridges *Alectoris rufa* and other game species (Faddoul *et al.*, 1966; Takata *et al.*, 2003; Millán, 2009; Díaz-Sánchez *et al.*, 2012). Consumption of contaminated meats may provoke severe gastroenteritis in humans (Paulsen *et al.*, 2012). The West Nile virus (Flaviviridae), the aetiological agent of the human West Nile fever, was detected in captive chukar partridges, which have been widely introduced in North America (Wünschmann and Ziegler, 2006).

Disease transmission by Columbiformes to humans has been studied widely. For instance, in the USA, fouling with faecal droppings from introduced pigeons facilitates the transmission of mycoses (*Aspergillus fumigatus*, *Candida albicans*, *Cryptococcus neoformans*, *Hystoplasma capsulatum* and *Blastomyces dermatidus*), bacterial (*Salmonella typhimurium* and *S. pullorum*, *Chlamydophila psittaci*, *Coxiella burnetti*, *Listeria monocytogenes*, *Erysipelothrus insidiosis*, *Pasteurella multicida*, *Yersinia pseudotuberculosis*, *Y. enterocolitica*), viral (Eastern/Western equine encephalitis, St Louis encephalitis, meningitis and West Nile disease) and protozoan (*Trypanosoma cruzi*, *Trichomonas gallinae* and *Toxoplasma gondii*) diseases (Weber, 1979; Long, 1981; González-Acuña *et al.*, 2007). It should be noted that other birds can also transmit some of the pathogens mentioned above. For instance, Indian house crows *Corvus splendens* introduced in Yemen are reported to be carriers of intestinal parasites (Enterobacteriaceae including *Salmonella*, and *Shigella* serotypes and *Proteus* strains as well as members of Vibrionaceae and Pseudomonads, *Giardia lamblia* and *Hymenolepis nana*), provoking diarrheal diseases in humans through aerosols (e.g. surrounding roosts: Al-Sallami, 1991). Cysts of *Giardia* spp., oocysts of *Cryptosporidium parvum* and other virulence factors have been detected in the faeces of Canada geese (Graczyk *et al.*, 1998; Kullas *et al.*, 2002). Similarly, introduced mynas pollute the environment by dropping their faecal material under nocturnal roosts and they may accordingly be related to important public health problems, spreading psittacosis and salmonellosis (Yap *et al.*, 2002; Lim *et al.*, 2003). Nonetheless, in the introduced range, mynas steal food from restaurant plates and scavenge food from human settlements (IUCN ISSG data, http://www.cabi.org/isc/datasheet/2994), both behaviours representing potential risks for contamination and disease transmission. *Campylobacter jejuni* is a bacterium responsible for food-borne diseases, human gastroenteritis, headaches and depression, and may be found in a number of passerine birds, e.g. Estrildidae (Boseret *et al.*, 2013); a number of species belonging to this order are currently established worldwide (Evans *et al.*, 2016), but no data are available on their associated pathogens. Parasitic worms present in rock pigeons that are potentially harmful to humans include cestodes (*Taenia saginata* and *Schistosoma* spp.) and trematodes (*Echinoparyphium paraulum*, *Echinoparyphium recurvatum*, *Echinostoma revolutum*, *Haplorchis pulimio* and *Hypoderaeum conoideum*: Weber, 1979).

9.2.2 Vector-borne transmission

Although many non-native species may lose their original pathogens and parasites once introduced in new environments ('enemy release hypothesis': Prenter *et al.*, 2004; Lymbery *et al.*, 2014), ticks, mites, fleas, flat-flies and lice brought by alien birds can transmit a number of diseases to native species and man (i.e. spill-over: Daszak *et al.*, 2000).

The tropical blood-eating mite *Ornithonyssus bursa* is frequent on chicks of monk parakeet, as well as on adult common mynas in their native range, and it has been recorded in many bird species from tropical areas (Argentina: Aramburù *et al.*, 2003). This parasite has also been occasionally found in Europe (Viviano and Bongiorno, 2014), where several alien populations of monk parakeet species occur (Edelaar *et al.*, 2015), and on alien mynas in South-east Asia (e.g. Lin Neo, Singapore, 2017, personal communication). Bites of *O. bursa* may provoke dermatitis, asthma, severe irritations and skin rashes in humans (Orton *et al.*, 2000). Within the native range, mynas also host Arboviruses, which can be transmitted to humans through bites from arthropod parasites (e.g. *Dermanyssus gallinae*: IUCN ISSG data, http://www.cabi.org/isc/datasheet/2994). Trypanosomiasis can be transmitted by the kissing bug *Triatoma rubrofasciata*, a parasite of introduced rock pigeons (Weber, 1979).

Incidence and spread of native parasites increase when they successfully attack alien birds ('spill-back': Daszak *et al.*, 2000; Kelly *et al.*, 2009). Among native ectoparasites detected in free-ranging alien parakeets in Europe, the pigeon tick *Argas reflexus*, detected in a ring-necked parakeet in northern Italy (Pavia: Mori *et al.*, 2015), is the only one potentially representing a threat to human health (Khouri and Maroli, 2004). Bites by this soft tick provoke papular erythematous injuries, urticarial skin rashes, mucocutaneous, cardiovascular or gastroenteric manifestations, up to anaphylactic shock. This species (as well as other mites and ticks present on the same species) may also represent a hazard for human health where rock pigeons have been introduced, whereas human infestations by other pigeon parasites are asymptomatic (Nelson and Murray, 1971).

The red mite *Dermanyssus gallinae*, native to Mediterranean countries, may attack introduced common mynas (IUCN ISSG data, http://www.cabi.org/isc/datasheet/2994): bites by this mite cause dermatitis with pruritus and skin lesions in humans (Bellanger *et al.*, 2008; Abdigoudarzi *et al.*, 2013).

Ticks belonging to the genus *Ixodes* are commonly found on free-ranging alien birds (e.g. *I. ricinus* on common pheasants: Hoodless *et al.*, 2003; *I. ricinus* and *I. frontalis* on ring-necked parakeets: Franz, 2010) and they have the ability to transmit viral pathogens (e.g. Flavoviruses) and *Borrelia burgdoferi*, the causative agent of Lyme disease in humans (Boseret *et al.*, 2013). The importance of alien Passeriformes as long-term reservoirs of borreliosis seems to be negligible (Boseret *et al.*, 2013). By contrast, this problem is emphasized when involving migrating alien birds (e.g. feral mallards), which may contribute to long-distance dispersal of vectors and spirochetes (Boseret *et al.*, 2013).

9.3 Other Effects of Introduced Birds on Human Health and Well-being

Introduced waterfowls may increase eutrophication of lakes (e.g. Canada goose: Allan and Feare, 1994), which may in turn affect human health (e.g. intoxication and human poisoning: Carmichael, 2001). Noise pollution is provoked by excessive sound, which may harm human activities and well-being (Slabbekoorn and Peet, 2003; Singh and Davar, 2004). While quantitative and experimental studies are still lacking, introduced loud birds, such as ring-necked parakeets, Indian house crows and common mynas rapidly become associated with noise pollution (Rahman *et al.*, 2014), as their urban nocturnal roosts may hold over 10,000 individuals (e.g. Stoner, 1923; Markula *et al.*, 2009; Van Kleunen *et al.*, 2010; Parau *et al.*, 2016). These species are particularly vocal while flying and resting, especially at dusk and dawn (Stoner, 1923; Menchetti *et al.*, 2016). Noise pollution from introduced birds has also been noted for monk parakeets in North America (Stafford, 2003) and for rainbow lorikeets in Western Australia (Chapman, 2005).

In addition, several alien bird species introduced to the UK have been involved in bird-strikes, i.e. aircraft–bird collision events, representing serious air hazards (Shirley and Kark, 2009). This is particularly dangerous when large bird species are involved (e.g. Canada geese: Allan and Feare, 1994; Watola *et al.*, 1996), whereas bird-strikes of introduced ring-necked parakeets roosting at Heathrow Airport (London) did not result in any damage to human health (Fletcher and Askew, 2007).

9.4 Conclusions

Our synthesis suggests that at least 42 species of alien birds out of 415 that have established self-sustaining non-native populations throughout the world (Menchetti and Mori, 2014; Evans *et al.*, 2016) may affect human health and well-being. Despite the wide distribution of alien free-ranging birds, impacts on biodiversity, economy or human well-being are only documented for a minority of species, and most documented impacts are negligible, in sharp contrast to alien mammals (Kumschick and Nentwig, 2010; Menchetti and Mori, 2014; Evans *et al.*, 2016). Among species with documented impacts, very few studies assessed the sanitary impacts of alien free-ranging birds, and most of them dealt with disease transmission to native wildlife (Mazza *et al.*, 2014) without considering human health. Yet, studies discussing impacts on human health often assign high (potential) risk scores to this impact category (Martin-Albarracin *et al.*, 2015). While reviewing the literature, we found that such high scores for human health risks are typically based on the fact that species may be reservoirs of pathogens potentially harmful to humans. By contrast, most empirical data on zoonoses are derived from captive birds living in zoological gardens or private aviaries, where close contacts with humans and overcrowding may favour transinfestations and disease transmission (Van Borm *et al.*, 2005; Circella *et al.*, 2011; Boseret *et al.*, 2013; Schmidt *et al.*, 2015; Nakamura *et al.*, 2016). Thus, information and impact scores for zoonoses that may be provoked by alien birds are typically based on a perceived risk that these birds may transmit pathogens and diseases; actual data showing that alien birds affect human health issues are rare to non-existent (only six documented cases).

When considering the potential risk listed in the literature, most knowledge is related to pathologies transmitted by Psittaciformes (47.6% of introduced bird species affecting human health), Anseriformes (28.6% species) and Columbiformes (9.5% species, with most impacts by *Columba livia*), which are widely traded (or introduced) and often synanthropic. Only a few species of introduced Passeriformes seem to have the potential to affect human health (Martin-Albarracin *et al.*, 2015). Galliformes are widely introduced in the northern hemisphere (mainly Europe, the Middle East and North America) for hunting purposes. Many potential pathogens are listed for this taxon, probably because these species are often bred for release in aviaries, where (animal and human) health issues are actively studied and monitored (Slota *et al.*, 2011). Disease impact assessments also consider the fact that invasive birds often have colonized urban areas as an extra risk factor (Andreotti *et al.*, 2001; Clergeau and Vergnes, 2011), but we note that further studies are also warranted for alien birds in other habitats. For example, species such as the sacred ibis (*Threskiornis aethiopicus*) and the reef heron (*Egretta gularis*), typical of wetlands, remain poorly known in terms of pathogens and parasites (cf. Andreotti *et al.*, 2001). Such species may also affect human health, as heronries are shared among hundreds of individuals, and such high population densities may increase pathogen transmission (Andreotti *et al.*, 2001).

After the discovery of pandemic risk due to avian influenza and other pathologies, a global consciousness on disease transmission risks associated with wildlife migration and international trade is growing, and bird origin traceability often required. While empirical data on the risk of disease transmission to humans by alien bird populations are still largely lacking, the long list of potential pathogens associated with non-native birds strongly suggests that sanitary monitoring of caged birds (e.g. pet shops, zoological gardens, aviaries) should be recommended, with a special regard to the most synanthropic species (e.g. ducks, parrots and songbirds) and to game species for human consumption (e.g. pheasants, quails, partridges and francolins). To conclude, further research and quantitative studies should be carried out on alien populations established within urban areas, to assess their parasite and pathogen load and to verify their role in affecting human health and well-being.

Acknowledgements

We would like to thank Lin Neo (Singapore) for the precious information provided on alien birds in Singapore, and Stefano Manfredini for the ring-necked parakeet photo. Vasco Sfondrini kindly took the time to revise the English of our early draft. We acknowledge the support provided by European Cooperation in Science and Technology COST Action ES1304 (ParrotNet) for the realization of this book chapter. The contents of the chapter are the authors' own responsibility and neither COST nor any person acting on its behalf is responsible for the use which might be made of the information contained in it. To conclude, we thank an anonymous reviewer for the useful comments, which greatly improved our manuscript.

References

Abdigoudarzi, M., Mirafzali, M.S. and Belgheiszadeh, H. (2013) Human infestation with *Dermanyssus gallinae* (Acari: Dermanyssidae) in a family referred with pruritus and skin lesions. *Journal of Arthropod Borne Diseases* 18, 119–123.

Al-Sallami, S. (1991) A possible role of crows in the spread of diarrhoeal diseases. *Journal of the Egyptian Public Health Association* 66, 441–449.

Allan, J.R. and Feare, C.J. (1994) Feral Canada geese (*Branta canadensis*) as a hazard to aircraft in Europe: options for management and control. *22nd Proceedings of the International Bird Strike Committee*, Wien, Austria, pp. 25–42.

Andreotti, A., Baccetti, N., Perfetti, A., Besa, M., Genovesi, P. and Guberti, V. (2001) *Mammiferi ed Uccelli esotici in Italia: analisi del fenomeno, impatto sulla biodiversità e linee guida gestionali.* Ministero dell'Ambiente, Istituto Nazionale Fauna Selvatica, Rome.

Aramburù, R.M., Calvo, S., Alzugaray, M.E. and Cicchino, A. (2003) Ectoparasitic load of monk parakeet (*Myiopsitta monachus*) nestlings. *Ornitologia Neotropical* 14, 415–418.

Bellanger, A.P., Bories, C., Foulet, F., Bretagne, S. and Botterel, F. (2008) Nosocomial dermatitis caused by *Dermanyssus gallinae*. *Infection Control and Hospital Epidemiology* 29, 282–283.

Boseret, G., Losson, B., Mainil, J.G., Thiry, E. and Saegerman, C. (2013) Zoonoses in pet birds: review and perspectives. *Veterinary Research* 44, 36.

Carmichael, W.W. (2001) Health effects of toxin-producing cyanobacteria: 'The CyanoHABs'. *Human and Ecological Risk Assessment: An International Journal* 7, 1393–1407.

Cassey, P., Blackburn, T.M., Sol, D., Duncan, R.P. and Lockwood, J.L. (2004) Global patterns of introduction effort and establishment success in birds. *Proceedings of the Royal Society of London B: Biological Sciences* 271, S405–S408.

Chapman, T. (2005) The status and impact of the rainbow lorikeet (*Trichoglossus haematodus moluccanus*) in South-Western Australia. Miscellaneous Publication, Department of Agriculture, Government of Western Australia.

Circella, E., Pugliese, N., Todisco, G., Cafiero, M.A., Sparagano, O.A.E. and Camarda, A. (2011) *Chlamydia psittaci* infection in canaries heavily infested by *Dermanyssus gallinae*. *Experimental and Applied Acarology* 55, 329–338.

Clergeau, P. and Vergnes, A. (2011) Bird feeders may sustain feral Rose-ringed parakeets *Psittacula krameri* in temperate Europe. *Wildlife Biology* 17, 248–252.

Daszak, P., Cunningham, A.A. and Hyatt, A.D. (2000) Emerging infectious diseases of wildlife-threats to biodiversity and human health. *Science* 287, 443–449.

Díaz-Sánchez, S., Moriones, A.M., Casas, F. and Höfle, U. (2012) Prevalence of *Escherichia coli*, *Salmonella* sp. and *Campylobacter* sp. in the intestinal flora of farm-reared, restocked and wild red-legged partridges (*Alectoris rufa*): is restocking using farm-reared birds a risk? *European Journal of Wildlife Research* 58, 99–105.

Duncan, R.P., Blackburn, T.M. and Sol, D. (2003) The ecology of bird introductions. *Annual Review of Ecology, Evolution, and Systematics* 1, 71–98.

Edelaar, P., Roques, S., Hobson, E.A., Gonçalves da Silva, A., Avery, M.L., Russello, M.A., Senar, J.C., Wright, T.F., Carrete, M. and Tella, J.L. (2015) Shared genetic diversity across the global invasive range of the monk parakeet suggests a common restricted geographic origin and the possibility of convergent selection. *Molecular Ecology* 24, 2164–2176.

Erbeck, D.H. and Nunn, S.A. (1999) Chlamydiosis in pen-raised bobwhite quail (*Colinus virginianus*) and chukar partridge (*Alectoris chukar*) with high mortality. *Avian Diseases* 43(4), 798–803.

Evans, T., Kumschick, S. and Blackburn, T.M. (2016) Application of the environmental impact classification for alien taxa (EICAT) to a global assessment of alien bird impacts. *Diversity and Distribution* 22(9), 919–931.

Faddoul, G.P., Fellows, G.W. and Baird, J. (1966) A survey on the incidence of salmonellae in wild birds. *Avian Diseases* 10, 89–94.

Fitzwater, W.D. (1988) Solutions to urban bird problems. In: Crabb, A.C. and Marsh, R.E (eds) *Proceedings of the 13th Vertebrate Pest Conference*. University of Nebraska, Lincoln, Nebraska, pp. 254–259.

Fletcher, M. and Askew, N. (2007) Review of the status, ecology and likely future spread of parakeets in England. CSL, York, UK. Available at: http://archive.defra.gov.uk/wildlife-pets/wildlife/management/non-native/documents/csl-parakeet-deskstudy.pdf (accessed 15 September 2016).

Franz, D. (2010) Zeckenbefall bei Halsbandsittichen. *Papageien* 3, 98–99.

Genovesi, P. and Shine, C. (2004) *European Strategy on Invasive Alien Species, Final Version*. Council of Europe, Strasbourg, France.

Gismondi, E. (1991) *Il grande libro degli Uccelli da gabbia e da voliera*. De Vecchi Editors, Milan, Italy.

Gonzalez, J.C. (2008) Impact of introduced birds in the Philippines. *Journal of Environmental Science and Management* 9, 66–79.

González-Acuña, D., Silva, F., Moreno, L., Cerda, F., Donoso, S., Cabello, J. and López, J. (2007) Detection of some zoonotic agents in the domestic pigeon (*Columba livia*) in the city of Chillan, Chile. *Revista chilena de infectología* 24, 199–203.

Graczyk, T.K., Fayer, R., Trout, J.M., Lewis, E.J., Farley, C.A., Sulaiman, I. and Lal, A.A. (1998) *Giardia* sp. cysts and infectious *Cryptosporidium parvum* oocysts in the feces of migratory Canada geese (*Branta canadensis*). *Applied and Environmental Microbiology* 64, 2736–2738.

Gregory, D.W. and Schaffner, W. (1997) Psittacosis. *Seminars in Respiratory Infections* 12, 7–11.

Hoodless, A.N., Kurtenbach, K., Nuttall, P.A. and Randolph, S.E. (2003) Effects of tick *Ixodes ricinus* infestation on pheasant *Phasianus colchicus* breeding success and survival. *Wildlife Biology* 9, 171–178.

Keawcharoen, J., Van Riel, D., van Amerongen, G., Bestebroer, T., Beyer, W., Van Lavieren, R., Osterhaus, A.D.M.E., Fouchier, R.A.M. and Kuiken, T. (2008) Wild ducks as long-distance vectors of highly pathogenic avian influenza virus (H5N1). *Emerging Infectious Diseases* 14(4), 600–607.

Kelly, D.W., Paterson, R.A., Townsend, C.R., Poulin, R. and Tompkins, D.M. (2009) Parasite spillback: a neglected concept in invasion ecology? *Ecology* 90, 2047–2056.

Keymer, I.F. (1974) Ornithosis in free-living and captive birds. *Proceedings of the Royal Society of Medicine* 67, 733.

Khouri, C. and Maroli, M. (2004) La zecca del piccione *Argas reflexus* (Acari: Argasidae) ed i rischi per la salute umana. *Annali dell'Istituto Superiore di Sanità* 40, 427–432.

Kilpatrick, A.M., Chmura, A.A., Gibbons, D.W., Fleischer, R.C., Marra, P.P. and Daszak, P. (2006) Predicting the global spread of H5N1 avian influenza. *Proceedings of the National Academy of Sciences* 103, 19368–19373.

Krawiec, M., Piasecki, T. and Wieliczko, A. (2015) Prevalence of *Chlamydia psittaci* and other *Chlamydia* species in wild birds in Poland. *Vector-Borne and Zoonotic Diseases* 15, 652–655.

Kullas, H., Coles, M., Rhyan, J. and Clark, L. (2002) Prevalence of *Escherichia coli* serogroups and human virulence factors in faeces of urban Canada geese (*Branta canadensis*). *International Journal of Environmental Health Research* 12, 153–162.

Kumschick, S. and Nentwig, W. (2010) Some alien birds have as severe an impact as the most effectual alien mammals in Europe. *Biological Conservation* 143, 2757–2762.

Kumschick, S., Bacher, S. and Blackburn, T.M. (2013) What determines the impact of alien birds and mammals in Europe? *Biological Invasions* 15, 785–797.

Kumschick, S., Blackburn, T.M. and Richardson, D.M. (2016) Managing alien bird species: time to move beyond '100 of the worst' lists? *Bird Conservation International* 26, 154–163.

Lever, C. (2005) *Naturalized Birds of the World.* Poyser Editors, London.

Lim, H., Sodhi, N.S., Brook, B.W. and Soh, M.C.K. (2003) Undesirable aliens: factors determining the distribution of three invasive bird species in Singapore. *Journal of Tropical Ecology* 19, 685–695.

Long, J.L. (1981) *Introduced Birds of the World.* Universe Books, New York.

Lymbery, A.J., Morine, M., Kanani, H.G., Beatty, S.J. and Morgan, D.L. (2014) Co-invaders: The effects of alien parasites on native hosts. *International Journal for Parasitology: Parasites and Wildlife* 3, 171–177.

Markula, A., Hannan-Jones, M. and Csurhes, S. (2009) Pest animal risk assessment: Indian myna *Acridotheres tristis*. Department of Employment, Economic Development and Innovation, The State of Queensland, Australia. Available at: http://www.dpi.qld.gov.au/documents/Biosecurity_Environmental-Pests/IPA-Indian-Myna-Risk-Assessment.pdf (accessed 15 September 2016).

Martin-Albarracin, V.L., Amico, G.C., Simberloff, D. and Nuñez, M.A. (2015) Impact of non-native birds on native ecosystems: a global analysis. *PLoS ONE* 10, e0143070.

Mase, M., Imada, T., Sanada, Y., Itoh, E., Sanada, N., Tsukamoto, K., Kawaoka, Y. and Yamaguchi, S. (2001) Imported parakeets harbor H9N2 influenza A viruses that are genetically closely related to those transmitted to humans in Hong Kong. *Journal of Virology* 75, 3490–3494.

Mazza, G., Tricarico, E., Genovesi, P. and Gherardi, F. (2014) Biological invaders are threats to human health: an overview. *Ethology Ecology and Evolution* 26, 112–129.

Menchetti, M. and Mori, E. (2014) Worldwide impact of alien parrots (Aves Psittaciformes) on native biodiversity and environment: a review. *Ethology Ecology and Evolution* 26, 172–194.

Menchetti, M., Mori, E. and Angelici, F.M. (2016) Effects of the recent world invasion by ring-necked parakeets *Psittacula krameri*. In: Angelici, F.M. (ed.) *Problematic Wildlife. A Cross-disciplinary Approach.* Springer, New York, pp. 253–266.

Millán, J. (2009) Diseases of the red-legged partridge (*Alectoris rufa* L.): a review. *Wildlife Biology in Practice* 5, 70–88.

Mori, E., Monaco, A., Sposimo, P. and Genovesi, P. (2014) Low establishment success of alien non-passerine birds in a Central Italy wetland (Selva di Paliano: Latium). *Italian Journal of Zoology* 81, 593–598.

Mori, E., Ancillotto, L., Groombridge, J., Howard, T., Smith, V.V. and Menchetti, M. (2015) Macroparasites of introduced parakeets in Italy: a possible role for parasite-mediated competition. *Parasitology Research* 114, 3277–3281.

Mustaffa-Babjee, A., Ibrahim, A.L. and Khim, T.S. (1976) A case of human infection with Newcastle disease virus. *The Southeast Asian Journal of Tropical Medicine and Public Health* 7, 622–624.

Nakamura, S., Hayashidani, H., Sotohira, Y. and Une, Y. (2016) Yersiniosis caused by *Yersinia pseudotuberculosis* in captive toucans (Ramphastidae) and a Japanese squirrel (*Sciurus lis*) in zoological gardens in Japan. *Journal of Veterinary Medical Science* 78, 297–299.

Nelson, B.C. and Murray, M.D. (1971) The distribution of Mallophaga on the domestic pigeon (*Columba livia*). *International Journal for Parasitology* 1, 21–29.

Nelson, C.B., Pomeroy, B.S., Schrall, K., Park, W.E. and Lindeman, R.J. (1952) An outbreak of conjunctivitis due to Newcastle disease virus (NDV) occurring in poultry workers. *American Journal of Public Health and the Nation Health* 42, 672–678.

Orton, D.I., Warren, L.J. and Wilkinson, J.D. (2000) Avian mite dermatitis. *Clinical and Experimental Dermatology* 25, 129–131.

Parau, L.G., Strubbe, D., Mori, E., Menchetti, M., Ancillotto, L., van Kleunen, A., White, R.L., Hernàndez-Brito, D., Le Louarn, M., Clergeau, P. *et al.* (2016) Rose-ringed parakeet *Psittacula krameri* populations and numbers in Europe: a complete overview. *The Open Ornithology Journal* 9, 1–13.

Paulsen, P., Smulders, F.J.M. and Hilbert, F. (2012) Salmonella in meat from hunted game: a central European perspective. *Food Research International* 45, 609–616.

Prenter, J., MacNeil, C., Dick, J.T. and Dunn, A.M. (2004) Roles of parasites in animal invasions. *Trends in Ecology and Evolution* 19, 385–390.

Rahman, N.A.A., Fadzly, N., Dzakwan, N.M. and Zulkifli, N.H. (2014) The numerical competency of two bird species (*Corvus splendens* and *Acridotheres tristis*). *Tropical Life Sciences Research* 25, 95–103.

Raidal, S.R. (1997) Bilateral necrotizing pectenitis causing blindness in a rainbow lorikeet (*Trichoglossus haematodus*). *Avian Pathology* 26, 871–876.

Runde, D.E., Pitt, W.C. and Foster, J.T. (2007) Population ecology and some potential impacts of emerging populations of exotic parrots. Managing Vertebrate Invasive Species. Paper 42. Available at: http://digitalcommons.unl.edu/nwrcinvasive/42 (accessed 15 September 2016).

Šatrović, E., Rešidbegović, E., Hadžiomerović, Z., Krkalić, L., Goletić, T., Kavazović, A., Džaja, P. and Vegara, M. (2010) Avian chlamydiosis in pheasants (*Phasianus colchicus*) in Bosnia and Herzegovina. *Veterinaria* 59, 29–36.

Schmidt, R.E., Reavill, D.R. and Phalen, D.N. (2015) *Pathology of Pet and Aviary Birds*. Wiley Blackwell Editions, New York.

Shirley, S.M. and Kark, S. (2009) The role of species traits and taxonomic patterns in alien bird impacts. *Global Ecology and Biogeography* 18, 450–459.

Singh, N. and Davar, S.C. (2004) Noise pollution: sources, effects and control. *Journal of Human Ecology* 16, 181–187.

Slabbekoorn, H. and Peet, M. (2003) Ecology: birds sing at a higher pitch in urban noise. *Nature* 424, 267.

Slota, K.E., Hill, A.E., Keefe, T.J., Bowen, R.A., Millers, R.S. and Pabilonia, K.L. (2011) Human–bird interactions in the United States upland gamebird industry and the potential for zoonotic disease transmission. *Vector Borne Zoonotic Diseases* 11(8), 1115–1123.

Stafford, T. (2003) Pest risk assessment for the monk parakeet in Oregon. Available at: http://www.oregon.gov/OISC/docs/pdf/monkpara.pdf (accessed 15 September 2016).

Stoner, D. (1923) The mynah. A study in adaptation. *Auk* 40, 328–330.

Takata, T., Liang, J., Nakano, H. and Yoshimura, Y. (2003) Invasion of *Salmonella enteritidis* in the tissues of reproductive organs in laying Japanese quail: an immunocytochemical study. *Poultry Science* 82, 1170–1173.

Van Borm, S., Thomas, I., Hanquet, G., Lambrecht, B., Boschmans, M., Dupont, G., Decaestecker, M., Snacken, R. and Van den Berg, T. (2005) Highly pathogenic H5N1 influenza virus in smuggled Thai eagles, Belgium. *Emerging Infectious Diseases* 11, 702–705.

Vangeluwe, D. (2014) *Belgishc Ringwerk. Overzicht van de in 2014. Uitgevoerde Activiteiten in België*. Koninklijk Belgisch Instituut Voor Naturwetenschappen, Vautierstraat, Brussels.

Van Kleunen, A., Van den Bremer, L., Lensink, R. and Wiersma, P. (2010) *De Halsbandparkiet, Monniksparkiet en Grote Alexanderparkiet in Nederland: risicoanalyse en beheer*. SOVON onderzoeksrapport 2010/10 Dit rapport is samengesteld in opdracht van Team Invasieve Exoten van het Ministerie van Landbouw, Natuur en Voedselkwaliteit.

Viviano, E. and Bongiorno, M.R. (2014) Avian mite dermatitis: an Italian case indicating the establishment and spread of *Ornithonyssus bursa* (Acari: Gamasida: Macronyssidae) (Berlese, 1888) in Europe. *International Journal of Dermatology* 54, 795–799.

Watola, G., Allan, J. and Feare, C. (1996) *Problems and Management of Naturalised Introduced Canada Geese Branta canadensis in Britain. The Introduction and Naturalisation of Birds*. HMSO, London.

Weber, W. (1979) Pigeon associated people diseases. *Bird Control Seminars Proceedings* 21, 156–158.

Wünschmann, A. and Ziegler, A. (2006) West Nile virus-associated mortality events in domestic chukar partridges (*Alectoris chukar*) and domestic Impeyan pheasants (*Lophophorus impeyanus*). *Avian Diseases* 50, 456–459.

Yap, C.A.M., Sodhi, N.S. and Brook, B.W. (2002) Roost characteristics of invasive mynas in Singapore. *Journal of Wildlife Management* 66, 1118–1127.

10

Impact of Alien Mammals on Human Health

Dario Capizzi[1]*, Andrea Monaco[1], Piero Genovesi[2], Riccardo Scalera[3] and Lucilla Carnevali[2]

[1]Latium Region, Environment and Natural Systems, Rome, Italy; [2]ISPRA Institute for Environmental Protection and Research, Rome, Italy and [3]IUCN SSC Invasive Species Specialist Group, Rome, Italy

Abstract

We provide an overview of the impact of wild invasive alien mammals on human health, focusing specifically on species acting as zoonotic hosts or pathogens, along with the diseases and mechanisms of disease transmission associated with mammals in terrestrial, freshwater and marine environments. We checked for published data on the impact on human health for 129 alien invasive mammals, reported in 123 different countries. The highest number of invasive alien mammals causing impacts on human health is reported in Japan (31 species), followed by Australia (24) and Argentina, New Zealand and Cuba (19). However, Australia, Brazil, Mexico, the Bahamas and the Czech Republic are characterized by the highest proportions of alien mammals impacting human health of the total number of invasive mammals reported in the country (range from 93% to 96%). Carnivores are the taxonomic group with the highest numbers of alien species impacting on human health, followed by ungulates and rodents. Our review highlights the important role of alien mammals in threatening human health and welfare, particularly through the transmission of zoonoses. Alien mammals can act as vectors of both alien and native pathogens, and as hosts of either native or alien parasites (which in turn can act as vectors of either native or alien pathogens). In this way, alien mammals can introduce new pathogens, alter the epidemiology of local pathogens, become reservoir hosts and increase disease risk for humans, along with other species. The increasing movements of humans and other species because of climate change and other factors could result in the sudden emergence of disease outbreaks, including new diseases and in new locations. This shows the urgent need for a better understanding of the parasite–vector–host trio and the environmental, climatic and socioeconomic factors involved, as well as the large potential for future zoonotic emergence.

10.1 Alien Mammals and Human Health

The role of wild mammals in the worldwide increase in frequency and prevalence of zoonotic diseases has been reviewed recently by Han *et al.* (2016). The potential public health risks posed by alien parasites, hosts and vectors introduced to Europe was also recently

---

* E-mail: dcapizzi@regione.lazio.it

summarized by Hulme (2014). In the present contribution, based on a comprehensive analysis of available data, we provide an overview focusing specifically on alien mammals acting as zoonotic hosts or pathogens, along with the diseases and mechanisms of disease transmission associated with mammals in terrestrial, freshwater and marine environments. We chose not to include domestic species in the analysis. We also discuss other types of impact of alien mammals on human health, including the risks of injuries and accidents linked to their behavioural traits. The results of this study can help in prioritizing prevention and management efforts targeting alien mammals.

As a first step, in order to compile available information on the impact of alien mammals on human health, we mined the Global Register of Introduced and Invasive Species (GRIIS: http://www.griis.org/), a comprehensive data source containing validated and verified inventories of introduced species for almost all countries of the world. The information provided by GRIIS includes: taxonomy, biological status at the country level (alien, cryptogenic, alien in part of the country) and invasiveness (meaning species causing any form of impact, based on published and unpublished data, including country editors' opinions) (Pagad et al., 2015). All the data available in GRIIS were verified by IAS experts identified for each relevant country (Pagad et al., 2018).

We considered all data referred to 'Animalia' stored in the GRIIS, using an offline version of the database provided by ISSG (version October 2016). We extracted 15,824 records concerning 7270 alien animals recorded in 201 countries or subnational regions (large islands, archipelagos, etc.) in all continents, obtaining an overall list of 247 alien mammals in 149 countries. Of these, 129 alien mammals, reported in 123 countries, are considered as invasive (52% of the total). Overall, all alien mammals may have a potential role as carriers of pathogens and diseases, even though they are not labelled as invasive. However, in our review we focused on the species whose impact on human health was known. The geographic distribution of data used for the review is shown in Fig. 10.1.

Japan is the country with the highest reported number of invasive alien mammals (38 species). Australia, New Zealand and Cuba recorded more than 25 invasive mammals, followed by Argentina (23), Italy (20), the Russian Federation (18), Spain (18) and South Africa (17). It must be noted that the distribution of data in GRIIS reflects the availability of information, and the efforts of the country editors. Therefore, while for regions such as Europe, Oceania and South America, a higher amount of data are available, there are regions that are not well represented, such as Africa. For the USA, we integrated the information in GRIIS with additional data extracted from Feldhamer et al. (2003) and Krausman and Bleich (2013).

As far as taxonomy is concerned, invasive alien mammals belong to 14 different groups: 33 Artiodactyla, 29 Carnivora, 29 Rodentia, 12 Primates, 7 Diprodontia, 5 Lagomorpha, 4 Perissodactyla, 4 Soricomorpha, 3 Erinaceomorpha, and 1 species for Chiroptera, Didelphimorphia, Cingulata, Afrosoricida and Eulipotyphla.

In order to identify alien mammals also impacting human health, we carried out a specific literature search and analysed the main global online databases, including the IUCN SSC Global Invasive Species Database (GISD: http://www.iucngisd.org/gisd/) and CABI's Invasive Species Compendium (http://www.cabi.org/isc/), both reporting impacts on human health. Because the present analysis focuses on alien mammals, we excluded from the review information on impacts caused by mammals on human health in their native range.

For 73 invasive alien mammals (57% of the total number of invasive alien mammals), impacts on human health are reported in the available sources. The distribution of data on invasive alien mammals impacting health reflects the general patterns of the distribution of data on alien mammals described above: the highest number of invasive alien mammals causing impacts on human health are reported in Japan (31 mammals), followed by Australia (24). In Argentina, New Zealand and Cuba 19, 17 and 15 invasive alien mammals with impacts on health are reported (Fig. 10.2).

D. Capizzi *et al.*

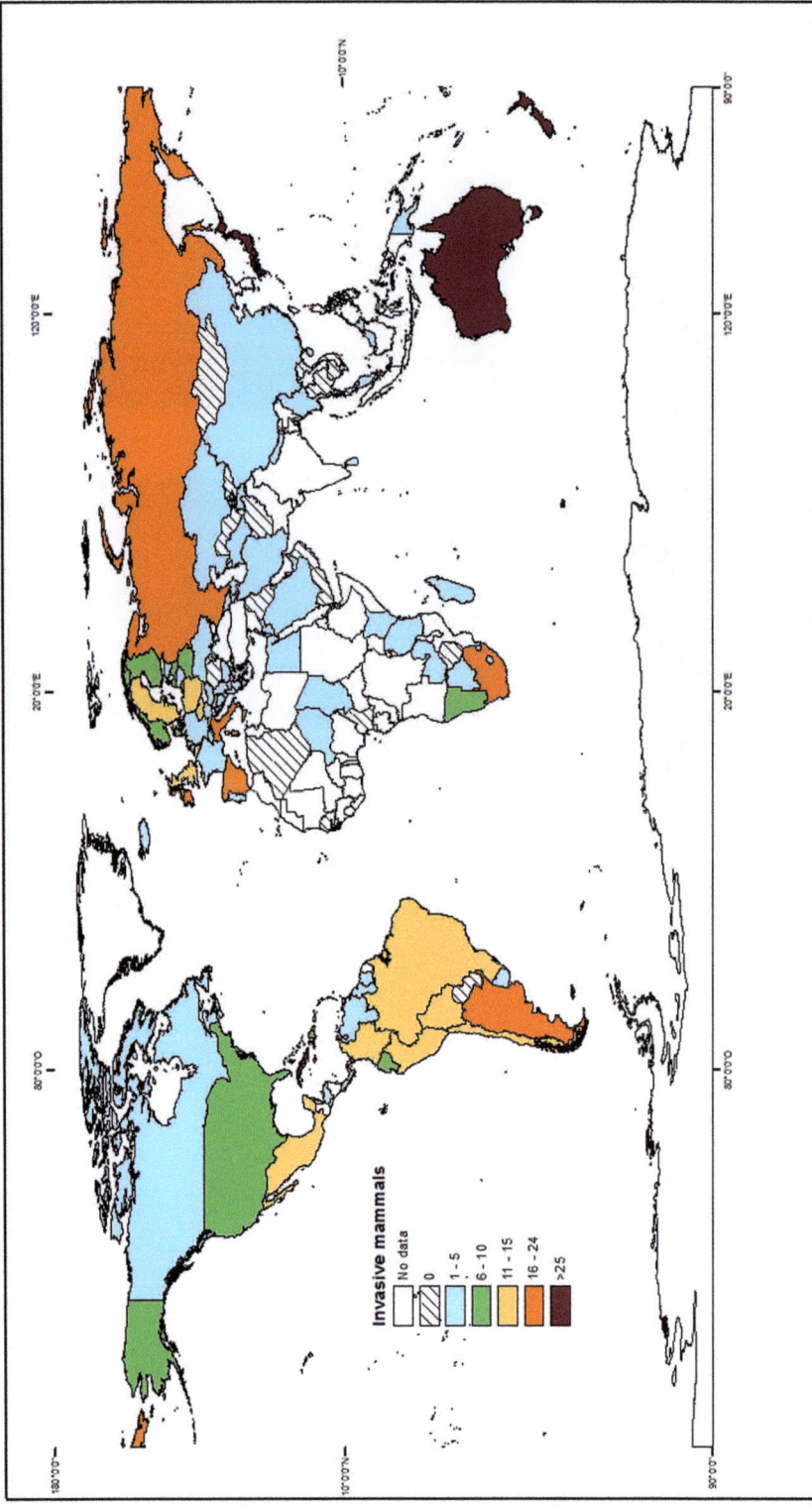

Fig. 10.1. Global distribution of data on invasive mammals used for the present review, based on GRIIS data and additional data for the USA. (From www.griis.org and USA data from Feldhamer *et al.* (2003) and Krausman and Bleich (2013).)

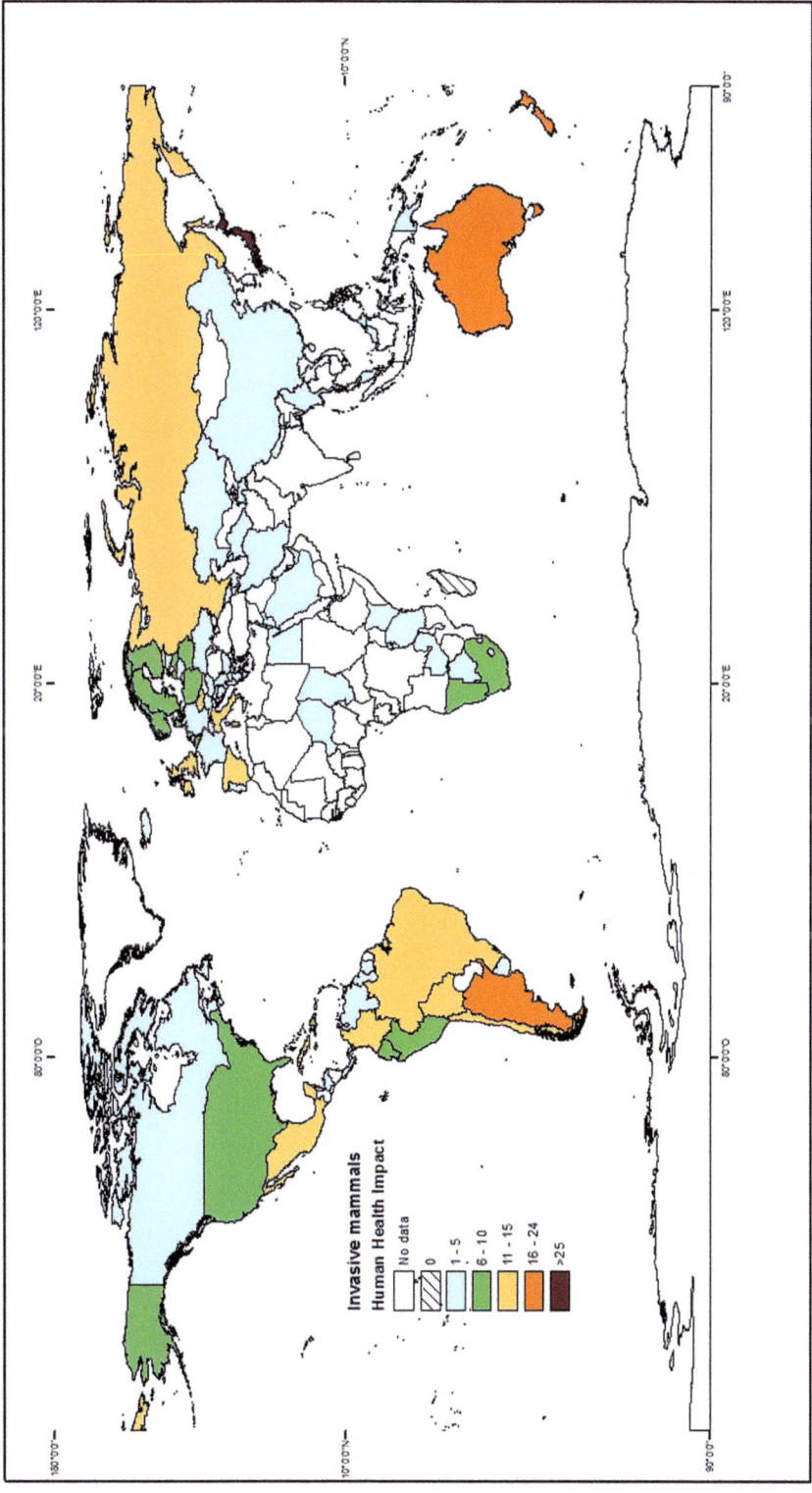

Fig. 10.2. Data on invasive alien mammals causing impacts on human health. (From www.griis.org and Feldhamer *et al.* (2003) and Krausman and Bleich (2013).)

Australia, Brazil, Mexico, the Bahamas and the Czech Republic are characterized by the highest proportions of alien mammals impacting human health out of the total number of invasive mammals reported in the country (range from 93% to 96%). On the contrary, South Africa, Cuba, New Zealand and the Russian Federation are characterized by lower percentages of invasive mammals with an impact on human health in comparison with the total (range from 53% to 67%).

Carnivores are the group with the highest numbers of alien species impacting on human health (20/29 species; 69% of the total number of invasive alien carnivores), followed by ungulates (19 species) and rodents (14 species), which are both characterized by lower percentages of invasive species with an impact on human health in comparison with the total (about 50%). Figure 10.3 shows the proportion of mammals with an impact on human health out of the total number of invasive alien mammals, by taxonomic order.

At the species level, not surprisingly synanthropic rats and mice (*Rattus rattus*, *Rattus norvegicus* and *Mus musculus*) are the most widespread invasive mammals causing impacts on human health (species reported as invasive in 85, 52 and 50 countries, respectively), followed by *Myocastor coypus* (30), *Herpestes javanicus* (29) and *Neovison vison* (28) (domestic species not considered in this analysis). Regarding wild ungulates, the most impacting species is the Axis deer (in six countries).

In Table 10.1, data on alien invasive mammals with an impact on human health, their taxonomic classification and the number of countries affected are reported.

10.2 Alien Mammals and Zoonoses

The main impact of alien mammals on human health is caused by the transmission of diseases. It is therefore essential to understand how pathogens are transmitted among hosts.

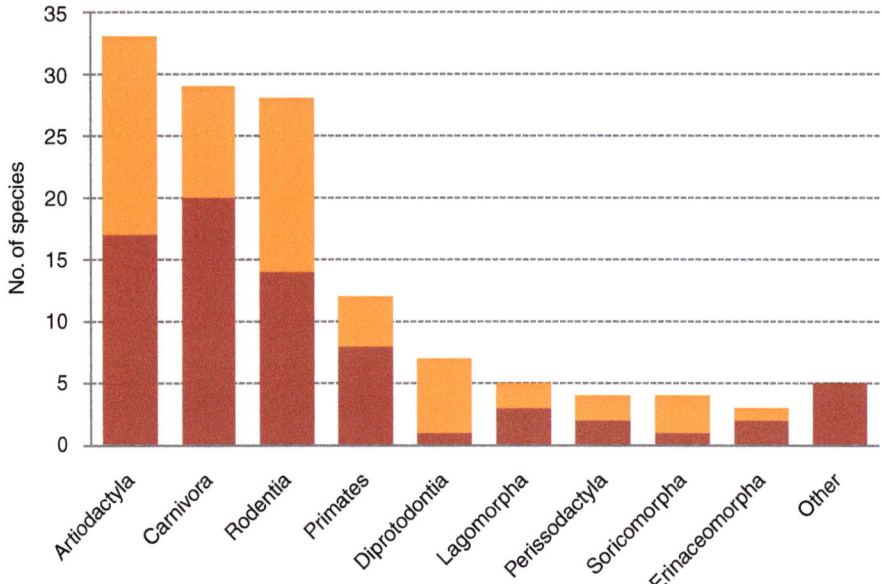

Fig. 10.3. Proportion of mammals with an impact on human health (red) out of the total number of invasive alien mammals (orange), by taxonomic order. (From www.griis.org and Feldhamer *et al.* (2003) and Krausman and Bleich (2013).)

Table 10.1. List of alien mammals with impact on human health, split by taxonomic order and number of countries where the species are reported as invasive. (From www.griis.org and Feldhamer *et al.* (2003) and Krausman and Bleich (2013).)

Order	Species name	No. of countries
Afrosoricida	*Tenrec ecaudatus*	3
Artiodactyla	*Ammotragus lervia*	2
	Axis axis	6
	Bubalus bubalis	1
	Camelus dromedarius	1
	Capra hircus	30
	Cervus elaphus	4
	Cervus nippon	1
	Dama dama	2
	Hippopotamus amphibius	1
	Hydropotes inermis	2
	Lama glama	2
	Lama pacos	1
	Ovis aries	17
	Rangifer tarandus	1
	Rusa timorensis	2
	Rusa unicolor	4
	Sus scrofa	48
Carnivora	*Canis latrans*	1
	Canis lupus	20
	Canis lupus dingo	1
	Canis familiaris	5
	Felis catus	2
	Felis silvestris	41
	Herpestes auropuntatus	1
	Herpestes javanicus	29
	Lycalopex griseus	1
	Martes zibellina	1
	Mustela furo	3
	Mustela itatsi	1
	Mustela sibirica	2
	Nasua nasua	1
	Neovison vison	28
	Nyctereutes procyonoides	11
	Paguma larvata	1
	Procyon lotor	14
	Vulpes vulpes	2
Chiroptera	*Rousettus egyptiacus*	1
Cingulata	*Chaetophractus villosus*	1
Didelphimorphia	*Didelphis marsupialis*	2
Diprotodontia	*Trichosurus vulpecula*	2
Erinaceomorpha	*Atelerix algirus*	3
	Erinaceus europaeus	9
Eulipotyphla	*Hemiechinus auritus*	1
Lagomorpha	*Lepus europaeus*	11
	Oryctolagus cuniculus	20
	Sylvilagus floridanus	1
Perissodactyla	*Equus asinus*	8
	Equus caballus	15

continued

Table 10.1. *continued.*

Order	Species name	No. of countries
Primates	*Callithrix geoffroyi*	1
	Callithrix jacchus	1
	Callithrix penicillata	5
	Chlorocebus aethiops	10
	Macaca cyclopis	1
	Macaca fascicularis	4
	Macaca mulatta	1
	Saimiri sciureus	2
Rodentia	*Callosciurus erythraeus*	1
	Glis glis	1
	Microtus levis	3
	Mus musculus	53
	Myocastor coypus	30
	Ondatra zibethicus	15
	Rattus exulans	15
	Rattus norvegicus	52
	Rattus rattus	85
	Sciurus carolinensis	5
	Sciurus vulgaris	2
	Sciurus yucatanensis	1
	Tamias sibiricus	4
Soricomorpha	*Crocidura russula*	1

At least three general routes by which pathogens can be transmitted among hosts can be identified: (i) close contact; (ii) environmental contamination; and (iii) through intermediate hosts. Such broad categories include many different routes of transmission. For example, pathogen transmission from mammals to humans may occur by injection, ingestion, inhalation or contact. In the case of transmission by inoculation, the infection is mediated by a vector, as in the case of diseases (e.g. plague or Lyme disease) transmitted by ectoparasitic arthropod vectors (ticks, mites, fleas). This is a very common mode of transmission in zoonoses involving rodents.

Pathogens can also be transmitted via inhalation of an aerosol, a suspension of liquid particles of small size, which can therefore be conveyed over long distances. The smaller the pathogen, the easier its dissemination via aerosol. Because of the small size of viruses, this is a common transmission mode for viral diseases.

The infection can also occur through the epidermis coming into contact with excreta, and the pathogen penetrating through small wounds or lesions. There are many pathogens that can be transmitted by contact, such as causative agents of rabies, leptospirosis or rat-bite fever.

Finally, the infection can be transmitted through ingestion of contaminated food or water, such as meat of the intermediate host containing the pathogen, as in the case of toxoplasmosis, or contaminated water, as may occur for leptospirosis.

10.2.1 Rodents

Species belonging to this particularly rich order (counting 2244 species, about 40% of mammals) have been associated with the transmission of many diseases to humans (e.g. Gratz, 2006; Meerburg *et al.*, 2009; Capizzi *et al.*, 2014). Throughout history, the human population has been decimated by plague, a rodent-borne zoonosis, as evidenced by historical accounts (despite some recent reinterpretations; Massad *et al.*, 2004), and the reputation of rodents as vectors of disease is still unchanged. This undeniable potential as carriers of zoonoses, combined with the great capacity of some species to establish themselves outside of their natural distribution area, makes rodents one of the invasive animal orders of greater health significance.

The potential invasiveness of rodents is indeed remarkable: in total, nine species are alien in at least one continent (Capizzi *et al.*, 2014), requiring expensive control measures, even when the control measures raise major concerns for their undesired environmental impacts. Three species (*Rattus rattus, Rattus norvegicus, Mus musculus*) are spread across all continents, impacting on agriculture, urban areas, public health and ecosystems in all of them.

In recent decades, rodent-borne zoonoses have often been defined as emerging, because some of them have been discovered or associated to rodents just recently (e.g. Mills and Childs, 1998). Some authors (e.g. Meerburg *et al.*, 2009) outline that the diseases involving small mammals in their pathogens' life cycles have increased.

Rodents may transmit zoonoses either directly or indirectly. Direct transmission can occur through a bite, as in the case of rat-bite fever, contamination of water or foodstuffs, as in the case of leptospirosis, or by aerosol, as in the case of hantaviruses. Indirect transmission is instead due to the intervention of an animal carrier, often an arthropod (especially fleas or ticks, but also other mites), thus constituting the role of reservoir of the disease. In addition, rodents can be responsible for livestock or pet animals ingesting the pathogen and passing it to humans, as in the case of alveolar echinococcosis or toxoplasmosis.

Rodents may carry different types of pathogens. Zoonoses, indeed, may be caused by viruses, bacteria, protozoa or other parasites (nematodes, cestodes, etc.).

Viral diseases

Alien rodents are involved in the cycle of several viral zoonoses, some of which are potentially relevant to human health. Among them, in recent decades there has been a rise in the importance attributed to hantaviruses, a vast group of the Bunyaviridae family, which may cause pulmonary or haemorrhagic fever with renal syndromes (HFRS; Schmaljohn and Hjelle, 1997; Meerburg *et al.*, 2009). However, although both syndromes are mainly linked to native rodents as reservoirs, HFRS may be spread by synanthropic *R. norvegicus* and *R. rattus*, which have been reported to be involved in the transmission of the Seoul virus (SEOV) (e.g. Wang *et al.*, 2000).

Another viral infection linked to commensal rodent species is lymphocytic choriomeningitis virus (LCMV), belonging to the family Arenaviridae (genus Arenavirus). Its impact on human health varies in severity, in the worst cases leading to meningitis and encephalitis (Lledó *et al.*, 2003). In several parts of the world, the virus is spread mainly by the house mouse (*M. musculus*) through infected aerosol.

Antibodies of cowpox virus (CPXV), a member of the genus Orthopoxvirus in the family Poxviridae usually associated to native rodents, have been found in the house mouse in the UK. Direct transmission from *Rattus norvegicus* to man has been demonstrated in at least one case (Wolfs *et al.*, 2002).

Bacterial diseases

Many bacterial diseases have been associated with rodents as reservoirs or carriers. The most important is perhaps leptospirosis, a zoonosis spread throughout the world caused by bacteria of the genus *Leptospira*. Its incidence in human populations is greater in humid areas than in temperate ones, especially in rural areas of Asia (Meerburg *et al.*, 2009), significantly increasing during natural disasters such as floods. It is associated with many species of rodents, including rats (both *R. rattus* and *R. norvegicus*; e.g. Matthias *et al.*, 2008), house mouse and coypu *Myocastor coypus*. Transmission occurs mainly through contaminated food or water, but also by contact between contaminated ground or vegetation and the skin.

Plague is a disease caused by the bacterium *Yersinia pestis*. Historically, it is the zoonosis that has claimed most lives through devastating pandemics over the millennia. Arthropod vectors transmit the disease, especially the rat flea *Xenopsylla cheopis*. Although the bacterium is present in many wild rodents, its transmission to humans had been associated with the black rat. Although its spread and lethality have greatly decreased in modern times, plague is still present in several countries of Asia, America and Africa (Meerburg *et al.*, 2009).

Lyme disease is an important zoonosis caused by the spirochetes *Borrelia burgdorferi sensu lato*, a bacterium transmitted by hard ticks, mainly *Ixodes ricinus* in Europe, which perpetuates in cycles involving rodents as reservoir hosts. Main reservoirs are wild mice and voles in America and Eurasia, but *R. norvegicus* and *R. rattus* may be locally important hosts to these pathogens in urban areas (Smith *et al.*, 1993; Matuschka *et al.*, 1996). In Europe, two alien squirrels have recently been found to be infected by the bacterium: in France, *Tamias sibiricus* has been found to have a higher infection rate than native wild voles and mice (Vourc'h *et al.*, 2007); in the UK, 11.9% of *Sciurus carolinensis* examined were found to be infected with *B. burgdorferi s.l.* (Millins *et al.*, 2015).

Commensal rats may be reservoirs of murine and scrub typhus, the former carried by *Rickettsia typhi*, and transmitted by the bite of the rat flea, *Xenopsylla cheopis*, the latter caused by the bacterium *Orientia tsutsugamushi*, transmitted to humans by trombiculid mites. Various rodent species are reservoirs of the pathogens, including *R. norvegicus* and *M. musculus*.

Rickettsialpox is a relatively mild bacterial disease, caused by *Rickettsia akari* and widespread worldwide, especially in urban areas. It is transmitted to humans by the bite of infected house mouse mites, *Liponyssoides sanguineus*. The pathogen is hosted by different rodent species, but its main host is the house mouse *M. Musculus* (Meerburg *et al.*, 2009).

The Norway rat may be the reservoir of some of the several *Bartonella* species (Boulouis *et al.*, 2005) transmitted to humans by fleas and lice, as well as of *Coxiella burnetii*, the causative agent of Q-Fever. In Japan a high seroprevalence of the latter bacterium has been found in *M. coypu* (Meerburg *et al.*, 2009). However, the actual role of rodents in the transmission of the illness is poorly known and is supposed to be of minor importance, because the main reservoirs of the pathogen are cattle and pets.

Although caused by the consumption of contaminated food, salmonellosis is spread by several rodent species, among them *M. musculus*, which is involved in the transmission of the disease to animals used by man for food, especially in chicken farms (Henzler and Opitz, 1992).

Rat-bite fever is caused by the bacterium *Streptobacillus moniliformis*, which can be transmitted to humans by bites of infected rodents as well as by holding pet rats. The disease is especially widespread in the USA, but Europe, Asia, Africa and Oceania have also had cases. The house mouse may also be implicated in the disease transmission (Meerburg *et al.*, 2009).

Finally, listeriosis, a zoonosis caused by the bacterium *Listeria monocytogenes*, is suspected to be transmitted to humans by rodents. Among these, commensal *R. rattus* has been found infected by this agent in Japan (Iida *et al.*, 1998).

Diseases caused by protozoa and other parasites

Alien rodents are associated with the transmission of several diseases caused by protozoa and other parasites. Toxoplasmosis is caused by *Toxoplasma gondii*, a pathogen with a rather complex life cycle. The asexual phase of the cycle is completed in a wide range of intermediate hosts, both birds and mammals. Among the most important are the Norway rat, the black rat and the house mouse (Gratz, 2006). The behavioural changes induced by the parasite in the intermediate host have been thoroughly investigated in *R. norvegicus*, showing an increase in activity and a lower neophobia in infected individuals, thus increasing susceptibility to predation by cats (Macdonald *et al.*, 1999).

In different parts of the world, alien rodents (*R. rattus*, *R. norvegicus*, *M. musculus*) are important reservoirs of protozoans of the genus *Cryptospridium*, the causative agent of cryptosporidiosis, a diarrheal disease of humans and livestock (e.g. Chalmers *et al.*, 1997; Torres *et al.*, 2000; Meerburg *et al.*, 2009).

Several rodent species, including *R. rattus*, *R. norvegicus* and *M. musculus*, are thought to be the reservoir of *Trypanosoma cruzi*, the causative agent of Chagas' disease, especially widespread in South America and transmitted by the bite from the hemipterous *Triatoma infestans*.

The Muskrat *Ondatra zibethicus* was found infected with several cestode species in The Netherlands, such as *Echinococcus multilocularis* and different species of the genus *Taenia* (Borgsteede *et al.*, 2003). The former parasite is especially important for human health, because it causes echinoccosis, a zoonosis widespread in several parts of the world, especially in the northern hemisphere. The main host of the pathogen is the fox, but rodents (especially *Arvicolinae*) are involved in the life cycle of the tapeworm. The disease is important also for domestic animals and pets, especially pigs and dogs.

Alien rodents may also be involved as reservoirs in the life cycle of two pathogens whose main hosts are water snails, i.e. worms of the genus *Schistosoma* and *Fasciola*. The former has been found in several rodent species, including *R. rattus* and *R. norvegicus* in Guadalupe (Denoya *et al.*, 1997); the latter is especially widespread in Eurasia. In France, high infection rates of *Fasciola hepatica* have been found in *R. rattus* (in Corsica) and *M. coypus* (Mas-Coma *et al.*, 1988; Ménard *et al.*, 2000).

Several rodent species, including *R. norvegicus*, may be involved in the cycle of the nematode *Angiostrongylus cantonensis*, the causative agent of angiostrongylosis, especially widespread in Asia, the Pacific region and Eastern Australia (Spratt, 2005), although cases in Egypt, Nigeria and the Caribbean region have been reported (Gratz, 1994).

Trichinosis is a zoonosis caused by *Trichinella spiralis*, and may be carried by a lot of rodent species. However, it is believed that commensal rats are responsible for the worldwide distribution of the disease (Meerburg *et al.*, 2009).

Other impacts of rodents to human health

Rodents may threaten human health in other ways besides the transmission of diseases to man, livestock or pets. Indeed, some rodent species may alter the water regime by obstructing channels and rivers, such as alien beavers *Castor canadensis* (Global Invasive Species Database, 2017), with indirect health effects on local communities. In Europe, damage to river ditches has been reported for coypu *M. coypus* and muskrat *O. zibethicus*, which in some cases is suspected to have caused flooding events (Panzacchi *et al.*, 2007; DAISIE, 2011). Finally, it is well known that commensal rats and mice may cause damage to electrical wires and gas pipes by gnawing them, risking serious accidents such as fires due to short circuits. In one case, alien squirrels *Callosciurus finlaysonii* were suspected to have caused an explosion in Southern Italy that had two victims (P. Genovesi, 2017, personal communication, based on news published in local newspapers).

10.2.2 Ungulates

A recent review stated that about 32% of wild species of ungulates (mostly Artiodactyla) can be zoonotic hosts (Han *et al.*, 2016) and play a crucial role in the epidemiology of zoonotic diseases, acting as vectors or reservoirs of pathogens or parasites (Kock *et al.*, 2010; Bekker *et al.*, 2012).

Of the 37 alien ungulates reported as invasive in the GRIIS database, 19 (51%) showed evidence of an impact on human health in at least one country in the world. Apart from the wild boar *Sus scrofa*, and its domesticated forms, the feral pig, the most widespread species, introduced in more than 50 non-native countries, cervids (especially fallow deer *Dama dama* and sika deer *Cervus nippon*) are the most introduced taxon worldwide, both for their hunting and aesthetical value.

After introduction, some populations of ungulates known as vectors of zoonoses sharply expanded their range and increased their population size. Arabian camels (*Camelus dromedarius*), known as host of *Mers coronavirus* (Azhar *et al.*, 2014), were imported from India between 1840 and 1907 by people exploring the semi-arid internal areas of Australia. Since the 1920s, a free-ranging population of feral camels has established in Australia and now totals about 1 million, the largest herd of camels in the wild (Saalfeld and Edwards, 2010).

Introduced outside of their native range mainly for hunting, farming or ornamental purposes, ungulates are currently present in a mixed assembly of native, non-native and feral species, often living in sympatry, in many countries worldwide. Of the nine species of free-living ungulates in the UK, six are feral or non-native (Putman, 2010). Considering both feral and alien species, 17 non-native ungulates are living and potentially exchanging diseases with 15 native species in the USA (Krausman and Bleich, 2013).

Biological intrinsic characteristics seem to predispose ungulates to be successful invaders. A study performed on 14 species introduced to New Zealand in the period 1851–1926 identified the high survival rate for adults as the most important biological traits that could explain the high success rate (Forsyth and Duncan, 2001).

Wild ungulates often share space and resources with domestic livestock (Chirichella *et al.*, 2014) and a pathogen's transmission can bi-directionally occur through the interface between wildlife and domesticated mammal species, most of which are ungulates (Bengis *et al.*, 2002; East *et al.*, 2011; Ferroglio *et al.*, 2011; Wiethoelter *et al.*, 2015). As a result of this contact, zoonoses carried by infected livestock can be transmitted to alien wild ungulates and spread in the environment (Riemann *et al.*, 1979), exposing hunters, hikers or people professionally linked with forestry or outdoor activities to disease (Ruiz-Fons, 2015). In Queensland (Australia), more than 90% of human brucellosis cases occurred in hunters exposed to carcasses of infected wild swine (Eales *et al.*, 2010). Zoonotic diseases transmitted by alien wild ungulates to livestock can enter the human population through the ingestion of contaminated meat, milk and water (Meng *et al.*, 2009). For example, a recent paper by Galetti *et al.* (2016) stressed that vampire bats (*Desmodus rotundus*), feeding on the constantly spreading feral pigs, may be viewed as a potential risk to wildlife, livestock and humans, in the light of their role as a major reservoir of rabies and other viruses, including hantavirus, coronavirus and adenovirus.

Alien ungulates can be vectors or reservoirs of a great number and variety of zoonoses (Ferroglio *et al.*, 2011). Böhm *et al.* (2007), summarizing the major diseases affecting deer species, found 19 zoonoses (9 bacterial, 4 viral, 5 parasitic and 1 prionic), most of which had as host a deer species established outside of their native range, as in the case of chronic wasting disease, suspected to be introduced to Korea by infected red deer from Canada (Lee *et al.*, 2013). Wild boars are well-known vectors or reservoirs for a number of parasitic and infectious zoonotic diseases (Meng *et al.*, 2009; Ruiz-Fons, 2015) affecting also populations living outside of the species' native range or in feral populations of wild pig (e.g. USA and Australia). The recent overall increase in numbers and geographical expansion of wild boar, also within urban and sub-urban contexts, leads to an increased exposure of humans to wild boar zoonotic pathogens, among which some viruses (hepatitis E, Japanese encephalitis, flu and Nipah) and bacteria (*Salmonella* spp., Shiga toxin producing *Escherichia coli*, *Campylobacter* and *Leptospira* spp.) have been recently identified as 'the most prone to be transmitted from wild swine to humans' (Ruiz-Fons, 2015, p. 68).

Alien ungulates can also impact human health by causing injuries or death due to their aggressive behaviour or, indirectly, causing traffic accidents. *Hippopotamus amphibious*, introduced to Colombia in 1981, attacked fishermen creating panic among local communities (Valderrama Vásquez, 2012). Even though very rare, wild boar or wild pig (*Sus scrofa*) attacks on humans have been reported in many countries of their non-native range, often involving a wounded animal during hunting. The consequences of such attacks can be very serious, causing severe injuries or even fatalities (Mayer, 2013).

Alien ungulates (in particular cervids and wild boar) are most often implicated in vehicle collisions, causing injuries and deaths to humans (Bruinderink and Hazebroek, 1996; Mayer and Johns, 2007; Beasley *et al.*, 2014).

10.2.3 Carnivores

Carnivores can harbour and spread a great number of pathogens or parasites to wildlife, livestock, pets and humans. The role of carnivores as zoonotic reservoirs is well known (Cleaveland *et al.*, 2001). Biological traits such as the high vagility and a generalist feeding ecology lead carnivores to colonize almost all natural habitats and urban areas, where mostly small- and medium-sized species can exploit anthropogenic resources, reach high densities and increase the interface with humans, pets and livestock (Bateman and Fleming, 2012). Raccoon *Procyon lotor* in Europe, red fox *Vulpes vulpes* or dingo *Canis lupus dingo* in Australia are examples of alien carnivores occurring in urban areas and hosting several zoonotic pathogens, such as rabies, tuberculosis or parasites (Bateman and Fleming, 2012; Mackenstedt *et al.*, 2015). Compared with other mammals, carnivores have the greatest fraction of species that can act as zoonotic host (about 49%) and carry the greatest diversity of zoonotic pathogens (Han *et al.*, 2016).

Of the 29 alien carnivores reported as invasive in the GRIIS database, 20 (69%) showed evidence of an impact on human health in at least one country. Most of them are Canidae or Mustelidae, and the other species belong to the families Viverridae, Herpestidae, Felidae and Procionidae. The feral domestic cat is the most widespread species, followed by American mink *Neovison vison*, raccoon dog *Nyctereutes procyonides* and small Indian mongoose *Herpestes auropunctatus*.

Carnivores have been introduced all over the world mostly as biocontrol agents, pets, for fur farming or food; alien populations of carnivores also often originate from individuals escaped from captivity.

Among Herpestidae, the small Indian mongoose *Herpestes auropunctatus* has been introduced worldwide for biological control of rats and snakes (Jennings and Veron, 2016). Listed in *100 of the World's Worst Invasive Alien Species* (Lowe *et al.*, 2000), the small Indian mongoose is a vector for rabies and *Leptospira* and was identified recently in Japan as a reservoir for Gram-negative bacteria *Bartonella henselae*, the causative agent of cat-scratch disease (Sato *et al.*, 2013).

Mustelidae is the best-represented family of alien carnivore with 14 species, 8 belonging to the genera *Mustela*. The Eurasian ferret *Mustela furo* has been introduced in many countries as a pet and to Australia and New Zealand for controlling rabbits. Ferrets also have been used for centuries for hunting rabbits, but escapees in some cases have led to established feral populations (Medina and Martin, 2010). Since the 1970s, their popularity as pets in the USA and around the world has increased to reach, in 2001, an estimated population of 1 million ferrets in 0.5% of all households in the USA (Chomel, 2015). The Eurasian ferret can act as a reservoir of many and various zoonoses, including salmonellosis, mycobacteriosis, flu type A and B, cryptosporidiosis and giardiasis (Pignon and Mayer, 2011).

The American mink *Neovison vison* is the most widespread wild carnivore outside its native range (34 countries) and one of the most invasive alien mammals in the world. Introduced as a fur animal in Russia and naturalized in many parts of Europe as a result of escapes and intentional releases (Bonesi and Palazon, 2007), the American mink can serve as a host for a number of zoonoses, such as as leptospirosis, trichinellosis, toxoplasmosis and hepatitis E (Barros *et al.*, 2014; Krog *et al.*, 2013; Niemczynowicz and Zalewski, 2016; Zheng *et al.*, 2016).

The global wildlife trade increases the interface between humans and wild animals, offering an ideal way for pathogens to spillover between species (Bengis *et al.*, 2002). The wildlife trade in small carnivores can play a crucial role in worldwide diffusion of infectious diseases to domestic and wild animals and to humans (Bell *et al.*, 2004). The case of Viverridae can be taken as an example of multiple transmission routes of zoonoses. Viverrids are generally omnivorous and opportunistic (Gittleman, 2013) and often prey on small rodents, primary vectors of zoonoses. A recent review identified a wide range of viruses, bacteria and parasites reported for species of civet or genet within this family (Wicker *et al.*, 2017). Viverrids are commonly exploited for many purposes, ranging from hunting for human consumption to farming for fur, scent glands or to produce the 'civet coffee', the world's most expensive coffee bean; viverrids are also kept as zoo animals or pets. Keeping viverrids in captivity poses the greatest risk for transmission of pathogens to humans by the faecal–oral route (Wicker *et al.*, 2017).

Carnivoran hosts of zoonoses can naturally spread after being accidentally or intentionally released in a country. This is the case of the raccoon dog *Nyctereures procyonides*, an important vector of many zoonoses. It was introduced for its fur to the European part of the former Soviet Union around 1929, from the Russian Far East, and it is now well established in a large part of central, eastern and northern Europe (Kauhala and Kowalczyk, 2011). A comparison of parasite fauna in its native and alien range (Laurimaa *et al.*, 2016) found 26 helminthic species (17 zoonotic) infecting the raccoon dog in its native range and 32 species (19 zoonotic) in its alien range, 9 of which were absent from the species in native range. The raccoon dog (together with the masked palm civet *Paguma larvata*, a Viverridae probably introduced to Japan for its fur) also has been identified as a possible source of SARS-like coronaviruses (Guan *et al.*, 2003).

Increasing human–carnivore interactions, especially in urban and sub-urban contexts, can result in attacks on people that can cause injuries (Gehrt *et al.*, 2010). Feral dogs and cats account for most animal bites to humans (Levy and Crawford, 2004). In Queensland (Australia), dingo attacks occur, causing bite wounds or more serious injuries and, in a case at Frazer Island, death (Thompson *et al.*, 2003; Schmidt and Timm, 2007; Smith, 2015).

10.2.4 Other mammals

Alien carnivores, ungulates and rodents are not the only groups of mammals that can contribute to the transmission of parasitic zoonoses. Although evidence of actual risks from other groups of alien mammals is rarely available, there are examples from the literature showing that monkeys, shrews and hedgehogs, hares and rabbits, bats, marsupials and marine mammals can also play a role. Not all examples below concern situations reported directly from the introduced range of the mentioned taxa, however all relevant species are known as being successful invaders, thus we assume that they can be used as an example of their potential to cause significant morbidity or mortality to humans. Additionally, since most taxa below often live in close contact with humans (including being traded as game, or as pests), this raises concerns about the potential for transmission of disease.

Not surprisingly, the risk of carrying zoonotic agents is very high in non-human primates (see also Goldsmid, 2005). Non-human primates are indeed considered among the most dangerous of all taxa with regard to the potential for disease transmission to humans (Fiennes, 1967). The relevant risks are only partially mitigated by the fact that the number of species of primates introduced outside their native range is very low (Long, 2003). Marsupials are not very frequently introduced outside their native range, with a few exceptions (Long, 2003); however, it is important to consider that they can also be important vectors of diseases. For example, the brushtail possum *Trichosurus vulpecula*, introduced in New Zealand, can carry leptospirosis (Hathaway *et al.*, 1978) and there are cases of salmonellosis in humans associated with exposure to sugar gliders (*Petaurus breviceps*) kept as pets (Schoemaker, 2008).

Bats are poorly represented among the catalogues of invasive alien species, and may not be a great concern considering that the few reported populations have been eradicated successfully (see Genovesi *et al.*, 2012). Nevertheless, the remarkable role of bats as natural reservoir hosts and sources of infection for several microorganisms, many of which cause severe human diseases (FAO, 2011; Allocati *et al.*, 2016), requires some attention. As pointed out by Luis *et al.* (2013), bats hold the record of hosting more zoonotic viruses per species than rodents (but the total number of zoonotic viruses identified in bats seems lower than in rodents). Indeed, interspecific transmission may be more prevalent in bats than in other groups (Luis *et al.*, 2013). Additionally, there are indications of greater numbers of host switches of viruses from bats to other mammals (i.e. in the case of paramyxoviruses) than from rodents, primates, carnivores, etc. (Luis *et al.*, 2013). The mechanisms of interspecific transfer of pathogens, particularly to humans, remain poorly understood (Luis *et al.*, 2013), but in some cases are complex and involve intermediate hosts (Allocati *et al.*, 2016). Their characteristic ecology undoubtedly influences the maintenance and transmission of microorganisms within the colony and directly or indirectly to humans (Allocati *et al.*, 2016). Thus, as pointed out by Luis *et al.* (2013), bats are receiving increasing attention as potential reservoirs for zoonotic diseases following recent identification of their involvement with severe acute respiratory syndrome-like coronaviruses, Ebola and Marburg filoviruses, as well as Hendra and Nipah paramyxoviruses (see also FAO, 2011). Additionally, many pathogens are not dangerous for bats and can therefore survive for a long time in the host without killing it, meaning that bats may control viral replication more efficiently than other mammals (Allocati *et al.*, 2016).

The role of insectivores (and soricomorphs) in the transmission and ecology of zoonoses is also relevant, although – in contrast to rodents – is largely unknown. In fact, as pointed out by Arai *et al.* (2007), because some soricomorphs share habitats with rodents, shrews might be involved in the maintenance of the enzootic cycle and contribute to the evolutionary history and genetic diversity of viruses. Examples are those viruses closely related antigenically to Hantaan virus, typically harboured by rodents but also isolated from species known to be introduced, such as the Asian house shrew (*Suncus murinus*) and the greater white-toothed shrew (*Crocidura russula*). This evidence challenges the long-accepted dogma that rodents are the sole reservoirs of hantaviruses, and may lead to the discovery of additional hantaviruses in soricids throughout Eurasia, Africa and the Americas (Arai *et al.*, 2007). According to Mackenstedt *et al.* (2015), small mammals may have an important role as 'bridge' hosts, because some species are competent vectors or reservoirs of parasites and pathogens that may be introduced to new habitats. For example, the European hedgehog (*Erinaceus europaeus*) is the maintenance host for different tick species, which means that pathogens can be exchanged between the different ticks (more or less generalist) with the result of maintaining stable pathogen populations, especially in urban areas. Exposure to hedgehogs kept as pets is also considered to be associated with cases of bacterial diseases in humans, such as salmonellosis (Schoemaker, 2008). Also shrews (*Crocidura russula*) were found to be a reservoir of tick-borne zoonotic bacteria, namely *Anaplasma phagocytophilum* (Barandika *et al.*, 2007). Outside its native range, i.e. in Ireland, *Crocidura russula* has been found infected by a novel serovar of *Leptospira* (Nally *et al.*, 2016).

Lagomorphs have been introduced all over the world and may be reservoirs of several zoonoses. The rabbit (*Oryctolagus cuniculus*), for example, was found to be infected with *Fasciola hepatica* in France (Ménard *et al.*, 2000). Rabbits are also thought to be responsible for the transmission to human of cryptosporidiosis and giardiasis (Zhang *et al.*, 2012). There are other infections commonly associated with disease in rabbits, like the microsporidial infection with *Encephalitozoon cuniculi* causing severe neurologic disease in humans (Schoemaker, 2008). Likewise, the European brown hare may harbour many zoonotic agents, including common diseases such as staphylococcosis, pasteurellosis, pseudotuberculosis and brucellosis, as well as tularemia, listeriosis, toxoplasmosis, leptospirosis and borreliosis, which are the most serious and dangerous ones (Treml *et al.*, 2007). Lagomorphs are also among the most important game animals. For this reason, Treml *et al.* (2007) suggest the need to be careful when handling affected hares, e.g. during hunting activities (their blood sera is used to survey the occurrence of natural nidi of zoonoses in hunting grounds).

Even marine mammals may play a role in the emergence of zoonoses. Some species of cetaceans and pinnipeds kept in coastal dolphinaria and oceanaria are known to have escaped from captivity to the sea, and in some cases were released deliberately, supposedly transmitting infections circulating in dolphinaria to wild populations (Birkun, 2002). Marine mammals can be infected with zoonotic pathogens and show clinical signs of disease or be asymptomatic carriers of such disease agents (Hunt *et al.*, 2008). Marine mammal work-related illnesses commonly reported included: 'seal finger' (*Mycoplasma* spp. or *Erysipelothrix rhusiopathiae*), conjunctivitis, viral dermatitis, bacterial dermatitis and non-specific contact dermatitis (Hunt *et al.*, 2008). Infections due to *Anisakis* spp. and *Pseudoterranova* spp. are known in marine mammals, whales and porpoises in particular (Goldsmid, 2005), which thus play a major role in the life cycle of the nematodes causing Anisakiasis. Reports in the literature of humans infected with strains of *Brucella* associated with marine mammal (e.g. both cetaceans and pinnipeds) are also available (McDonald *et al.*, 2006).

10.3 Conclusions and Perspectives

Our review highlights the important role of alien mammals in threatening human health and welfare, particularly through the transmission of zoonoses. Alien mammals can act as vectors of both alien and native pathogens, and as hosts of either native or alien parasites (which in turn can act as vectors of either native or alien pathogens). In this way, alien mammals can introduce new pathogens, alter the epidemiology of local pathogens, become reservoir hosts and increase disease risk for humans, along with other species (Prenter *et al.*, 2004; Dunn, 2009). The examples above stress the importance of looking at the various changes in vector–host–parasite relationships that may influence invasion success and impact, e.g. by looking at enemy release, parasite spillover and spillback in relation to the introduction and spread of alien mammals. The importance of alien mammals in relation to the rise of emerging disease is also pivotal (OIE, 2010). Additionally, some mechanisms are still poorly studied and understood and deserve greater attention. For example, while the emergence of mammalian pathogens is often accompanied by host switching in a zoonotic context from one vertebrate species to another, some range extension from invertebrates to vertebrates is also hypothesized (Vilela *et al.*, 2015), showing how complex the dynamics of pathogen and disease transmission at the human–animal interface can be.

Mammals, however, appear less studied than other groups, regardless of their crucial epidemiological role, highlighted in this and other studies. We point out that the issue is not dealt with in its entirety and specificity, but usually placed in general reviews on the impact of individual groups of mammals on human health. On the one hand, this does not allow the problem to be analysed in its peculiarities, on the other there is a fragmentation of knowledge and a lack of homogeneity among the various groups.

Such partial and fragmented knowledge does not allow proper appreciation and quantifying of the complexity of the problem, and affects the possibility of identifying priorities for managing and preventing the threat. This may affect the assessment of trends, dynamics and other general and peculiar patterns of invasion, impacts, etc., which in turn may prevent the establishment of sound surveillance and monitoring systems to address the problem and its impact adequately (for an attempt to prioritize pathogens with a focus on ruminants, see Ciliberti et al., 2015). Additionally, because the increasing movement of humans and other species (as a result of climate change and other factors) could result in the sudden emergence of disease outbreaks, including new diseases and in new locations, it is important to assess future scenarios to identify further unexpected threats, and strengthen the relevant policy and legislative frameworks, also considering interlinking health and environmental legislation when appropriate. This shows the urgent need for a better understanding of the parasite–vector–host trio, the environmental, climatic and socio-economic factors involved, and the large potential for future zoonotic emergence.

Among the specificities of mammals, there are also the introductions of domestic and semi-domestic species in the wild (for example, reindeer introduced to Scotland for breeding; see Whitaker, 1986), as well as species introduced for hunting purposes, as in the case of rabbits and wild boars all over the world. Such introductions may carry the inherent risk of a further spread of alien mammals with their hidden load of diseases and pathogens. This shows the crucial importance of raising awareness in the public and key stakeholders on the health risks associated with the translocation of wildlife.

Increased research efforts on relevant pathogen life histories are needed to enhance our understanding of threats, methods and knowledge gaps. Priority should be given to improving our understanding of the pathways for mammal introductions, an issue quite poorly studied despite the conspicuousness of mammals compared with other taxa of health interest (e.g. invertebrates), as well as to measures to prevent the introduction of the species with the greatest impacts on human health. See, for example, the key role of the international wildlife trade in the emergence of global disease (Karesh et al., 2005). It would also be desirable that field studies and restoration projects aimed at eradicating alien mammals for biodiversity conservation purposes include a focus on the subsequent decrease in the risk of transmission of zoonotic diseases, such as occurs with the removal of rats from semi-inhabited islands (Capizzi et al., 2016).

References

Allocati, N., Petrucci, A.G., Di Giovanni, P., Masulli, M., Di Ilio, C. and De Laurenzi, V (2016) Bat-man disease transmission: zoonotic pathogens from wildlife reservoirs to human populations. Cell Death Discovery 27, 16048.

Arai, S., Song, J.W., Sumibcay, L., Bennett, S.N., Nerurkar, V.R., Parmenter, C., Cook, J., Yates, T.L. and Yanagihara, R. (2007) Hantavirus in northern short-tailed shrew, United States. Emerging Infectious Diseases 13, 1420–1423.

Azhar, E.I., El-Kafrawy, S.A., Farraj, S.A., Hassan, A.M., Al-Saeed, M.S., Hashem, A.M. and Madani, T.A. (2014) Evidence for camel-to-human transmission of MERS coronavirus. New England Journal of Medicine 370, 2499–2505.

Barandika, J.F., Hurtado, A., García-Esteban, C., Gil, H., Escudero, R., Barral, M., Jado, I., Juste, R.A., Anda, P. and García-Pérez, A.L. (2007) Tick-borne zoonotic bacteria in wild and domestic small mammals in northern Spain. *Applied Environmental Microbiology* 73, 6166–6171.

Barros, M., Sáenz, L., Lapierre, L., Nuñez, C. and Medina-Vogel, G. (2014) High prevalence of pathogenic Leptospira in alien American mink (*Neovison vison*) in Patagonia. *Revista chilena de historia natural* 87, 19.

Bateman, P.W. and Fleming, P.A. (2012) Big city life: carnivores in urban environments. *Journal of Zoology* 287, 1–23.

Beasley, J.C., Grazia, T.E., Johns, P.E. and Mayer, J.J. (2014) Habitats associated with vehicle collisions with wild pigs. *Wildlife Research* 40, 654–660.

Bekker, J.L., Jooste, P.J. and Hoffman, L.C. (2012) Wildlife-associated zoonotic diseases in some southern African countries in relation to game meat safety: a review. *Onderstepoort Journal of Veterinary Research* 79, 1–12.

Bell, D., Roberton, S. and Hunter, P.R. (2004) Animal origins of SARS coronavirus: possible links with the international trade in small carnivores. *Philosophical Transactions of the Royal Society of London B: Biological Sciences* 359, 1107–1114.

Bengis, R.G., Kock, R.A. and Fischer, J. (2002) Infectious animal diseases: the wildlife/livestock interface. *Revue Scientifique et Technique* 21, 53–65.

Birkun, A. (2002) The current status of bottlenose dolphins (*Tursiops truncatus*) in the Black Sea. *AC18 Inf.2 ACCOBAMS, Agreement on the Conservation of Cetaceans of the Black Sea, Mediterranean Sea and Contiguous Atlantic Area. First Meeting of the Parties*, 28 February–2 March 2002, Monaco.

Böhm, M., White, P.C., Chambers, J., Smith, L. and Hutchings, M.R. (2007) Wild deer as a source of infection for livestock and humans in the UK. *The Veterinary Journal* 174, 260–276.

Bonesi, L. and Palazon, S. (2007) The American mink in Europe: status, impacts, and control. *Biological Conservation* 134, 470–483.

Borgsteede, F.H., Tibben, J.H. and van der Giessen, J.W. (2003) The musk rat (*Ondatra zibethicus*) as intermediate host of cestodes in the Netherlands. *Veterinary Parasitology* 117, 29–36.

Boulouis, H.J., Chang, C.C., Henn, J.B., Kasten, R.W. and Chomel, B.B. (2005) Factors associated with the rapid emergence of zoonotic Bartonella infections. *Veterinary Research* 36, 383–410.

Bruinderink, G.W.T.A. and Hazebroek, E. (1996) Ungulate traffic collisions in Europe. *Conservation Biology* 10, 1059–1067.

Capizzi, D., Bertolino, S. and Mortelliti, A. (2014) Rating the rat: global patterns and research priorities in impacts and management of rodent pests. *Mammal Review* 44, 148–162.

Capizzi, D., Baccetti, N. and Sposimo, P. (2016) Fifteen years of rat eradication on Italian islands. In: Angelici, F.M. (ed.) *Problematic Wildlife*. Springer International Publishing, Cham, Switzerland, pp. 205–227.

Chalmers, R.M., Sturdee, A.P., Bull, S.A., Miller, A. and Wright, S.E. (1997) The prevalence of *Cryptosporidium parvum* and *C. muris* in *Mus domesticus*, *Apodemus sylvaticus* and *Clethrionomys glareolus* in an agricultural system. *Parasitology Research* 83, 478–482.

Chirichella, R., Apollonio, M. and Putman, R.J. (2014) Competition between domestic and wild ungulates. In: Putman, R. and Apollonio, M. (eds) *Behaviour and Management of European Ungulates*. Whittle, Dunbeath, UK, pp. 110–123.

Chomel, B.B. (2015) Diseases transmitted by less common house pets. *Microbiology Spectrum* 3, IOL5-0012-2015.

Ciliberti, A., Gavier-Widén, D., Yon, L., Hutchings, M.R. and Artois, M. (2015) Prioritisation of wildlife pathogens to be targeted in European surveillance programmes: expert-based risk analysis focus on ruminants. *Preventive Veterinary Medicine* 118, 271–284.

Cleaveland, S., Laurenson, M.K. and Taylor, L.H. (2001) Diseases of humans and their domestic mammals: pathogen characteristics, host range and the risk of emergence. *Philosophical Transactions of the Royal Society of London B: Biological Sciences* 356, 991–999.

DAISIE (2011) European Invasive Alien Species Gateway. Available at: http://www.europe-aliens.org/ (accessed December 2017).

Denoya, V.A., Pointer, J.P., Colmenares, C., Theron, A., Balzan, C., Cesari, I.M., Gonzalez, S. and Noya, O. (1997) Natural *Schistosoma mansoni* infection in wild rats from Guadeloupe – parasitological and immunological aspects. *Acta Tropica* 68, 11–21.

Dunn, A.M. (2009) Parasites and biological invasions. *Advances in Parasitology* 68, 161–184.

Eales, K.M., Norton, R.E. and Ketheesan, N. (2010) Brucellosis in northern Australia. *The American Journal of Tropical Medicine and Hygiene* 83, 876–878.

East, M.L., Bassano, B. and Ytreus, B. (2011) The role of pathogens in the population dynamics of European ungulates. In Apollonio, M., Andersen, R. and Putman, R.J. (eds) *European Ungulates and their Management in the 21 Century*. Cambridge University Press, Cambridge, pp. 319–348.

FAO (Food and Agriculture Organization of the United Nations) (2011) Investigating the role of bats in emerging zoonoses: Balancing ecology, conservation and public health interests. In: Newman, S.H., Field, H.E., de Jong, C.E. and Epstein, J.H. (eds) *FAO Animal Production and Health Manual No. 12*. FAO, Rome.

Feldhamer, G.A., Thompson, B.C. and Chapman, J.A. (2003) *Wild Mammals of North America: Biology, Management, and Conservation*. Johns Hopkins University Press, Baltimore, Maryland.

Ferroglio, E., Gortazar, C. and Vicente, J. (2011) Wild ungulate diseases and the risk for live-stock and public health. In: Apollonio, M., Andersen, R. and Putman, R.J. (eds) *European Ungulates and their Management in the 21 Century*. Cambridge University Press, Cambridge, UK, pp. 192–214.

Fiennes, R. (1967) *Zoonoses of Primates*. Cornell University Press, Ithaca, New York.

Forsyth, D.M. and Duncan, R.P. (2001) Propagule size and the relative success of exotic ungulate and bird introductions to New Zealand. *The American Naturalist* 157, 583–595.

Galetti, M., Pedrosa, F., Keuroghlian, A. and Sazima, I. (2016) Liquid lunch – vampire bats feed on invasive feral pigs and other ungulates. *Frontiers in Ecology and the Environment* 14, 505–506.

Gehrt, S.D., Riley, S.P. and Cypher, B.L. (2010) *Urban Carnivores: Ecology, Conflict, and Conservation*. Johns Hopkins University Press, Baltimore, Maryland, USA.

Genovesi, P., Carnevali, L., Alonzi, A. and Scalera, R. (2012) Alien mammals in Europe: updated numbers and trends, and assessment of the effects on biodiversity. *Integrative Zoology* 7, 247–253.

Gittleman, J.L. (2013) *Carnivore Behavior, Ecology, and Evolution*. Springer Science & Business Media, Dordrecht, the Netherlands.

Global Invasive Species Database (2017) Species profile: *Castor canadensis*. Available at: http://www.iucngisd.org/gisd/species.php?sc=981 (accessed 9 March 2017).

Goldsmid, J.M. (2005) Zoonotic infections: an overview. In: *Primer of Tropical Medicine* 14. Australasian College of Tropical Medicine, Redhill, Queensland, Australia, pp. 1–13.

Gratz, N.G. (1994) Rodents as carriers of disease. In: Buckle, A.P. and Smith, R.H. (eds.) *Rodent Pests and Their Control*. CAB International, Wallingford, UK, pp. 85–108.

Gratz, N.G. (2006) *Vector- and Rodent-borne Diseases in Europe and North America: Distribution, Public Health Burden, and Control*. Cambridge University Press, Cambridge, UK.

Guan, Y., Zheng, B.J., He, Y.Q., Liu, X.L., Zhuang, Z.X., Cheung, C.L., Luo, S.W., Li, P.H., Zhang, L.J., Guan, Y.J., Butt, K.M., Wong, K.L., Chan, K.W., Lim, W., Shortridge, K.F., Yuen, K.Y., Peiris, J.S. and Poon, L.L. (2003) Isolation and characterization of viruses related to the SARS coronavirus from animals in southern China. *Science* 302, 276–278.

Han, B.A., Kramer, A.M. and Drake, J.M. (2016) Global patterns of zoonotic disease in mammals. *Trends in Parasitology* 32, 565–577.

Hathaway, S.C., Blackmore, D.K. and Marshall, R.B. (1978) The serologic and cultural prevalence of *Leptosira interrogans balcanica* in possums *Trichosurus vulpecula* in New Zealand. *Journal of Wildlife Diseases* 14, 345–350.

Henzler, D.J. and Opitz, H.M. (1992) The role of mice in the epizootiology of *Salmonella enteritidis* infection on chicken layer farms. *Avian Diseases* 36, 625–631.

Hulme, P.E. (2014) Invasive species challenge the global response to emerging diseases. *Trends in Parasitology* 30, 267–270.

Hunt, T.D., Ziccardi, M.H., Gulland, F.M., Yochem, P.K., Hird, D.W. and Rowles, T. (2008) Health risks for marine mammal workers. *Diseases of Aquatic Organisms* 81, 81–92.

Iida, T., Kanzaki, M., Nakama, A., Kokubo, Y., Maruyama, T. and Kaneuchi, C. (1998) Detection of *Listeria monocytogenes* in humans, animals and foods. *The Journal of Veterinary Medical Science* 60, 1341.

Jennings, A. and Veron, G. (2016) *Herpestes auropunctatus*. In: *The IUCN Red List of Threatened Species 2016*. Available at: http://www.iucnredlist.org/details/70204120/0 (accessed 17 February 2017).

Karesh, W.B., Cook, R.A., Bennett, E.L. and Newcomb, J. (2005) Wildlife trade and global disease emergence. *Emerging Infectious Disease* 11, 1000–1002.

Kauhala, K. and Kowalczyk, R. (2011) Invasion of the raccoon dog *Nyctereutes procyonoides* in Europe: history of colonization, features behind its success, and threats to native fauna. *Current Zoology* 57, 584–598.

Kock, R.A., Woodford, M.H. and Rossiter, P.B. (2010) Disease risks associated with the translocation of wildlife. *Revue scientifique et technique* 29, 329.

Krausman, P.R. and Bleich, V.C. (2013) Conservation and management of ungulates in North America. *International Journal of Environmental Studies* 70, 372–382.

Krog, J.S., Breum, S.Ø., Jensen, T.H. and Larsen, L.E. (2013) Hepatitis E virus variant in farmed mink, Denmark. *Emerging Infectious Diseases* 19, 2028–2030.

Laurimaa, L., Süld, K., Davison, J., Moks, E., Valdmann, H. and Saarma, U. (2016) Alien species and their zoonotic parasites in native and introduced ranges: the raccoon dog example. *Veterinary Parasitology* 219, 24–33.

Lee, J., Kim, S.Y., Hwang, K.J., Ju, Y.R. and Woo, H.J. (2013) Prion diseases as transmissible zoonotic diseases. *Osong Public Health and Research Perspectives* 4, 57–66.

Levy, J.K. and Crawford, P.C. (2004) Humane strategies for controlling feral cat populations. *Journal of the American Veterinary Medical Association* 225, 1354–1360.

Lledó, L., Gegúndez, M.I., Saz, J.V., Bahamontes, N. and Beltrán, M. (2003) Lymphocytic choriomeningitis virus infection in a province of Spain: analysis of sera from the general population and wild rodents. *Journal of Medical Virology* 70, 273–275.

Long, J.L. (2003) *Introduced Mammals of the World.* CAB International and CSIRO Publishing, Melbourne, Australia.

Lowe, S., Browne, M., Boudjelas, S. and De Poorter, M. (2000) *100 of the World's Worst Invasive Alien Species. A Selection from the Global Invasive Species Database.* World Conservation Union (IUCN), Gland, Switzerland.

Luis, A.D., Hayman, D.T.S., Fooks, A.R., Rupprecht, C.E., Wood, J.L.N., Webb, C.T., O'Shea, T.J., Cryan, P.M., Gilbert, A.T., Pulliam, J.R.C., Mills, J.N., Timonin, M.E., Willis, C. and Cunningham, A.A. (2013) A comparison of bats and rodents as reservoirs of zoonotic viruses: are bats special? *Proceedings of the Royal Society – Biological Series* 280, 2012–2753.

Macdonald, D.W., Mathews, F. and Berdoy, M. (1999) The behaviour and ecology of *Rattus norvegicus*: from opportunism to kamikaze tendencies. In: Singleton, G.R., Hinds, L.A., Leirs, H. and Zhang, Z. (eds) *Ecologically Based Rodent Management.* Australian Centre for International Agricultural Research, Canberra, Australia, pp. 49–80.

Mackenstedt, U., Jenkins, D. and Romig, T. (2015) The role of wildlife in the transmission of parasitic zoonoses in peri-urban and urban areas. *International Journal for Parasitology: Parasites and Wildlife* 4, 71–99.

Mas-Coma, S., Fons, R., Feliu, C., Bargues, M.D., Valero, M.A. and Galan-Puchades, T. (1988) Small mammals as natural definitive hosts of the liver fluke *Fasciola hepatica* (Linnaeus, 1758) (Trematoda: Fasciolidae); a review and two new records of epidemiologic interest on the island of Corsica. *Rivista di Parassitologia* 5, 73.

Massad, E., Coutinho, F.A.B., Burattini, M.N. and Lopez, L.F. (2004) The Eyam plague revisited: did the village isolation change transmission from fleas to pulmonary? *Medical Hypotheses* 63, 911–915.

Matthias, M.A., Ricaldi, J.N., Cespedes, M., Diaz, M.M., Galloway, R.L., Saito, M., Steigerwalt, A.G., Patra, K.P., Ore, C.V., Gotuzzo, E., Gilman, R.H., Levett, P.N. and Vinetz, J.M. (2008) Human Leptospirosis caused by a new, antigenically unique *Leptospira* associated with a *Rattus* species reservoir in the Peruvian Amazon. *PLoS Neglected Tropical Diseases* 2, e213.

Matuschka, F.-R., Endepols, S., Richter, D., Ohlenbusch, A., Eiffert, H. and Spielman, A. (1996) Risk of urban Lyme disease enhanced by the presence of rats. *Journal of Infectious Disease* 174, 1108–1111.

Mayer, J.J. (2013) Wild pig attacks on humans. In: Armstrong, J.B. and Gallagher, G.R. (eds) *Proceedings of the 15th Wildlife Damage Management Conference*, p. 151.

Mayer, J.J. and Johns, P.E. (2007) Characterization of wild pig–vehicle collisions. *Proceedings of the Wildlife Damage Management Conference* 12, 175–187.

McDonald, W.L., Jamaludin, R., Mackereth, G., Hansen, M., Humphrey, S., Short, P., Taylor, T., Swingler, J., Dawson, C.E., Whatmore, A.M., Stubberfield, E., Perrett, L.L. and Simmons, G. (2006) Characterization of a *Brucella* sp. strain as a marine-mammal type despite isolation from a patient with spinal osteomyelitis in New Zealand. *Journal of Clinical Microbiology* 44, 4363–4370.

Medina, F.M. and Martín, A. (2010) A new invasive species in the Canary Islands: a naturalized population of ferrets *Mustela furo* in La Palma Biosphere Reserve. *Oryx* 44, 41–44.

Meerburg, B.G., Singleton, G.R. and Kijlstra, A. (2009) Rodent-borne diseases and their risks for public health. *Critical Reviews in Microbiology* 35, 221–270.

Ménard, A., L'Hostis, M., Leray, G., Marchandeau, S., Pascal, M., Roudot, N., Michel, V. and Chauvin, A. (2000) Inventory of wild rodents and lagomorphs as natural hosts of *Fasciola hepatica* on a farm located in a humid area in Loire Atlantique (France). *Parasite* 7, 77–82.

Meng, X.J., Lindsay, D.S. and Sriranganathan, N. (2009) Wild boars as sources for infectious diseases in livestock and humans. *Philosophical Transactions of the Royal Society B: Biological Sciences* 364, 2697–2707.

Millins, C., Magierecka, A., Gilbert, L., Edoff, A., Brereton, A., Kilbride, E., Denwood, M., Birtles, R. and Biek, R. (2015) An invasive mammal (the gray squirrel, *Sciurus carolinensis*) commonly hosts diverse and atypical genotypes of the zoonotic pathogen *Borrelia burgdorferi sensu lato*. *Applied Environmental Microbiology* 81, 4236–4245.

Mills, J.N. and Childs, J.E. (1998) Ecologic studies of rodent reservoirs: their relevance for human health. *Emerging Infectious Diseases* 4, 529–537.

Nally, J.E., Arent, Z., Bayles, D.O., Hornsby, R.L., Gilmore, C., Regan, S., McDevitt, A.D., Yearsley, J., Fanning, S. and McMahon, B.J. (2016) Emerging infectious disease implications of invasive mammalian species: the greater white-toothed shrew (*Crocidura russula*) is associated with a novel serovar of pathogenic Leptospira in Ireland. *PLoS Neglected Tropical Diseases* 10, e0005174.

Niemczynowicz, A. and Zalewski, A. (2016) An invasive species as an additional parasite reservoir: *Trichinella* in introduced American mink (*Neovison vison*). *Veterinary Parasitology* 231, 106–110.

OIE (2010) *Training Manual on Wildlife Diseases and Surveillance.* Workshop for OIE National Focal Points for Wildlife. World Organisation for Animal Health, Paris.

Pagad, S., Genovesi, P., Carnevali, L., Scalera, R. and Clout, M. (2015) IUCN SSC Invasive Species Specialist Group: invasive alien species information management supporting practitioners, policy makers and decision takers. *Management of Biological Invasions* 6, 127–135.

Pagad, S., Genovesi, P., Carnevali, L., Schigel, D. and McGeoch, M.A. (2018) Introducing the Global Register of Introduced and Invasive Species. *Scientific Data* 5, 170202 doi: 10.1038/sdata.2017.202.

Panzacchi, M., Cocchi, R., Genovesi, P. and Bertolino, S. (2007) Population control of coypu *Myocastor coypus* in Italy compared to eradication in UK: a cost-benefit analysis. *Wildlife Biology* 13, 159–171.

Pignon, C. and Mayer, J. (2011) Zoonoses of ferrets, hedgehogs, and sugar gliders. *Veterinary Clinics of North America: Exotic Animal Practice* 14, 533–549.

Prenter, J., MacNeil, C., Dick, J.T. and Dunn, A.M. (2004) Roles of parasites in animal invasions. *Trends in Ecology & Evolution* 19, 385–390.

Putman, R.J. (2010) Ungulates and their management in Great Britain and Ireland. In: Apollonio, M., Andersen, R. and Putman, R.J. (eds) *European Ungulates and their Management in the 21 Century.* Cambridge University Press, Cambridge, UK, pp. 129–164.

Riemann, H., Zaman, M.R., Ruppaner, R., Aalund, O., Jorgensen, J.B., Worsae, H. and Behymer, D. (1979) Paratuberculosis in cattle and free-living exotic deer. *Journal of the American Veterinary Medical Association* 174, 841–843.

Ruiz-Fons, F. (2015) A review of the current status of relevant zoonotic pathogens in wild swine (*Sus scrofa*) populations: changes modulating the risk of transmission to humans. *Transboundary and Emerging Diseases* 64, 68–88.

Saalfeld, W.K. and Edwards, G.P. (2010) Distribution and abundance of the feral camel (*Camelus dromedarius*) in Australia. *The Rangeland Journal* 32, 1–9.

Sato, S., Kabeya, H., Shigematsu, Y., Sentsui, H., Une, Y., Minami, M., Murata, K., Ogura, G. and Maruyama, S. (2013) Small Indian mongooses and masked palm civets serve as new reservoirs of *Bartonella henselae* and potential sources of infection for humans. *Clinical Microbiology and Infection* 19, 1181–1187.

Schmaljohn, C. and Hjelle, B. (1997) Hantaviruses: a global disease problem. *Emerging Infectious Diseases* 3, 95–104.

Schmidt, R.H. and Timm, R.M. (2007) Bad dogs: why do coyotes and other canids become unruly? *Proceedings of the Wildlife Damage Management Conference* 12, 287–302.

Schoemaker, N.J. (2008) Exotic companion mammal zoonoses small animals can have big consequences. *Proceedings of the North American Veterinary Conference*, Orlando, Florida, pp. 1877–1879.

Smith, B. (2015) *The Dingo Debate: Origins, Behaviour and Conservation.* Csiro Publishing, Clayton South, Victoria.

Smith, R.P., Rand, P.W., Lacombe, E.H., Telford, S.R., Rich, S.M., Piesman, J. and Spielman, A. (1993) Norway rats as reservoir hosts for Lyme disease spirochetes on Monhegan Island, Maine. *Journal of Infectious Diseases* 168, 687–691.

Spratt, D.M. (2005) Australian ecosystems, capricious food chains and parasitic consequences for people. *International Journal of Parasitology* 35, 717–724.

Thompson, J., Shirreffs, L. and McPhail, I. (2003) Dingoes on Fraser Island: tourism dream or management nightmare. *Human Dimensions of Wildlife* 8, 37–47.

Torres, J., Gracenea, M., Gomez, M.S., Arrizabalaga, A. and Gonzalez-Moreno, O. (2000) The occurrence of *Cryptosporidium parvum* and *C. muris* in wild rodents and insectivores in Spain. *Veterinary Parasitology* 92, 253.

Treml, F., Pikula, J., Bandouchova, H. and Horakova, J. (2007) European brown hare as a potential source of zoonotic agents. *Veterinarni Medicina Praha* 52, 451–456.

Valderrama Vásquez, C.A. (2012) Wild hippos in Colombia. *Aliens: The Invasive Species Bulletin* 32, 8–12.

Vilela, R., Taylor, J.W., Walker, E.D. and Mendoza, L. (2015) *Lagenidium giganteum* pathogenicity in mammals. *Emerging Infectious Diseases* 21, 290–297.

Vourc'h, G., Marmet, J., Chassagne, M., Bord, S. and Chapuis, J.L. (2007) *Borrelia burgdorferi sensu lato* in Siberian chipmunks (*Tamias sibiricus*) introduced in suburban forests in France. *Vector-Borne and Zoonotic Diseases* 7, 637–642.

Wang, H., Yoshimatsu, K., Ebihara, H., Ogino, M., Araki, K., Kariwa, H., Wang, Z., Luo, Z., Li, D., Hang, C. and Arikawa, J. (2000) Genetic diversity of hantaviruses isolated in China and characterization of novel hantaviruses isolated from *Niviventer confucianus* and *Rattus rattus*. *Virology* 278, 332–345.

Whitaker, I. (1986) The survival of feral reindeer in northern Scotland. *Archives of Natural History* 13, 11–18.

Wicker, L.V., Canfield, P.J. and Higgins, D.P. (2017) Potential pathogens reported in species of the family Viverridae and their implications for human and animal health. *Zoonoses and Public Health* 64, 75–93.

Wiethoelter, A.K., Beltrán-Alcrudo, D., Kock, R. and Mor, S. (2015) Global trends in infectious diseases at the wildlife–livestock interface. *Proceedings of the National Academy of Sciences* 112, 9662–9667.

Wolfs, T.F., Wagenaar, J.A., Niesters, H.G. and Osterhaus, A.D. (2002) Rat-to-human transmission of cowpox infection. *Emerging Infectious Diseases* 8, 1495–1496.

Zhang, W., Shen, Y., Wang, R., Liu, A., Ling, H., Li, Y., Cao, J., Zhang, X., Shu, J. and Zhang, L. (2012) *Cryptosporidium cuniculus* and *Giardia duodenalis* in rabbits: genetic diversity and possible zoonotic transmission. *PLoS One* 7, e31262.

Zheng, W.B., Cong, W., Meng, Q.F., Ma, J.G., Wang, C.F., Zhu, X.Q. and Qian, A.D. (2016) Seroprevalence and risk factors of *Toxoplasma gondii* infection in farmed minks (*Neovison vison*) in northeastern and eastern China. *Vector-Borne and Zoonotic Diseases* 16, 485–488.

11

Climate Change and Increase of Impacts on Human Health by Alien Species

Stefan Schindler[1]*, Wolfgang Rabitsch[1] and Franz Essl[1,2]

[1]Environment Agency Austria, Vienna, Austria and [2]University of Vienna, Austria

Abstract

There is mounting evidence that climate change will facilitate biological invasions. Regarding alien species relevant for human health, climate change can modify their impacts by altering the likelihood of their introduction, establishment, distribution and abundance, the scale of impacts and management. In this chapter, we summarize climate change impacts on health-relevant alien species with a focus on Europe. Climate change can support introductions of alien species impacting human health, but its role is moderate compared with the other aspects of globalization, such as increased trade and people's increased mobility. There is strong evidence that changing climate characteristics, notably temperature, are crucial for the establishment and spread of human health-relevant alien species. Resulting increases and shifts in distribution might cause increasing health impacts, particularly in cold regions such as temperate and polar regions or areas of higher elevation. In Europe, changes in health impacts caused by alien species are mainly related to arthropod vectors and plants with allergenic pollen. As numbers of alien species will probably increase, preventive management needs to be strengthened. Further, a better understanding of drivers and management options of alien diseases and vectors, as well as joint efforts in education and outreach to the public and decision-makers, is required. There is a need for action and research, particularly in the fields of pathway management, epidemiology, modelling and vector monitoring.

11.1 Introduction

Climate change brings about changes in climate characteristics, such as temperature and precipitation, and the frequency and intensity of extreme events, as well as affecting sea levels, sea ice, the extent of glaciers and ocean acidification (IPCC, 2014). Impacts of climate change on the biotic environment are plentiful and act from genes to biomes to people (Scheffers et al., 2016; Wiens, 2016). There is mounting evidence that climate change will facilitate the risk of biological invasions (Diez et al., 2012; Hulme, 2017), in particular in cold environments at high elevations and latitudes that until recently have been viewed as

* E-mail: Stefan.Schindler@umweltbundesamt.at

© CAB International 2018. *Invasive Species and Human Health*
(eds G. Mazza and E. Tricarico)

more resistant to biological invasions (Alsos *et al.*, 2015; Pauchard *et al.*, 2016), but also in aquatic systems (Rahel and Olden, 2008; Gallardo and Aldridge, 2013) and more generally in regions that become more climatically hospitable (Sorte *et al.*, 2013). However, all species have upper and lower limits of their ecophysiological tolerances, and thus climate change-induced range losses at the trailing range limit may partly offset range gains at the leading edge (Lafferty, 2009; Harter *et al.*, 2015).

Alien species can impact human health directly by biting and stinging, poisoning, causing allergies and ailment, as a pathogen causing disease or infection, and indirectly as a vector or reservoir of pathogens (Pyšek and Richardson, 2010; Mazza *et al.*, 2014; Hulme, 2014; Schindler *et al.*, 2015; Bayliss *et al.*, 2017; Rabitsch *et al.*, 2017; see Chapters 4 and 5, this volume). Climate change can modify these impacts in various ways, such as by altering the likelihood of alien species' introduction or establishment, their geographic distribution, the magnitude of the impact and management of alien species (Fig. 11.1) (e.g. Walther *et al.*, 2009; Fischer *et al.*, 2011; Chapman *et al.*, 2016; Hulme, 2017). However, a recent review revealed that on a European scale, the evidence on climate change-driven responses of human health-relevant alien species is largely limited to predicting their spread under climate change, while it is scarce for other relevant aspects (Schindler *et al.*, 2015).

Here, we present and discuss evidence (i.e. climate change effects in recent decades) and projections (i.e. how climate change is predicted to affect the future) for climate change effects on alien species relevant to human health. We focus on temperate and cool climates

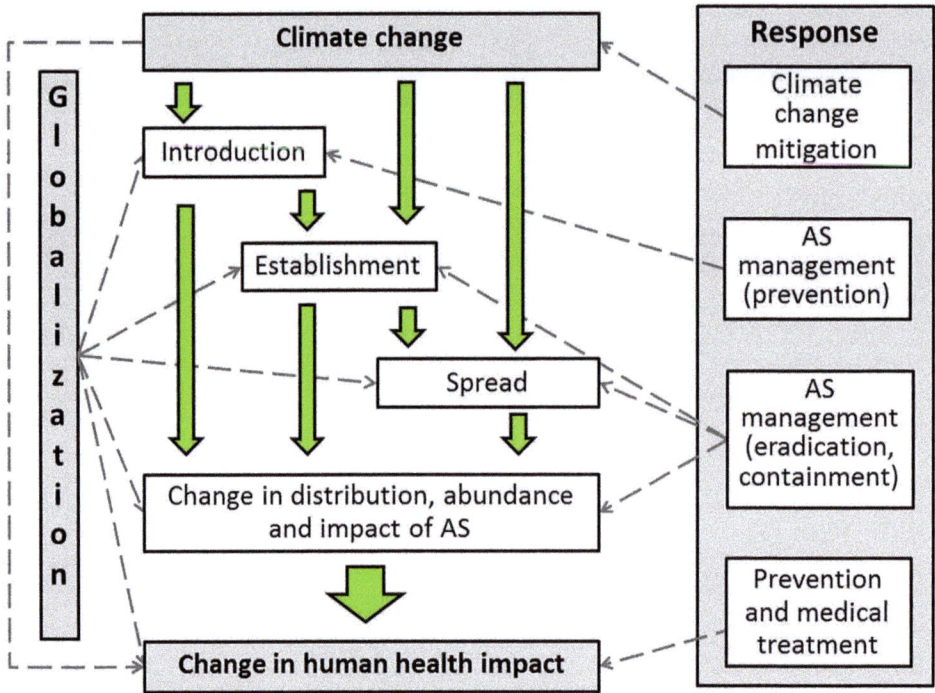

Fig. 11.1. Climate change impacts on introduction, establishment and spread of human health-relevant alien species (AS) and management responses of society and policy aiming at the mitigation of climate change and the management of AS. The current level of responses are insufficient to halt the introduction of new alien species (Tittensor *et al.*, 2014; Seebens *et al.*, 2017). Thus, prevention and medical treatment of human health impacts are of increasing importance. Green arrows indicate processes that are subjects of this chapter. (Modified from Bayliss *et al.* (2015) and Blackburn *et al.* (2011).)

such as Europe and the Arctic (Bellard *et al.*, 2013), but also include examples from other regions. The evidence was mainly obtained from recent reviews on health-relevant alien species (e.g. Hellmann *et al.*, 2008; Semenza and Menne, 2009; Dudley *et al.*, 2015; Essl *et al.*, 2015b; Schindler *et al.*, 2015, 2016; Bayliss *et al.*, 2017) and additional literature searches. The chapter is structured along the invasion stages introduction, establishment, spread and impact (Blackburn *et al.*, 2011; Jeschke *et al.*, 2013). We apply for alien species the definitions of Richardson *et al.* (2000) and Blackburn *et al.* (2011), excluding species that shift their distribution range in response to climate change only, without any further human assistance. We also exclude indirect impacts of alien species on human health, such as those expected due to increased fire risk and subsequent respiratory problems associated with the devastation of large forests by alien pine beetles (Embrey *et al.*, 2012; see Chapter 5, this volume) or yield losses and subsequent food security risks caused by range shifts of pests (Bebber, 2015).

11.2 Climate Change Impacts on Human Health-Relevant Alien Species

11.2.1 Climate change impacts on introduction rates

Introduction rates of alien species may be modified by climate change in several ways. Native ranges may shift and subsequently pathways that were not available before may become relevant. For instance, marine species often show fast range shifts tracking changing climates (Poloczanska *et al.*, 2013) and so might end up in ballast water tanks of ships that travel along routes that were previously outside the climatic niches of potentially invasive marine species. Climate change may also precipitate changes in trading patterns or shipping routes, such as the opening of Arctic passages (Hellmann *et al.*, 2008; Chan *et al.*, 2013), and new pathways may emerge in sectors such as agriculture, forestry (Felton *et al.*, 2016), energy provision (Raghu *et al.*, 2006; Chimera *et al.*, 2010; Barney, 2014) or recreation and tourism (Rahel and Olden, 2008; Hulme, 2015).

Evidence for human health-relevant alien species

Evidence for climate change impacts on introduction rates of human health-relevant alien species is scarce. First, the association of alien species with pathways often is insufficiently known, and partly inferred from species ecology and anecdotal information (Essl *et al.*, 2015a). Second, quantitative data on propagule pressure and introduction rates of specific pathways and sources of origin are hardly known. For instance, West Nile Virus (WNV) is a mosquito-borne flavivirus and human and avian neuropathogen that is native to parts of Africa, Asia, Europe and Australia and maintained in nature in a mosquito–bird–mosquito transmission cycle (Campbell *et al.*, 2002; Kilpatrick, 2011). Substantial outbreaks of West Nile neuroinvasive disease in humans have recently been recorded in several European countries, but the relative importance of human-aided dispersal of WNV through infected mosquitoes or birds is unknown (Rizzoli *et al.*, 2015). A large epidemic in North and South America started in New York City in 1999. While it is evident that the origin of the introduced strain was the Middle East, the mode of introduction is uncertain and might have occurred naturally by a migrating bird or been human assisted by a stowaway mosquito or a captive bird (Lanciotti *et al.*, 1999; Giladi *et al.*, 2001; Kilpatrick, 2011).

For the marine environment, however, Raitsos *et al.* (2010) demonstrated a strong increase in the rate of introduced warm water (mainly Lessepsian) species after a sudden rise in sea water temperatures in the eastern Mediterranean around 1998. For species with health impacts (cf. Streftaris and Zenetos, 2006), the algae *Ostreopsis ovata*, the jellyfish *Rhopilema nomadica* and the fishes *Lagocephalus sceleratus* and *Seriola fasciata* were recorded first in the eastern Mediterranean after 1998, whereas the fishes *Siganus luridus*, *Siganus rivulatus* and *Sphoeroides pachygaster* had already been recorded earlier (Raitsos *et al.*, 2010).

Predictions for human health-relevant alien species

According to a survey among experts conducted by Gale *et al.* (2010), climate change was predicted to increase the risk of incursion into the EU through entry of vectors on a qualitative scale from low to medium for WNV and the Crimean-Congo haemorrhagic fever virus until 2080.

11.2.2 Climate change impacts on establishment of alien species

Establishment rates of alien species may be affected by climate change as the climatic suitability in the potential alien regions changes. In response, alien species that were present as casuals only, that survived only in sheltered environments such as greenhouses, or that were completely absent, may be able to establish self-sustaining permanent populations (Hellmann *et al.*, 2008; Raitsos *et al.*, 2010; Roques *et al.*, 2016; Hulme, 2017).

Evidence for human health-relevant alien species

There is evidence that climate change increases establishment risks for subtropical and tropical health-relevant species in Europe (Roy *et al.*, 2009). Such species include the poisonous common lionfish (*Pterois miles*), a Lessepsian migrant that is spreading in the eastern Mediterranean likely due to rising sea temperatures (Kletou *et al.*, 2016). An example for Mediterranean health-relevant species expanding to temperate Europe is the pine processionary moth *Thaumetopoea pityocampa*, whose larvae carry urticating setae (Battisti *et al.*, 2017). In response to climate change since the 1990s, this species has colonized higher altitudes and expanded its range from the Mediterranean northwards, supported by human-mediated introductions far beyond the moth expansion front edge (by pathways such as the trade of mature pine trees transplanted with soil and including moth pupae and often found in recent plantations of pine trees, e.g. along highways, on roundabouts, near urban buildings and in recreation parks) (see Chapter 5, this volume). Natural range expansion is caused by greater nocturnal dispersal of females during unusually warm night temperatures in summer and increased winter survival of larvae during warmer winters (Battisti *et al.*, 2006; Chapter 5, this volume). Thus, moth establishment became possible in large parts of previously inhospitable Western Europe (Battisti *et al.*, 2017).

Further evidence comes from vectors of tropical diseases established in countries with temperate climate (e.g. Rezza *et al.*, 2007), but how far their establishment can be attributed to climate change is often uncertain (Schindler *et al.*, 2015). Where alien ranges were limited by climatic suitability, local establishment of species in adjacent areas should clearly be supported by warmer climates, as shown for the Asian Tiger mosquito *Aedes albopictus* in Northern Italy (Neteler *et al.*, 2011). This insect is an aggressive, daytime biting mosquito that is emerging throughout the world as a public health threat due to its relevance in (among others) West Nile virus, dengue virus and chikungunya virus outbreaks (Bonizzoni *et al.*, 2013). It was introduced with the used tyre and ornamental plant trade from its native range in East Asia and has established on every continent except Antarctica during the last decades, including Europe (1979), the continental USA (1985) and Central and South America (1980–1990s) (Medlock *et al.*, 2012; Bonizzoni *et al.*, 2013). The alien mosquitoes *Aedes atropalpus*, *Ae. japonicus* and *Ae. koreicus* have also established locally in Europe, but their role in disease transmission is less clear and the role of climate change in their establishment was probably negligible (Medlock *et al.*, 2012).

Predictions for human health-relevant alien species

Under warming climate, it is highly likely that range expansion of the pine processionary moth *T. pityocampa* will continue during the next decades (Battisti *et al.*, 2005, 2006, 2017). Medlock *et al.* (2006) predict the establishment of *Aedes albopictus* throughout large parts of lowland UK, with highest risk in the urban centres around London and the south. Climate plays the main role in these predictions, as well as in others for Japan and the USA (Kobayashi *et al.*, 2002), and it is safe to assume that climate warming will ease establishment of the mosquito in many temperate areas (Benedict *et al.*, 2007). Since WNV is present in large parts of Europe, the facilitated establishment of one of its vectors in new areas will increase its public health relevance (Paz and Semenza, 2013).

11.2.3 Climate change effects on spread

Climate change may allow alien species to move to higher altitudes and latitudes after their introduction and establishment (Neteler *et al.*, 2011; Atkinson *et al.*, 2014; Pauchard *et al.*, 2016). Spread of alien insects might also be supported because a more suitable climate can increase fitness, reproduction or dispersal capacities of established species (Stireman *et al.*, 2005; Hellmann *et al.*, 2008; Hulme, 2017).

Evidence for human health-relevant alien species

In the marine environment, spread and increase in abundance of stinging jellyfish species are evident in many seas (Purcell *et al.*, 2007; Brotz *et al.*, 2012). The relative importance of the causative factors vary among species, but human mediation might have played a role along with climate change because the geographic spread of ballast water has been increasing (Seebens *et al.*, 2013). The majority of well-studied jellyfish species of temperate seas are favoured by rising temperatures (Purcell *et al.*, 2007), such as in the case of the Lessepsian migrant *Rhopilema nomadica* (Öztürk and İşinibilir, 2010; Raitsos *et al.*, 2010).

However, it often remains uncertain how far observed increases in range and abundance can be attributed to recent climatic changes (e.g. Aspöck, 2008; Peters *et al.*, 2014). This is the case for *Dirofilaria immitis* and *D. repens*, parasitic nematodes that are currently expanding in many parts of the world and that cause human dirofilariasis that is hosted by dogs and other carnivores and transmitted by mosquitoes (Simón *et al.*, 2009). Although the cause of this spreading is multifactorial, climate warming likely plays a role because vector populations can find more suitable conditions for development and increase their distribution and abundance (Morchón *et al.*, 2012; Fuehrer *et al.*, 2016).

Similarly, climate plays a decisive role in the spread of allergenic common ragweed *Ambrosia artemisiifolia* in Europe (Essl *et al.*, 2009, 2015b; Richter *et al.*, 2013; Chapman *et al.*, 2016). *Ambrosia artemisiifolia* is an Asteraceae native to North America that has colonized temperate regions worldwide since the mid-20th century (Essl *et al.*, 2015b). It is a major weed of agricultural crops and has strongly allergenic pollen that causes severe public health problems in Central and Eastern Europe (Essl *et al.*, 2015b). Further allergenic alien species that are currently rare (e.g. *Ambrosia trifida*, *A. psilostachya*, *Iva xanthiifolia*) are expanding in European countries (Follak *et al.*, 2013). Giant hogweed *Heracleum mantegazzianum* on the other hand, a successful invader in many European countries that poses a health hazard due to its phototoxic sap, prefers humid and cool montane regions in the Czech Republic (Pyšek *et al.*, 1998). It is assumed that the lack of low winter temperatures contributes to the southern distributional limit and its scarcity in parts of south-eastern Europe (Pyšek *et al.*, 1998).

For the Arctic, a review by Dudley *et al.* (2015) highlighted that the indigenous human population of about 400,000 persons is highly susceptible to introduced human pathogens. Further, the speed of climate change in polar regions is particularly high, with ramifications for the establishment of new parasites and pathogens (Dudley *et al.*, 2015). Zoonotic pathogens of circumpolar concern include *Brucella* spp., *Toxoplasma gondii*, *Trichinella* spp., *Clostridium botulinum*, *Francisella tularensis*, *Borrelia burgdorferi*, *Bacillus anthracis*, *Echinococcus* spp., *Leptospira* spp., *Giardia* spp., *Cryptosporida* spp., *Coxiella burnetti*, rabies virus, WNV, hantaviruses and tick-borne encephalitis viruses (Parkinson *et al.*, 2014).

Predictions for human health-relevant alien species

Using climatic niche models, Hales *et al.* (2002) projected for 2085 an increasing risk of dengue fever at a global scale. For Europe, projections of *Ae. albopictus* reveal that a shift from Mediterranean strongholds of distribution to temperate regions will occur due to climate change in the 21st century (Fischer *et al.*, 2011; Caminade *et al.*, 2012). Different models agree on increasing suitability for parts of temperate Europe, but partly disagree about the extent to which the Mediterranean would become unsuitable (Fischer *et al.*, 2011). Further, *Ae. albopictus* will spread northwards and into the valleys of the Alps (Neteler *et al.*, 2011). The distribution of *Ae. aegypti*, the principal vector of dengue virus, is projected to expand in temperate climates where human behaviour and culture increases exposure to this mosquito (Weaver and Reisen, 2010).

Regarding health-relevant hymenopterans, predictions for the yellow-legged hornet *Vespa velutina nigrithorax* show an increase in the climatic suitability for the species in the northern hemisphere until 2100, especially close to the already invaded range in Europe up to southern Sweden and in the USA (Barbet-Massin *et al.*, 2013). Predictions for 2020, 2050 and 2080 by Bertelsmeier *et al.* (2013) show trend reversals of potential future climatically suitable habitat relative to the current range for 6 out of 15 partly health-relevant invasive ant species, including the red imported fire ant *Solenopsis invicta* and the black imported fire ant *S. richteri*.

Also for *Ambrosia artemisiifolia*, spread related to increasing climatic suitability is predicted for Europe (Vogl *et al.*, 2008; Essl *et al.*, 2009) and globally (Essl *et al.*, 2015b), because warmer summers and later autumn frosts will expand the climatically suitable regions as far north as southern Scandinavia and the British Isles (Essl *et al.*, 2015b). Changing climate is also predicted to cause shifts in habitat composition and increased habitat suitability for *Ambrosia artemisiifolia* (Richter *et al.*, 2013).

In the Arctic, climate warming is expected to lead to changes in geographic and altitudinal distributions of host and pathogen populations and exposures of historically native host populations to endemic or exotic pathogens (Dudley *et al.*, 2015). Significant northward expansion due to climate warming is predicted for the alien WNV and its primary reservoir *Culex tarsalis* in the Canadian provinces Manitoba, Saskatchewan and Alberta (Chen *et al.*, 2013). A northward spread driven by climate change has been predicted for the alien plant *Heracleum mantegazzianum* in Alaska by predictive modelling (US Fish and Wildlife Service, 2009).

11.2.4 Climate change effects on human health impacts and their magnitude

Under changing climates, health-relevant alien species are expected to cause more impact, because they become more widespread and abundant (Munson *et al.*, 2008) or they cause stronger impact per capita or per unit biomass, for instance due to increased vector competence, host preference and host exposition (Pfeffer and Dobler, 2009; Weaver and Reisen,

2010; Aspöck and Walochnik, 2014; Randall and van Woesik, 2015). Furthermore, alien species impacts likely coincide with urban heat islands, causing cumulative health impacts in cities, which are known invasion hotspots (La Sorte et al., 2014). However, the opposite might also be the case, i.e. that alien species move with climate change from more developed areas towards less developed ones, where they cause fewer relative impacts (Williams et al., 2010).

Evidence for climate change effects on human health impacts

Evidence that increasing spread and abundance are related to increased health impact has recently been synthesized by Bayliss et al. (2017) and include first reports of transmission of exotic diseases such as chikungunya virus and dengue fever by established alien mosquito species (Aedes albopictus and A. aegypti), harmful blooms of alien unicellular algae such as Ostreopsis spp. and spread and impacts of the poisonous jellyfish Rhopilema nomadica in the Mediterranean.

In Europe, there is clear evidence for the increasing health impact of diseases transmitted by alien vectors (La Ruche et al., 2010; Schmidt-Chanasit et al., 2010; Gjenero-Margan et al., 2011; ECDC, 2012; Morchón et al., 2012). Climate change might have set the stage for the emergence of WNV in Europe and the New World (Weaver and Reisen, 2010). However, it is not conclusively demonstrated that recent climate change has resulted in increased disease incidence at the pan-European level (Semenza and Menne, 2009; Bayliss et al., 2017).

Semenza and Menne (2009) and Paz and Semenza (2013) summarized evidence for climate change effects on WNV impacts, which probably increase under warmer conditions. Evidence from the Camargue (France) shows that the aggressiveness of the vector Culex modestus is positively correlated with rising temperature, humidity coupled with rainfall and sunshine, and that modelled vector aggressiveness was particularly high during a WNV outbreak in the year 2000 (Ludwig et al., 2005). Also in Israel, it was found that a WNV outbreak was preceded by a summer heat wave with minimum temperature being the most important climatic factor, resembling climatic patterns before outbreaks in Romania and New York (Paz, 2006). Paz et al. (2013) analysed more than 900 cases of WNV infection from several European countries and found significant positive correlations between the number of West Nile Fever cases and temperature and weak correlations with relative humidity. For its alien range, El Adlouni et al. (2007) assessed mild winters and hot summers as main climatic characteristics of the WNV outbreak in eastern North America during 2002. Because transmission in mosquitoes increases non-linearly with incubation temperature, minor temperature increases may have strong impacts (Kilpatrick et al., 2008). Paz and Semenza (2013) present in their review evidence that ambient temperature: (i) plays an important role in viral replication rates and transmission of WNV; (ii) affects the length of extrinsic incubation, seasonal phenology of mosquito host populations and geographical variation in human case incidence; (iii) increases growth rates of vector populations; and (iv) decreases the interval between blood meals and accelerates the rate of virus evolution. Most of the WNV cases occurred in dense cities in central Israel. It is therefore assumed that the urban heat island could cause increased impact of WNV, because it enhances vector competency, replication, abundance and biting capabilities, and weakens elderly people, the main cohort affected by West Nile Fever (Paz, 2006).

Effects of climate change on the health impact of alien species include further evidence on rodents in northern latitudes (Dudley et al., 2015). Warm winters coupled with wet springs increase adult survival, forage availability and juvenile survival, and drive the increase in the magnitude of zoonotic disease outbreaks such as human plague in the western USA (Ben Ari et al., 2008) and Puumala hantavirus in northern Sweden (Pettersson et al., 2008). A strong Puumala hantavirus epidemic in northern Sweden occurred during a particularly warm winter and was related to increased human exposure to infected voles because of the absence of snow cover (Evander and Ahlm, 2009).

Predictions for climate change effects on human health impacts

Climate change has the potential to enhance pathogen transmission in temperate and polar climates by extending transmission seasons, increasing over-winter survival, increasing host–vector contact by shortening vector reproduction cycles, and shortening extrinsic incubation times (Jenkins *et al.*, 2006, 2013; Weaver and Reisen, 2010; Paz and Semenza, 2013). The ability of some arboviruses to adopt anthroponotic or enzootic transmission cycles in urban environments involving highly efficient and anthropophilic vectors such as *Ae. aegypti*, *Ae. albopictus* and *Culex* spp. populations is of particular concern (Weaver and Reisen, 2010).

Predictions by Thomas *et al.* (2011) for changes in the impact of dengue fever for Europe, based on the IPCC emission scenarios B1 and A1B, showed that the southwest of the Iberian peninsula and Sicily are exposed to high risks until 2040, the southwest of Europe and parts of Greece and Turkey until 2070 and considerable parts of Italy until 2100. Applying weaker assumptions for developmental constraints of dengue virus in *Ae. aegypti*, more regions of Europe, including parts of central Europe, would offer climatically suitable conditions until the late 21st century (Thomas *et al.*, 2011). Jetten and Focks (1997) projected an increase in the dengue transmission season in temperate locations under climate change, alongside increases in the latitudinal and altitudinal range of dengue. Hales *et al.* (2002) showed that almost 5–6 billion people would be at risk of dengue by 2085, instead of 3.5 billion in a scenario without climate change.

It is also assumed that European citizens will be exposed more to the allergenic pollen of *Ambrosia artemisiifolia* as a consequence of climate change (Leiblein and Lösch, 2011; Chapman *et al.*, 2014; Essl *et al.*, 2015b). Hamaoui-Laguel *et al.* (2015) demonstrated that future ragweed pollen loads might increase substantially due to climate change effects on pollen production, release and transport. These climate change effects are responsible for two-thirds of projected future increases of pollen loads. These increases can be attributed primarily to rising pollen production due to enhanced photosynthesis at higher CO_2 levels, and also to an extension of the growing season and pollen production under warmer climates. In addition, allergenicity of ragweed pollen will likely increase under elevated CO_2 and drought stress (El Kelish *et al.*, 2014).

11.3 Discussion

Climate change facilitates introductions of alien species impacting human health in many ways, but its role is moderate compared with the various aspects of globalization such as people's increased mobility and international trade (Becker, 2008). In Europe, there is evidence for natural introductions due to migrating birds, while other emerging pathogens and vectors have likely been present but undiscovered during substantial periods before they expanded (Aspöck and Walochnik, 2014). There is strong evidence that changes in climatic parameters, particularly in temperature, are of crucial importance for the establishment and spread of human health-relevant alien species (Smith *et al.*, 2012; Paz and Semenza, 2013; Kaffenberger *et al.*, 2017; Rabitsch *et al.*, 2017). The most relevant alien species causing human health impacts in Europe are arthropod vectors and plants with allergenic pollen (Schindler *et al.*, 2015; Bayliss *et al.*, 2017). In the Arctic and high mountains, climate change-driven invasions of health-relevant species will likely accelerate particularly quickly in the future (Dudley *et al.*, 2015).

Certainly, climate change will shift the distribution of health-relevant native species, too (e.g. Smith *et al.*, 2012; Aspöck and Walochnik, 2014; König *et al.*, 2014). In Europe, it is predicted that *Phlebotomus* sandflies, the vectors of leishmaniasis, will increase their distribution range northwards (Dujardin *et al.*, 2008; Stark *et al.*, 2009; Ready, 2010; Dobler and

Aspöck, 2010) and that tick-borne diseases will increase in altitude and latitude (Holzmann et al., 2009; Jaenson and Lindgren, 2011; Aspöck and Walochnik, 2014; see also Kaffenberger et al., 2017 for North America). European vectors of malaria might also benefit from climate and land-use changes (Holy et al., 2011; Boccolini et al., 2012).

Uncertainties of the climate change effects on introduction, establishment, spread and impacts of alien species are plentiful because pathogens, vectors and reservoir hosts may interact in many different native/alien combinations (Rabitsch et al., 2017). These are related to: (i) the general difficulties of predicting invasions (Leung et al., 2004; Seebens et al., 2013); (ii) the inherent ecological complexities of multi-species interactions (Lafferty, 2009; Hatcher et al., 2012; Smith et al., 2012); (iii) the difficulties of disentangling climate change impacts from other factors (La Ruche et al., 2010); (iv) climatic niche shifts during the invasion process (Medley, 2010); (v) uncertainty of future human actions (Fischer et al., 2011; Thomas et al., 2011; Smith et al., 2012); (vi) climate change affecting relevant species in a non-linear fashion (Lafferty, 2009); (vii) the underestimation of the relevance of stochastic climatic extreme events (Parmesan et al., 2000; Battisti et al., 2006; Diez et al., 2012); (viii) the difficulties of disentangling human-mediated movements and natural migration processes (Walther et al., 2009); and (ix) lack of data or limitations in data quality (Lobo et al., 2010; Smith et al., 2012).

11.4 Recommendations

Due to phenomena such as invasion debt (Essl et al., 2011), increasing current invasion trends (Seebens et al., 2017) and increase of the species pool reached by vectors (Seebens et al., 2018), the number of introduced species will increase further and preventive management needs to be strengthened. However, the uncertainties of climate change impacts pose challenges for decision making for alien species management and nature conservation (Battisti et al., 2006). In particular, a better understanding of drivers and management options of alien diseases and vectors is required as well as joint efforts in education and outreach to the public and decision makers (Rabitsch et al., 2017). Improved understanding of the ecology of pathogens and vectors, the reservoirs and transmission cycles, and the role of human activities under different climates is required to predict effects of climate change on infectious diseases (Hulme, 2003; Lafferty, 2009; Kilpatrick, 2011; Ostfeld and Brunner, 2015).

There is a need for action and research, particularly in the fields of pathway management, epidemiology, modelling and vector monitoring and control (Becker, 2008; Dujardin et al., 2008; Stark et al., 2009; Grandadam et al., 2011; Essl et al., 2015a; Medlock and Leach, 2015). Semenza and Menne (2009) report 45 specific management and policy measures to reduce infectious diseases under climate change. These measures include monitoring, outbreak investigation, information, education and empowering, interagency and community partnerships, policy development, law enforcement, health care access, workforce competence, evaluation and research. We additionally recommend enforcing multidisciplinary efforts in research and implementation to link environmental changes with environmental, animal and human health, as promoted by the 'One Health' initiative (Ready, 2010; Rabinowitz and Conti, 2013; Dudley et al., 2015; Rabitsch et al., 2017; Roy et al., 2017).

Acknowledgements

We acknowledge helpful comments by Wolfgang Nentwig on a previous draft of the manuscript and funding by the Austrian Climate and Energy Fund within the framework of the Austrian Climate Research Program (ACRP; project number KR13AC6K11141).

References

Alsos, I.G., Ware, C. and Elven, R. (2015) Past Arctic aliens have passed away, current ones may stay. *Biological Invasions* 17, 3113–3123.

Aspöck, H. (2008) Durch Arthropoden übertragene Erreger von Infektionen des Menschen in Mitteleuropa – ein Update. *Mitteilungen der Deutschen Gesellschaft zur allgemeinen angewandten Entomologie* 16, 371–392.

Aspöck, H. and Walochnik, J. (2014) Durch blutsaugende Insekten und Zecken übertragene Krankheitserreger des Menschen in Mitteleuropa aus der Sicht von Klimawandel und Globalisierung. *Gredleriana* 14, 61–98.

Atkinson, C.T., Utzurrum, R.B., Lapointe, D.A., Camp, R.J., Crampton, L.H., Foster, J.T. and Giambelluca, T.W. (2014) Changing climate and the altitudinal range of avian malaria in the Hawaiian Islands – an ongoing conservation crisis on the island of Kaua'i. *Global Change Biology* 20, 2426–2436.

Barbet-Massin, M., Rome, Q., Muller, F., Perrard, A., Villemant, C. and Jiguet, F. (2013) Climate change increases the risk of invasion by the Yellow-legged hornet. *Biological Conservation* 157, 4–10.

Barney, J.N. (2014) Bioenergy and invasive plants: quantifying and mitigating future risks. *Invasive Plant Science and Management* 7, 199–209.

Battisti, A., Stastny, M., Netherer, S., Robinet, C., Schopf, A., Roques, A. and Larsson, S. (2005) Expansion of geographic range in the pine processionary moth caused by increased winter temperatures. *Ecological Applications* 15, 2084–2096.

Battisti, A., Stastny, M., Buffo, E. and Larsson, S. (2006) A rapid altitudinal range expansion in the pine processionary moth produced by the 2003 climatic anomaly. *Global Change Biology* 12, 662–671.

Battisti, A., Larsson, S. and Roques, A. (2017) Processionary moths and associated urtication risk: global-change driven effects. *Annual Review of Entomology* 62, 323–342.

Bayliss, H., Schindler, S., Essl, F., Rabitsch, W. and Pullin, A.S. (2015) What evidence exists for changes in the occurrence, frequency or severity of human health impacts resulting from exposure to alien invasive species in Europe? A systematic map protocol. *Environmental Evidence* 4, 10.

Bayliss, H., Schindler, S., Adam, M., Essl, F., Rabitsch, W. and Pullin, A.S. (2017) Evidence for changes in the occurrence, frequency or severity of human health impacts resulting from exposure to species alien to Europe: a systematic map. *Environmental Evidence* 6, 21.

Bebber, D.P. (2015) Range-expanding pests and pathogens in a warming world. *Annual Review of Phytopathology* 53, 335–356.

Becker, N. (2008) Influence of climate change on mosquito development and mosquito-borne diseases in Europe. *Parasitology Research* 103, 19–28.

Bellard, C., Thuiller, W., Leroy, B., Genovesi, P., Bakkenes, M. and Courchamp, F. (2013) Will climate change promote future invasions? *Global Change Biology* 19, 3740–3748.

Ben Ari, T., Gershunov, A., Gage, K.L., Snäll, T., Ettestad, P., Kausrud, K.L. and Stenseth, N.C. (2008) Human plague in the USA: the importance of regional and local climate. *Biology Letters* 4, 737–740.

Benedict, M.Q., Levine, R.S., Hawley, W.A. and Lounibos, L.P. (2007) Spread of the tiger: global risk of invasion by the mosquito *Aedes albopictus*. *Vector-borne and Zoonotic Diseases* 7, 76–85.

Bertelsmeier, C., Luque, G.M. and Courchamp, F. (2013) The impact of climate change changes over time. *Biological Conservation* 167, 107–115.

Blackburn, T.M., Pyšek, P., Bacher, S., Carlton, J.T., Duncan, R.P., Jarošík, V., Wilson, J.R.U. and Richardson, D.M. (2011) A proposed unified framework for biological invasions. *Trends in Ecology and Evolution* 26, 333–339.

Boccolini, D., Toma, L., Di Luca, M., Severini, F., Cocchi, M., Bella, A., Massa, A., Mancini Barbieri, F., Bongiorno, G., Angeli, L. *et al.* (2012) Impact of environmental changes and human-related factors on the potential malaria vector, *Anopheles labranchiae* (Diptera: Culicidae), in Maremma, Central Italy. *Journal of Medical Entomology* 49, 833–842.

Bonizzoni, M., Gasperi, G., Chen, X. and James, A.A. (2013) The invasive mosquito species *Aedes albopictus*: current knowledge and future perspectives. *Trends in Parasitology* 29, 460–468.

Brotz, L., Cheung, W.W., Kleisner, K., Pakhomov, E. and Pauly, D. (2012) Increasing jellyfish populations: trends in large marine ecosystems. *Hydrobiologia* 690, 3–20.

Caminade, C., Medlock, J.M., Ducheyne, E., McIntyre, K.M., Leach, S., Baylis, M. and Morse, A.P. (2012) Suitability of European climate for the Asian tiger mosquito *Aedes albopictus*: recent trends and future scenarios. *Journal of the Royal Society Interface* DOI: 10.1098/rsif.2012.0138.

Campbell, G.L., Marfin, A.A., Lanciotti, R.S. and Gubler, D.J. (2002) West nile virus. *The Lancet Infectious Diseases* 2, 519–529.

Chan, F.T., Bailey, S.A., Wiley, C.J. and MacIsaac, H.J. (2013) Relative risk assessment for ballast-mediated invasions at Canadian Arctic ports. *Biological Invasions* 15, 295–308.

Chapman, D.S., Haynes, D., Beal, S., Essl, F. and Bullock, J. (2014) Phenology predicts the native and invasive range limits of common ragweed. *Global Change Biology* 20, 192–202.

Chapman, D.S., Makra, L., Albertini, R., Bonini, M., Páldy, A., Rodinkova, V., Šikoparija, B., Weryszko-Chmielewska, E. and Bullock, J.M. (2016) Modelling the introduction and spread of non-native species: international trade and climate change drive ragweed invasion. *Global Change Biology* 22, 3067–3079.

Chen, C.-C., Jenkins, E., Epp, T., Waldner, C., Curry, P.S. and Soos, C. (2013) Climate change and West Nile virus in a highly endemic region of North America. *International Journal of Environmental Research and Public Health* 10, 3052–3071.

Chimera, C.G., Buddenhagen, C.E. and Clifford, P.M. (2010) Biofuels: the risks and dangers of introducing invasive species. *Biofuels* 1, 785–796.

Diez, J.M., D'Antonio, C.M., Dukes, J.S., Grosholz, E.D., Olden, J.D., Sorte, C.J., Blumenthal, D.M., Bradley, B.A., Early, R., Ibáñez, I. *et al.* (2012) Will extreme climatic events facilitate biological invasions? *Frontiers in Ecology and the Environment* 10, 249–257.

Dobler, G. and Aspöck, H. (2010) Durch Sandmücken und Gnitzen übertragene Arboviren als Erreger von Infektionen des Menschen. *Denisia* 30, 555–563.

Dudley, J.P., Hoberg, E.P., Jenkins, E.J. and Parkinson, A.J. (2015) Climate change in the North American Arctic: a one health perspective. *EcoHealth* 12, 713.

Dujardin, J.-C., Campino, L., Cañavate, C., Dedet, J.-P., Gradoni, L., Soteriadou, K., Mazeris, A., Ozbel, Y. and Boelaert, M. (2008) Spread of vector-borne diseases and neglect of Leishmaniasis, Europe. *Emerging Infectious Diseases* 14, 1013–1018.

ECDC (2012) *Annual Epidemiological Report. Reporting on 2010 Surveillance Data and 2011 Epidemic Intelligence Data.* European Centre for Disease Prevention and Control, Stockholm.

El Adlouni, S., Beaulieu, C., Ouarda, T.B., Gosselin, P.L. and Saint-Hilaire, A. (2007) Effects of climate on West Nile Virus transmission risk used for public health decision-making in Quebec. *International Journal of Health Geographics* 6, 1.

El Kelish, A., Zhao, F., Heller, W., Durner, J., Winkler, J.B., Behrendt, H., Traidl-Hoffmann, C., Horres, R., Pfeifer, M., Frank, U. and Ernst, D. (2014) Ragweed (*Ambrosia artemisiifolia*) pollen allergenicity: SuperSAGE transcriptomic analysis upon elevated CO_2 and drought stress. *BMC Plant Biology* 14, 1.

Embrey, S., Remais, J.V. and Hess, J. (2012) Climate change and ecosystem disruption: the health impacts of the North American Rocky Mountain pine beetle infestation. *American Journal of Public Health* 102, 818–827.

Essl, F., Dullinger, S. and Kleinbauer, I. (2009) Changes in the spatio-temporal patterns and habitat preferences of *Ambrosia artemisiifolia* during its invasion of Austria. *Preslia* 81, 119–133.

Essl, F., Dullinger, S., Rabitsch, W., Hulme, P.E., Hülber, K., Jarošík, V., Kleinbauer, I., Krausmann, F., Kühn, I., Nentwig, W. *et al.* (2011) Socioeconomic legacy yields an invasion debt. *Proceedings of the National Academy of Sciences USA* 108, 203–207.

Essl, F., Bacher, S., Blackburn, T.M., Booy, O., Brundu, G., Brunel, S., Cardoso, A.-C., Eschen, R., Gallardo, B., Galil, B. *et al.* (2015a) Crossing frontiers in tackling pathways of biological invasions. *Bioscience* biv082.

Essl, F., Bíró, K., Brandes, D., Broennimann, O., Bullock, J.M., Chapman, D.S., Chauvel, B., Dullinger, S., Fumanal, B., Guisan, A. *et al.* (2015b) Biological flora of the British Isles: *Ambrosia artemisiifolia*. *Journal of Ecology* 103, 1069–1098.

Evander, M. and Ahlm, C. (2009) Milder winters in northern Scandinavia may contribute to larger outbreaks of haemorrhagic fever virus. *Global Health Action* doi: 10.3402/gha.v2i0.2020.

Felton, A., Gustafsson, L., Roberge, J.-M., Ranius, T., Hjältén, J., Rudolphi, J., Lindbladh, M., Weslien, J., Rist, L., Brunet, J. and Felton, A.M. (2016) How climate change adaptation and mitigation strategies can threaten or enhance the biodiversity of production forests: insights from Sweden. *Biological Conservation* 194, 11–20.

Fischer, D., Thomas, S.M., Niemitz, F., Reineking, B. and Beierkuhnlein, C. (2011) Projection of climatic suitability for *Aedes albopictus* Skuse (Culicidae) in Europe under climate change conditions. *Global and Planetary Change* 78, 54–64.

Follak, S., Dullinger, S., Kleinbauer, S., Moser, D. and Essl, F. (2013) Invasion dynamics of three allergenic invasive Asteraceae (*Ambrosia trifida, Artemisia annua, Iva xanthiifolia*) in central and eastern Europe. *Preslia* 85, 41–61.

Fuehrer, H.-P., Auer, H., Leschnik, M., Silbermayr, K., Duscher, G. and Joachim, A. (2016) *Dirofilaria* in humans, dogs, and vectors in Austria (1978–2014). From imported pathogens to the endemicity of *Dirofilaria repens. PLoS Neglected Tropical Diseases* 10: e0004547.

Gale, P., Brouwer, A., Ramnial, V., Kelly, L., Kosmider, R., Fooks, A.R. and Snary, E.L. (2010) Assessing the impact of climate change on vector-borne viruses in the EU through the elicitation of expert opinion. *Epidemiology and Infection* 138, 214–225.

Gallardo, B. and Aldridge, D.C. (2013) Evaluating the combined threat of climate change and biological invasions on endangered species. *Biological Conservation* 160, 225–233.

Giladi, M., Metzkor-Cotter, E., Martin, D.A., Siegman-Igra, Y., Korczyn, A.D., Rosso, R., Berger, S.A., Campbell, G.L. and Lanciotti, R.S. (2001) West Nile encephalitis in Israel, 1999: the New York connection. *Emerging Infectious Diseases* 7, 659–661.

Gjenero-Margan, I., Aleraj, B., Krajcar, D., Lesnikar, V., Klobučar, A. and Pem-Novosel, I. (2011) Autochthonous dengue fever in Croatia, August–September 2010. *EuroSurveillance* 16, pii=19805.

Grandadam, M., Caro, V., Plumet, S., Thiberge, J.M., Souarès, Y. and Failloux, A.B. (2011) Chikungunya virus, southeastern France. *Emerging Infectious Diseases* 17, 910–914.

Hales, S., De Wet, N., Maindonald, J. and Woodward, A. (2002) Potential effect of population and climate changes on global distribution of dengue fever: an empirical model. *The Lancet* 360, 830–834.

Hamaoui-Laguel, L., Vautard, R., Liu, L., Solmon, F., Viovy, N., Khvorostyanov, D., Essl, F., Chuine, I., Colette, A. and Semenov, M.A. (2015) Effects of climate change and seed dispersal on airborne ragweed pollen loads in Europe. *Nature Climate Change* 5, 766–771.

Harter, D.E., Irl, S.D., Seo, B., Steinbauer, M.J., Gillespie, R., Triantis, K.A., Fernández-Palacios, J.-M. and Beierkuhnlein, C. (2015) Impacts of global climate change on the floras of oceanic islands. Projections, implications and current knowledge. *Perspectives in Plant Ecology Evolution and Systematics* 17, 160–183.

Hatcher, M.J., Dick, J.T. and Dunn, A.M. (2012) Diverse effects of parasites in ecosystems: linking interdependent processes. *Frontiers in Ecology and the Environment* 10, 186–194.

Hellmann, J.J., Byers, J.E., Bierwagen, B.G. and Dukes, J.S. (2008) Five potential consequences of climate change for invasive species. *Conservation Biology* 22, 534–543.

Holy, M., Schmidt, G. and Schröder, W. (2011) Potential malaria outbreak in Germany due to climate warming: risk modelling based on temperature measurements and regional climate models. *Environmental Science and Pollution Research* 18, 428–435.

Holzmann, H., Aberle, S.W., Stiasny, K., Werner, P., Mischak, A., Zainer, B., Netzer, M., Koppi, S., Bechter, E. and Heinz, F.X. (2009) Tick-borne encephalitis from eating goat cheese in a mountain region of Austria. *Emerging Infectious Diseases* 15, 1671–1673.

Hulme, P.E. (2003) Biological invasions: winning the science battles but losing the conservation war? *Oryx* 37, 178–193.

Hulme, P.E. (2014) Invasive species challenge the global response to emerging diseases. *Trends in Parasitology* 30, 267–270.

Hulme, P.E. (2015) Invasion pathways at a crossroad: policy and research challenges for managing alien species introductions. *Journal of Applied Ecology* 52, 1418–1424.

Hulme, P.E. (2017) Climate change and biological invasions: evidence, expectations, and response options. *Biological Reviews* 92, 1297–1313.

IPCC (2014) Climate Change 2014: Synthesis Report. In: Core Writing Team, Pachauri, R.K. and Meyer, L.A. (eds) *Contribution of Working Groups I, II and III to the Fifth Assessment Report of the Intergovernmental Panel on Climate Change*. IPCC, Geneva, Switzerland.

Jaenson, T.G. and Lindgren, E. (2011) The range of *Ixodes ricinus* and the risk of contracting Lyme borreliosis will increase northwards when the vegetation period becomes longer. *Ticks and Tick-borne Diseases* 2, 44–49.

Jenkins, E.J., Veitch, A.M., Kutz, S.J., Hoberg, E.P. and Polley, L. (2006) Climate change and the epidemiology of protostrongylid nematodes in northern ecosystems: *Parelaphostrongylus odocoilei* and *Protostrongylus stilesi* in Dall's sheep (*Ovis d. dalli*). *Parasitology* 132, 387–401.

Jenkins, E.J., Castrodale, L.J., de Rosemond, S.J.C., Dixon, B.R., Elmore, S.A., Gesy, K.M., Hoberg, E.P., Polley, L., Schurer, J.M., Simard, M. and Thompson, R.C.A. (2013) Tradition and transition: parasitic zoonoses of people and animals in Alaska, Northern Canada, and Greenland. In: Rollinson, D. (ed.) *Advances in Parasitology*. Academic Press, New York, pp. 33–204.

Jeschke, J.M., Keesing, F. and Ostfeld, R.S. (2013) Novel organisms: comparing invasive species, GMOs, and emerging pathogens. *Ambio* 42, 541–548.

Jetten, T.H. and Focks, D.A. (1997) Potential changes in the distribution of dengue transmission under climate warming. *American Journal of Tropical Medicine and Hygiene* 57, 285–297.

Kaffenberger, B.H., Shetlar, D., Norton, S. and Rosenbach, M. (2017) The effect of climate change on skin disease in North America. *Journal of the American Academy of Dermatology* 76, 140–147.

Kilpatrick, A.M. (2011) Globalization, land use, and the invasion of West Nile Virus. *Science* 334, 323–327.

Kilpatrick, A.M., Meola, M.A., Moudy, R.M. and Kramer, L.D. (2008) Temperature, viral genetics, and the transmission of West Nile Virus by *Culex pipiens* mosquitoes. *PLoS Pathogens* 4, e1000092.

Kletou, D., Hall-Spencer, J.M. and Kleitou, P. (2016) A lionfish (*Pterois miles*) invasion has begun in the Mediterranean Sea. *Marine Biodiversity Records* 9, 46.

Kobayashi, M., Nihei, N. and Kurihara, T. (2002) Analysis of northern distribution of *Aedes albopictus* (Diptera: Culicidae) in Japan by geographical information system. *Journal of Medical Entomology* 39, 4–11.

König, M., Loibl, W., Steiger, R., Aspöck, H., Bednar-Friedl, B., Brunner, K.-M., Haas, W., Höferl, K.-M., Huttenlau, M., Walochnik, J. and Weisz, U. (2014) Der Einfluss des Klimawandels auf die Anthroposphäre. In: Kromp-Kolb, H., Nakicenovic, N., Steininger, K., Gobiet, A., Formayer, H., Köppl, A., Prettenthaler, F., Stötter, J. and Schneider, J. (eds) *Austrian Assessment Report 2014*. Austrian Panel on Climate Change, Verlag der Österreichischen Akademie der Wissenschaften, Vienna, pp. 641–704.

La Ruche, G., Souarès, Y., Armengaud, A., Peloux-Petiot, F., Delaunay, P., Desprès, P., Lenglet, A., Jourdain, F., Leparc-Goffart, I., Charlet, F. *et al.* (2010) First two autochthonous dengue virus infections in metropolitan France, September 2010. *EuroSurveillance* 15, 19676.

La Sorte, F.A., Aronson, M.F.J., Williams, N.S.G., Clarkson, B., Celesti-Grapow, L., Cilliers, S., Dolan, R.W., Hipp, A., Klotz, S., Kühn, I. *et al.* (2014) Beta diversity of urban floras within and among European and non-European cities. *Global Ecology and Biogeography* 23, 769–779.

Lafferty, K.D. (2009) The ecology of climate change and infectious diseases. *Ecology* 90, 888–900.

Lanciotti, R.S., Roehrig, J.T., Deubel, V., Smith, J., Parker, M., Steele, K., Crise, B., Volpe, K.E., Crabtree, M.B., Scherret, J.H. *et al.* (1999) Origin of the West Nile virus responsible for an outbreak of encephalitis in the northeastern United States. *Science* 286, 2333–2337.

Leiblein, M.C. and Lösch, R. (2011) Biomass development and CO_2 gas exchange of *Ambrosia artemisiifolia* L. under different soil moisture conditions. *Flora* 206, 511–516.

Leung, B., Drake, J.M. and Lodge, D.M. (2004) Predicting invasions: propagule pressure and the gravity of Allee effects. *Ecology* 85, 1651–1660.

Lobo, J.M., Jiménez-Valverde, A. and Hortal, J. (2010) The uncertain nature of absences and their importance in species distribution modelling. *Ecography* 33, 103–114.

Ludwig, A., Bicout, D., Chalvet-Monfray, K. and Sabatier, P. (2005) Modelling the aggressiveness of *Culex modestus*, possible vector of West Nile fever in Camargue, as a function of meteorological data. *Environnement, Risques & Santé* 4, 109–113.

Mazza, G., Tricarico, E., Genovesi, P. and Gherardi, F. (2014) Biological invaders are threats to human health: an overview. *Ethology Ecology & Evolution* 26, 112–129.

Medley, K.A. (2010) Niche shifts during the global invasion of the Asian tiger mosquito, *Aedes albopictus* Skuse (Culicidae), revealed by reciprocal distribution models. *Global Ecology and Biogeography* 19, 122–133.

Medlock, J.M. and Leach, S.A. (2015) Effect of climate change on vector-borne disease risk in the UK. *Lancet Infectious Diseases* 15, 721–730.

Medlock, J.M., Avenell, D., Barrass, I. and Leach, S. (2006) Analysis of the potential for survival and seasonal activity of *Aedes albopictus* (Diptera: Culicidae) in the United Kingdom. *Journal of Vector Ecology* 31, 292–304.

Medlock, J.M., Hansford, K.M., Schaffner, F., Versteirt, V., Hendrickx, G., Zeller, H. and Bortel, W.V. (2012) A review of the invasive mosquitoes in Europe: ecology, public health risks, and control options. *Vector-borne and Zoonotic Diseases* 12, 435–447.

Morchón, R., Carretón, E., González-Miguel, J. and Mellado-Hernández, I. (2012) Heartworm disease (*Dirofilaria immitis*) and their vectors in Europe: new distribution trends. *Frontiers in Physiology* 3, 75–85.

Munson, L., Terio, K.A., Kock, R., Mlengeya, T., Roelke, M.E., Dubovi, E., Summers, B., Sinclair, A.R.E. and Packer, C. (2008) Climate extremes promote fatal co-infections during canine distemper epidemics in African lions. *PLoS One* 3, e2545.

Neteler, M., Roiz, D., Rocchini, D., Castellani, C. and Rizzoli, A. (2011) Terra and Aqua satellites track tiger mosquito invasion: modelling the potential distribution of *Aedes albopictus* in north-eastern Italy. *International Journal of Health Geographics* 10, 1.

Ostfeld, R.S. and Brunner, J.L. (2015) Climate change and Ixodes tick-borne diseases of humans. *Philosophical Transactions of the Royal Society B* 370, 20140051.

Öztürk, B. and İşinibilir, M. (2010) An alien jellyfish *Rhopilema nomadica* and its impacts to the Eastern Mediterranean part of Turkey. *Journal of the Black Sea/Mediterranean Environment* 16, 149–156.

Parkinson, A.J., Evengard, B., Semenza, J.C., Ogden, N., Børresen, M.L., Berner, J., Brubaker, M., Sjöstedt, A., Evander, M., Hondula, D.M. *et al.* (2014) Climate change and infectious diseases in the Arctic: establishment of a circumpolar working group. *International Journal of Circumpolar Health* 73, 25163.

Parmesan, C., Root, T.L. and Willig, M.R. (2000) Impacts of extreme weather and climate on terrestrial biota. *Bulletin of the American Meteorological Society* 81, 443–450.

Pauchard, A., Milbau, A., Albihn, A., Alexander, J., Burgess, T., Daehler, C., Englund, G., Essl, F., Evengård, B., Greenwood, G.B. *et al.* (2016) Non-native and native organisms moving into high elevation and high latitude ecosystems in an era of climate change: new challenges for ecology and conservation. *Biological Invasions* 18, 345–353.

Paz, S. (2006) The West Nile Virus outbreak in Israel (2000) from a new perspective: the regional impact of climate change. *International Journal of Environmental Health Research* 16, 1–13.

Paz, S. and Semenza, J.C. (2013) Environmental drivers of West Nile fever epidemiology in Europe and Western Asia: a review. *International Journal of Environmental Research and Public Health* 10, 3543–3562.

Paz, S., Malkinson, D., Green, M.S., Tsioni, G., Papa, A., Danis, K., Sirbu, A., Ceianu, C., Krisztalovics, K., Ferenczi, E. *et al.* (2013) Permissive summer temperatures of the 2010 European West Nile fever upsurge. *PLoS One* 8, e56398.

Peters, K., Breitsameter, L. and Gerowitt, B. (2014) Impact of climate change on weeds in agriculture: a review. *Agronomy for Sustainable Development* 34, 707–721.

Pettersson, L., Boman, J., Juto, P., Evander, M. and Ahlm, C. (2008) Outbreak of Puumala virus infection, Sweden. *Emerging Infectious Diseases* 14, 808–810.

Pfeffer, M. and Dobler, G. (2009) What comes after bluetongue? Europe as target for exotic arboviruses. *Berliner und MünchenerTierärztliche Wochenschrift* 122, 458–466.

Poloczanska, E.S., Brown, C.J., Sydeman, W.J., Kiessling, W., Schoeman, D.S., Moore, P.J., Brander, K., Bruno, J.F., Buckley, L.B., Burrows, M.T. *et al.* (2013) Global imprint of climate change on marine life. *Nature Climate Change* 3, 919–925.

Purcell, J.E., Uye, S.I. and Lo, W.T. (2007) Anthropogenic causes of jellyfish blooms and their direct consequences for humans: a review. *Marine Ecology Progress Series* 350, 153–174.

Pyšek, P. and Richardson, D.M. (2010) Invasive species, environmental change and management, and health. *Annual Review of Environment and Resources* 35, 25–55.

Pyšek, P., Kopecky, M., Jarosik, V. and Kotkova, P. (1998) The role of human density and climate in the spread of *Heracleum mantegazzianum* in the Central European landscape. *Diversity and Distributions* 4, 9–16.

Rabinowitz, P. and Conti, L. (2013) Links among human health, animal health, and ecosystem health. *Annual Review of Public Health* 34, 189–204.

Rabitsch, W., Essl, F. and Schindler, S. (2017) The rise of non-native vectors and reservoirs of human diseases. In: Vilà, M. and Hulme, P.E. (eds) *Impact of Biological Invasions on Ecosystem Services.* Springer, Berlin, pp. 263–275.

Raghu, S., Anderson, R.C., Daehler, C.C., Davis, A.S., Wiedenmann, R.N., Simberloff, D. and Mack, R.N. (2006) Adding biofuels to the invasive species fire? *Science* 313, 1742.

Rahel, F.J. and Olden, J.D. (2008) Assessing the effects of climate change on aquatic invasive species. *Conservation Biology* 22, 531–533.

Raitsos, D.E., Beaugrand, G., Georgopoulos, D., Zenetos, A., Pancucci-Papadopoulou, A.M., Theocharis, A. and Papathanassiou, E. (2010) Global climate change amplifies the entry of tropical species into the Eastern Mediterranean Sea. *Limnology and Oceanography* 55, 1478–1484.

Randall, C.J. and van Woesik, R. (2015) Contemporary white-band disease in Caribbean corals driven by climate change. *Nature Climate Change* 5, 375–379.

Ready, P.D. (2010) Leishmaniasis emergence in Europe. *EuroSurveillance* 15, 19505.

Rezza, G., Nicoletti, L., Angelini, R., Romi, R., Finarelli, A.C., Panning, M., Cordioli, P., Fortuna, C., Boros, S., Magurano, F. *et al.* (2007) Infection with chikungunya virus in Italy: an outbreak in a temperate region. *The Lancet* 370, 1840–1846.

Richardson, D.M., Pyšek, P., Rejmánek, M., Barbour, M.G., Panetta, F.D. and West, C.J. (2000) Naturalization and invasion of alien plants: concepts and definitions. *Diversity and Distributions* 6, 93–107.

Richter, R., Berger, U.E., Dullinger, S., Essl, F., Leitner, M., Smith, M. and Vogl, G. (2013) Spread of invasive ragweed: climate change, management and how to reduce allergy costs. *Journal of Applied Ecology* 50, 1422–1430.

Rizzoli, A., Jiménez-Clavero, M.A., Barzon, L., Cordioli, P., Figuerola, J., Koraka, P., Martina, B., Moreno, A., Nowotny, N., Pardigon, N. *et al.* (2015) The challenge of West Nile virus in Europe: knowledge gaps and research priorities. *EuroSurveillance* 20, pii=21135.

Roques, A., Auger-Rozenberg, M.-A., Blackburn, T.M., Garnas, J., Pyšek, P., Rabitsch, W., Richardson, D.M., Wingfield, M.J., Liebhold, A.M. and Duncan, R.P. (2016) Temporal and interspecific variation in rates of spread for insect species invading Europe during the last 200 years. *Biological Invasions* 18, 907–920.

Roy, H.E., Beckmann, B.C., Comont, R.F., Hails, R.S., Harrington, R., Medlock, J., Purse, B. and Shortall, C.R. (2009) *An Investigation into the Potential for New and Existing Species of Insect with the Potential to Cause Statutory Nuisance to Occur in the UK as a Result of Current and Predicted Climate Change.* DEFRA, Centre for Ecology and Hydrology, Wallingford, UK.

Roy, H.E., Hesketh, H., Purse, B.V., Eilenberg, J., Santini, A., Scalera, R., Stentiford, G.D., Adriaens, T., Amtoft Wynns, A., Bacela-Spychalska, K. *et al.* (2017) Alien pathogens on the horizon: opportunities for predicting their threat to wildlife. *Conservation Letters* 10, 477–484.

Scheffers, B.R., De Meester, L., Bridge, T.C., Hoffmann, A.A., Pandolfi, J.M., Corlett, R.T., Butchart, S.H.M., Pearce-Kelly, P., Kovacs, K.M., Dudgeon, D. *et al.* (2016) The broad footprint of climate change from genes to biomes to people. *Science* 354, aaf7671.

Schindler, S., Staska, B., Adam, M., Rabitsch, W. and Essl, F. (2015) Alien species and public health impacts in Europe: a literature review. *NeoBiota* 27, 1–23.

Schindler, S., Bayliss, H., Essl, F., Rabitsch, W., Follak, S. and Pullin, A.S. (2016) Management effectiveness of invasive common ragweed *Ambrosia artemisiifolia*: a systematic review protocol. *Environmental Evidence* 5, 11.

Schmidt-Chanasit, J., Haditch, M., Schöneberg, I., Günther, S., Stark, K. and Frank, C. (2010) Dengue virus infection in a traveller returning from Croatia to Germany. *EuroSurveillance* 15, pii=19677.

Seebens, H., Gastner, M.T. and Blasius, B. (2013) The risk of marine bioinvasion caused by global shipping. *Ecology Letters* 16, 782–790.

Seebens, H., Blackburn, T.M., Dyer, E.E., Genovesi, P., Hulme, P.E., Jeschke, J.M., Pagad, S., Pyšek, P., Winter, M., Arianoutsou, M. *et al.* (2017) No saturation in the accumulation of alien species worldwide. *Nature Communications* 8, 14435.

Seebens, H., Blackburn, T.M., Dyer, E.E., Genovesi, P., Hulme, P.E., Jeschke, J.M., Pagad, S., Pyšek, P., van Kleunen, M., Winter, M., *et al.* (2018) Global rise in emerging alien species results from increased accessibility of new source pools. *Proceedings of the National Academy of Sciences USA*, in press.

Semenza, J.C. and Menne, B. (2009) Climate change and infectious diseases in Europe. *The Lancet Infectious Diseases* 9, 365–375.

Simón, F., Morchón, R., González-Miguel, J., Marcos-Atxutegi, C. and Siles-Lucas, M. (2009) What is new about animal and human dirofilariosis? *Trends in Parasitology* 25, 404–409.

Smith, A.L., Hewitt, N., Klenk, N., Bazely, D.R., Yan, N., Wood, S., Henriques, I., MacLellan, J.I. and Lipsig-Mummé, C. (2012) Effects of climate change on the distribution of invasive alien species in Canada: a knowledge synthesis of range change projections in a warming world. *Environmental Reviews* 20, 1–16.

Sorte, C.J., Ibáñez, I., Blumenthal, D.M., Molinari, N.A., Miller, L.P., Grosholz, E.D., Diez, J.M., D'Antonio, C.M., Olden, J.D., Jones, S.J. and Dukes, J.S. (2013) Poised to prosper? A cross-system comparison of climate change effects on native and non-native species performance. *Ecology Letters* 16, 261–270.

Stark, K., Niedrig, M., Biederbick, W., Merkert, H. and Hacker, J. (2009) Die Auswirkungen des Klimawandels. Welche neuen Infektionskrankheiten und gesundheitlichen Probleme sind zu erwarten? *Bundesgesundheitsblatt* 52, 699–714.

Stireman, J.O., Dyer, L.A., Janzen, D.H., Singer, M.S., Lill, J.T., Marquis, R.J., Ricklefs, R.E. Gentry, G.L., Hallwachs, W., Coley, P.D. *et al.* (2005) Climatic unpredictability and parasitism of caterpillars: implications of global warming. *Proceedings National Academy of Sciences USA* 102, 17384–17387.

Streftaris, N. and Zenetos, A. (2006) Alien marine species in the Mediterranean – the 100 'Worst Invasives' and their impact. *Mediterranean Marine Science* 7, 87–118.

Thomas, S.M., Fischer, D., Fleischmann, S., Bittner, T. and Beierkuhnlein, C. (2011) Risk assessment of dengue virus amplification in Europe based on spatio-temporal high resolution climate change projections. *Erdkunde* 65, 137–150.

Tittensor, D.P., Walpole, M., Hill, S.L.L., Boyce, D.G., Britten, G.L., Burgess, N.D., Butchart, S.H.M., Leadley, P.W., Regan, E.C., Alkemade, R. *et al.* (2014) A mid-term analysis of progress toward international biodiversity targets. *Science* 346, 241–244.

US Fish and Wildlife Service (2009) Invasive plant species response to climate change in Alaska. Available at: https://www.fws.gov/alaska/climate/pdf/lecture/Report_final.pdf (accessed 29 December 2002).

Vogl, G., Smolik, M., Stadler, L.M., Leitner, M., Essl, F., Dullinger, S., Kleinbauer, I. and Peterseil, J. (2008) Modelling the spread of ragweed: effects of habitat, climate change and diffusion. *The European Physical Journal Special Topics* 161, 167–173.

Walther, G.-R., Roques, A., Hulme, P.E., Sykes, M.T., Pyšek, P., Kühn, I., Zobel, M., Bacher, S., Botta-Dukát, Z., Bugmann, H. *et al.* (2009) Alien species in a warmer world: risks and opportunities. *Trends in Ecology and Evolution* 24, 686–693.

Weaver, S.C. and Reisen, W.K. (2010) Present and future arboviral threats. *Antiviral Research* 85, 328–345.

Wiens, J.J. (2016) Climate-related local extinctions are already widespread among plant and animal species. *PLoS Biology* 14, e2001104.

Williams, F., Eschen, R., Harris, A., Djeddour, D., Pratt, C., Shaw, R.S., Varia, S., Lamontagne-Godwin, J., Thomas, S.E. and Murphy, S.T. (2010) *The Economic Cost of Invasive Non-native Species on Great Britain.* CAB International, Wallingford, UK.

Conclusions

Giuseppe Mazza[1,2]* and Elena Tricarico[2]

[1]CREA Research Centre for Plant Protection and Certification, Florence, Italy and [2]Department of Biology, University of Florence, Italy

Invasive alien species are renowned for their negative ecological and economic effects. However, despite the potential hazards they present to human health, until now their impacts on human well-being have rarely been analysed, except for arthropod vectors and allergenic plants.

In this book, we have provided a general overview of this important issue with the final goals of identifying future research priorities and suggesting possible ways forward. The rate of introduction of alien species is increasing with no evidence of saturation. The effects of invasive alien species on human health vary, ranging from psychological effects, phobias, discomfort and nuisance to allergies, poisoning, bites, disease and even death. Apart from the alien invasive species that cause serious risks to human health directly, other species may inflict displaced or deferred impacts. Humans are even menaced by alien invasive species affecting the services provided by ecosystems. These services are vital to our well-being: changes may decrease the availability of drinking water and of products from fisheries, agriculture and forestry; alter pollination; and impoverish culture and recreation. Moreover, climate change can facilitate biological invasions and can modify the impacts of alien invasive species by altering the likelihood of their introduction, establishment, distribution and abundance, as well as the scale of impacts and management.

Arthropods and allergenic plants have been the most-studied taxonomical groups: humans are still afflicted by mosquito-borne diseases, newly emerging in some areas. The recent continental-wide outbreaks of chikungunya and Zika viruses, also in Europe, testify to the success of mosquitoes, facilitated by climate change, and the limits of management procedures in controlling the pathogens they spread. Concerning aquatic environments, fresh waters are better investigated than marine waters. Alien as well as native crustaceans and molluscs are mostly well-known pathogen vectors and bioaccumulators. However, some other taxa are neglected (e.g. fishes, reptiles, birds), but it is expected that they will elicit more concern for public health in the future, considering the increase of ornamental trade involving these groups. Despite being the most numerous group, the number of insects causing impacts on human health is probably underestimated. In the future, attention should be devoted to the new cross-combinations between native/alien insects and pathogens (e.g. alien species being a further vector for a native pathogen).

Overall, the lack of quantitative data, as well as the ignorance of the extent and the severity of the health hazards posed by alien invasive species and their treatment, is worrying. This is particularly evident for marine species that are spreading into the Mediterranean Sea from the Red Sea and causing severe hazard. There is thus a need for action and research,

* E-mail: giuseppe.mazza@unifi.it

particularly in the fields of pathway management, epidemiology, modelling and vector monitoring. A better understanding of drivers and management options of alien diseases and vectors as well as joint efforts in educating the public and decision-makers is crucial since most health consequences of global change, biological invasions included, affect the weakest sectors of the society, i.e. developing countries and children.

Index

Note: Page numbers in **bold** type refer to **figures**
Page numbers in *italic* type refer to *tables*